Microwave Effects on DNA and Proteins

Chris D. Geddes
Editor

Microwave Effects on DNA and Proteins

Springer

Editor
Chris D. Geddes
Institute of Fluorescence
University of Maryland Baltimore County
Baltimore, MD, USA

ISBN 978-3-319-50288-5 ISBN 978-3-319-50289-2 (eBook)
DOI 10.1007/978-3-319-50289-2

Library of Congress Control Number: 2017932393

Printed on acid-free paper

This Springer imprint is published by Springer Nature
The registered company is Springer International Publishing AG
The registered company address is: Gewerbestrasse 11, 6330 Cham, Switzerland

Preface

For many years, the scientific community has studied ionizing and nonionizing electromagnetic field (EMF) radiation, either stand-alone or in combination with other agents, with regard to possible adverse health effects on biological systems. In particular, the roles of microwaves have received much attention with significant discussion focused on the underlying mechanisms of the so-called *thermal* or *nonthermal* effects, i.e., the reaction-promoting effects that occur with microwaves, with notable attention also on the role of both microwave-generated reactive oxygen and nitrogen species. In addition, these adverse health effects can also be taken in light of the fact that microwaves are indeed used for biological sample preparation/destruction, such as in histopathology to reduce the preparation time of tissue sections for microscopy, or used in the microwave-based Lyse-it™ technology (www.lyse-it.com), which can rapidly lyse cells and fragment DNA (and RNA/proteins) into desired base pair sizes.

Given the increasing use of devices which continuously emit polarized fields, which are believed to be significantly more bioactive than natural unpolarized ones, I have subsequently invited a group of well-known internationally recognized scientists to contribute leading-edge articles on the roles and effects that microwaves have on biological systems, particularly with regard to their effects on DNA and proteins.

At this time, I would like to thank the contributors for their excellent and timely contributions to this notable volume.

Many thanks.

Baltimore, MD, USA
November 11, 2016

Chris D. Geddes

Contents

About the Authors

Dr. Jitendra Behari is Ph.D. in physics (1971) from the Indian Institute of Technology Delhi. He joined the Department of Biomedical Engineering of All India Institute of Medical Sciences, in 1973. Dr. Behari was professor in the School of Environmental Sciences of Jawaharlal Nehru University, New Delhi. He was conferred the honorary D.Sc. degree in bioelectromagnetics (Ansted University, UK). He has been a Fulbright fellow and adjunct faculty at Michigan State University, USA. He is a DAAD fellow and recipient of fellowship from the Japanese Association of Cultivation Sciences and British Council and has been a career awardee (University Grants Commission). He is a fellow of the Institution of Electronics and Telecommunication Engineers and the Ultrasonics Society of India and a senior member of the IEEE. He has been a member of Commission K of the URSI (engineering in medicine and biology society). Dr. Behari has been a past chairman of the IEEE, ED MTT Chapter. Presently, he is researching on the health implications of mobile phone frequency exposure.

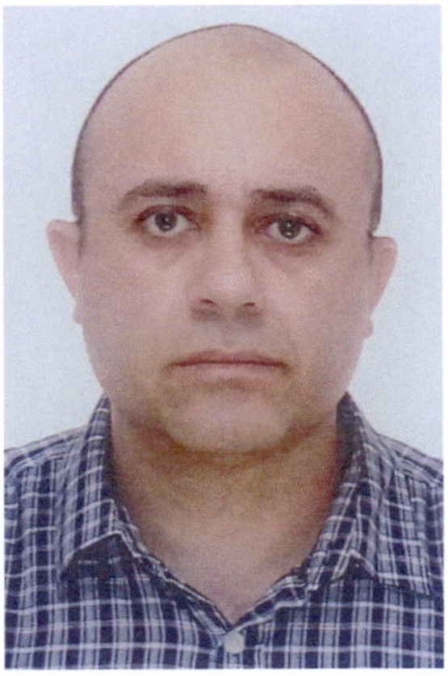

Dr. Orlando Vieira Furtado-Filho graduated in biological sciences from the Educational Foundation of Alegrete, Brazil (Current University Campaign Region), and with a master's degree in biological sciences (molecular biology) from the University of Brasilia, Brazil (UNB), and a Ph.D. in biological sciences (cell and molecular biology) from the Federal University of Rio Grande do Sul, Brazil (UFRGS). He is major in Brazilian Army and a research collaborator of Biotechnology Center and Department of Biophysics, Institute of Biosciences, UFRGS. He has experience in molecular biology, biochemistry, and biophysics with emphasis on biochemistry of free radicals and metabolism of fatty acid, acting on the following topics: oxidants, antioxidants, oxidative damage to macromolecules, oxidative stress, free radicals, bioelectromagnetism, electromagnetic field, aging, and plant extract.

Dr. Chris D. Geddes, Ph.D., FRSC, and professor (www.chrisgeddes.com), has extensive experience in fluorescence spectroscopy, particularly in fluorescence sen sing, microwave-accelerated plasmonics, and metal-fluorophore interactions, publishing over 275 peer-reviewed papers (h-index: 43) and more than 30 books.

Dr. Geddes is internationally known in fluorescence and plasmonics, and his laboratory is widely attributed to the development of the metal-enhanced fluorescence (MEF) and related plasmon-fluorescence technologies, securing in excess of $25 million in recent years to pursue his research aspirations. He is the editor in chief of the *Journal of Fluorescence* and founding editor in chief of the *Who's Who in Fluorescence*, *Annual Reviews in Fluorescence*, and the *Annual Reviews in Plasmonics* volumes. In addition, due to the labs' pioneering efforts in the fields of metallic nanoparticle-fluorophore interactions, Dr. Geddes launched a Springer journal *Plasmonics* in 2005, which is a leading journal in the field today.

Dr. Geddes is director of the Institute of Fluorescence, a department within UMBC (University of Maryland, Baltimore County), which focuses on the nano-bio-technological applications of fluorescence. Dr. Geddes has been a permanent member of the NIH's EBIT R01 study section (2007–2012) and chaired the NIH's Analytical and BioAnalytical SBIR study section from ~ 2004 to 2009. Dr. Geddes is a fellow of both the Royal Society of Chemistry (FRSC) and the Institute of Physics. Dr. Geddes holds more than 100 patents in the fields of fluorescence and plasmonics, and his roles and interactions with industry have created enterprise value in excess of $100 million today. In 2011, Professor Geddes was honored by the Maryland House of Delegates in Annapolis, (House Resolution #326), for his outstanding contributions to education, biotechnology, economic development, and innovation.

Dr. Kazuhiko Imaizumi, Ph.D., D.V.M., received his bachelor of veterinary science from Tokyo University of Agriculture and Technology and Ph.D. in forensic medicine from Saitama Medical University. He is currently the chief of forensic anthropology laboratory in National Research Institute of Police Science, Japan. He has engaged in forensic identification of skeletal remains over 20 years and applied DNA identification technique to bone specimens at early stage of its development with studies on DNA extraction from compact bones. His current interest is three-dimensional morphological analyses in bone and face shapes that provide highly sophisticated scientific certainty in forensic anthropology. He is a fellow of the Japanese Association of Forensic Science and Technology.

Dr. Ronald N. Kostoff received a Ph.D. in aerospace and mechanical sciences from Princeton University in 1967. He has worked for Bell Laboratories, Department of Energy, Office of Naval Research, and MITRE Corp. He has published over 200 peer-reviewed articles, served as guest editor of four journal special issues since 1994, obtained two text mining system patents, and presently is a research affiliate at Georgia Institute of Technology. He has published on numerous medical topics in the peer-reviewed literature, including (1) potential treatments for multiple sclerosis, Parkinson's disease, Raynaud's phenomenon, cataracts, SARS, vitreous restoration, and chronic kidney disease, (2) potential causes of chronic kidney disease, and (3) potential impacts of electromagnetic fields on health.

Radiation biophysicist **Dr. Dimitris J. Panagopoulos** was born in Athens, Greece (December 6, 1961). He has a degree in physics and Ph.D. in biology from the University of Athens. He completed his Ph.D. on the biological effects of electromagnetic fields (EMFs) in 2001 and postdoctoral research on cell death induction by microwave radiation in 2006. Since 2002, he has been working at the section of cell biology and biophysics and has been giving lectures on radiation biophysics (physical properties and biological effects of ionizing and nonionizing radiations). Since 2014, he has been working at the National Center for Scientific Research "Demokritos," Laboratory of Health Physics, Radiobiology and Cytogenetics, researching effects of ionizing and nonionizing radiations on human cells. His experiments were among the first that showed effects of mobile telephony radiation on DNA, and his theory on the mechanism of action of EMFs on cells is considered the most valid.

Dr. Luc Verschaeve received Ph.D. in and is a professor of the University of Antwerp (faculty of biomedical sciences). He is responsible for the toxicology laboratory of the Scientific Institute of Public Health in Brussels and a present or past president and/or member of different national and international working groups on nonionizing radiations. Activities related to nonionizing radiation comprise research on the (genetic) toxicology of extreme low-frequency and radiofrequency fields (e.g., in the framework of the Belgian BioElectroMagnetics Group, cf. http://www.bbemg.ulg.ac.be/); the performing of many national projects involving the writing of information booklets and reports; communications for the general public, professionals, and authorities; etc. Research or involvements in activities related to nonionizing radiations started in 1992 and were since conducted without interruption. Another scientific research is mainly on natural products (e.g., mutagenicity and antimutagenicity of traditional medicinal plant extracts) (Scientific publications: more than 200 (from which more than 100 are referenced in PubMed)).

Dr. Takeo Yoshimura received his Dr. of Engineering degree from Kyushu Institute of Technology in 2009. He joined Kitakyushu Foundation for the Advancement of Industry Science and Technology as a researcher in 2009, and he transferred to Kyushu Institute of Technology as a research staff. From 2011 to 2014, he was an assistant professor of Tokyo University of Science. Since 2015, he joined Tokyo Institute of Technology as a researcher. Currently he joined SAIDA FDS INC., where he has been engaged in development of the microwave flow-reactor.

Chapter 1
Mobile Telephony EMFs Effects on Insect Ovarian Cells. The Necessity for Real Exposures Bioactivity Assessment. The Key Role of Polarization, and the "Ion Forced-Oscillation Mechanism"

Dimitris J. Panagopoulos

Abstract Exposure of *Drosophila melanogaster* young adult insects to Electromagnetic Fields (EMFs)/Radiation (EMR) emitted by an active GSM (Global System for Mobile telecommunications) mobile phone handset during a usual "talk" operation for a few minutes daily for 2–5 days, revealed an impressive decrease (up to 57%) in reproductive capacity (fecundity) (Panagopoulos et al. 2004). That effect directed us to focus our next studies on the effects of this type of EMF/EMR on the DNA and proteins of the insect's reproductive cells (gametes). More specifically, we focused on the effects on the female ovarian cells. We used the TUNEL (Terminal deoxynucleotide transferase dUTP Nick End Labeling) assay, to detect fragmented DNA in the ovarian cells. Moreover, we used the Rhodamine-conjugated Phalloidin staining assay, to detect possible damage in the actin cytoskeleton of the ovarian cells. We found a high degree of DNA fragmentation in the nuclei of ovarian cells of the exposed insects (up to +55% compared to the sham-exposed insects) (Panagopoulos et al. 2007a). The DNA fragmentation was highly dependent on the intensity of radiation (distance from the handset) and was found to be maximum for intensities higher than 250 μW/cm^2 (in close proximity with the handset) and within a "window" around 10 μW/cm^2 (at 20–30 cm distance from the handset) (Panagopoulos et al. 2010). The DNA fragmentation in the nuclei of the exposed ovarian cells was found to be accompanied by actin cytoskeleton damage (Chavdoula et al. 2010). These effects caused a destruction of a significant percentage of egg chambers in the ovaries of the

---

The original version of this chapter was revised.
An erratum to this chapter can be found at 10.1007/978-3-319-50289-2_9

D.J. Panagopoulos (✉)
National Center for Scientific Research "Demokritos", Athens, Greece

Radiation and Environmental Biophysics Research Centre, Athens, Greece

Department of Biology, University of Athens, Athens, Greece
e-mail: dpanagop@biophysics.gr

C.D. Geddes (ed.), *Microwave Effects on DNA and Proteins*,
DOI 10.1007/978-3-319-50289-2_1

exposed females (Panagopoulos 2012a). New data (Panagopoulos et al. 2015a, b) suggest that the continuous and unpredictable variability of the mobile telephony signals, in combination with the fact that they are totally polarized (just like every type of man-made EMF), and the inclusion of Extremely Low Frequencies (ELF) - due to pulsing and modulation of the microwave carrier - in all modern mobile telecommunication microwave signals, constitute the main reasons for their intense bioactivity. A significant opposition is found between the results of experimental studies employing real exposures of biological samples from commercially available mobile phones, and the results of studies employing simulated exposures from generators or "test" phones as suggested by health authorities (Health Protection Agency 2012; IARC 2013). While experimental studies employing simulated EMF-emissions present a strong inconsistency among their results with nearly 50% of them reporting no effects, studies employing real-life emissions demonstrate an almost 100% consistency in showing adverse effects (Panagopoulos et al. 2015a). Finally, in the present chapter we show why polarized (man-made) EMFs are significantly more bioactive than natural (unpolarized) ones, and we describe the "Ion Forced-Oscillation Mechanism" for the action of polarized EMFs on biological systems.

Keywords Electromagnetic fields • Microwave radiation • Mobile phones • Real exposures • Biological effects • DNA damage • Protein damage • Reproduction • Ovarian cells • Insects • *Drosophila melanogaster* • Polarization • Mechanism of action

1.1 Introduction

Microwaves are called the electromagnetic waves produced by electronic oscillators of human technology, with frequencies higher than those which can be reflected by the ionosphere (thus higher than ~ 200 MHz) and up to the low limit of infrared (~ 300 GHz). Microwaves occupy the higher frequency part of a wider category called Radio-Frequency (RF) electromagnetic waves and start from frequencies around 10 kHz. In other words, microwaves are not reflected by the ionosphere unlike the electromagnetic waves of lower frequencies, and thus the receiving and the emitting microwave antennas need to have optical contact between them. The continuous demand for increasing the volume of transmitted information by microwave antennas leads to the continuous increase in the microwave frequencies, and the consequent approximation closer to the low limit of infrared (Lioliousis 1979).

In addition to the artificially produced microwaves which constitute the vast majority in our modern environment, there are natural microwaves of broad spectrum (10 MHz – 10 GHz) and cosmic origin which reach the earth's surface at very low intensities (~10^{-17} mW/cm^2/MHz) (Presman 1977). These are called cosmic microwaves. An important difference between man-made and natural (cosmic) microwaves is that the first are totally polarized like every form of man-made electromagnetic radiation (EMR)/field (EMF), while the second are unpolarized just like every form of natural EMR/EMF. This difference is of decisive importance when it comes to the issue of the mechanism of how microwaves (and EMFs in general) interact with living matter, as we shall explain later on.

One of the main applications of man-made microwaves is in modern telecommunications, such as digital mobile telephony, domestic cordless phones (DECT-Digitally Enhanced Cordless Technology), or internet connection wireless devices and local wireless networks (Wi-Fi), or satellite communications. Radars employ microwaves as well.

With the tremendous development of telecommunications during the past few decades, the levels of RF/microwave radiation are exponentially increased especially in modern urban environments, and there is consequent increasing concern among the scientists and the public about the possible adverse health effects of these radiation types.

It is very important to emphasize that all (wireless) telecommunication devices emitting microwaves, emit extremely low frequencies (ELF) as well since this is absolutely necessary for pulsing and modulating the RF carrier signal in order to be able to transmit increasing amount of information (Tisal 1998; Hyland 2000; Hillebrand 2002; Tuor et al. 2005; Curwen and Whalley 2008). In other words, exposure to EMFs from mobile or cordless phones, Wi-Fi and other wireless communication devices, is simultaneous exposure to several frequencies both RF and ELF. It is suspected that the ELF emissions included in all telecommunication signals are those more responsible for the biological/health effects, and not the carrier RF signal itself (Panagopoulos et al. 2015b).

Hundreds of biological, clinical, and statistical/epidemiological studies investigating the potential health effects of microwave radiation are carried out so far. A first look shows contradicting results (Verschaeve 2009; Verschaeve et al. 2010; Bourthoumieu et al. 2010; Vijayalaxmi 2012; Health Protection Agency 2012; Ingole and Ghosh 2012; Ros-Llor et al. 2012; Karaca et al. 2012; Vignera et al. 2012; Waldmann et al. 2013; Cucurachi et al. 2013; IARC 2013; Balmori 2014; Paul et al. 2015; Lerchl et al. 2015; Roggeveen et al. 2015; Morgan et al. 2015; Singh et al. 2016; Manna and Ghosh 2016; Shahin-Jafari et al. 2016). But a closer look shows that there is a significant difference between studies employing simulated mobile phone signals from generators or "test" phones programmed to emit constant signals of fixed frequency, waveform and output power, and studies employing real signals from commercially available mobile phones or other microwave devices. While experimental studies employing simulated EMF-emissions present a strong inconsistency among their results with nearly 50% of them reporting no effects (Health Protection Agency 2012; Vijayalaxmi 2012; IARC 2013), studies employing real mobile phone exposures demonstrate an almost 100% consistency in showing adverse effects (Panagopoulos et al. 2015a). This consistency is in addition supported by studies showing association with brain tumors, symptoms of un-wellness, and declines in animal populations. These statistical/epidemiological studies concern exposures to real emissions as well, mostly from mobile phone handsets and base station antennas (Navarro et al. 2003; Salama and Abou El Naga 2004; Kundi 2004; Hutter et al. 2006; Balmori

2005, 2010; Balmori and Hallberg 2007; Everaert and Bauwens 2007; Khurana et al. 2009; Blettner et al. 2009; Kundi and Hutter 2009; Viel et al. 2009; Hardell et al. 2007, 2013; Bhattacharya and Roy 2014; Singh et al. 2016).

Determination of realistic exposures from mobile phones and other wireless telecommunication devices is one of the most important issues in scientific studies examining the biological/health effects of microwaves, since it is key to defining public health protection. The situation becomes confusing by the divergent results reported in the literature which can very well be due to unrealistic exposures, which in turn leads to wrong conclusions and ineffective and misdirected regulations (Panagopoulos et al. 2015a).

While the International Agency for Research on Cancer (IARC) has classified both ELF and RF EMFs as possibly carcinogenic to humans (IARC 2002, 2013), in its 2013 report IARC criticized and excluded from consideration experimental studies that used commercially available mobile phone handsets in exposing biological samples, as having "unreliable dosimetry" without further scientific explanation (IARC 2013). Similarly the Health Protection Agency criticized this exposure methodology reporting that the exposure is "highly variable" with "lack of control" due to network reasons (number of subscribers each moment) and movement of the animals within the vials/boxes in case of freely moving animals, but recognizes that restriction of the animals during the exposures will result in additional stress. Their critique recommended that exposures should be performed by devices or handsets set to produce emissions at fixed frequency and output power by use of engineering or hardware controls (Health Protection Agency 2012). In both reports the critique was based on the fact that real mobile phone emissions always include large variations in their intensity, frequency, etc., especially in the near-field of the antenna.

But billions of mobile phone users are daily exposed for increasing periods to real emissions from their handsets in the near-field of the antenna in contact with their ears/bodies, not to any simulated emissions with fixed parameters. Is it then scientifically correct to study the effects of a "highly variable" field by using fields with fixed parameters? In our opinion, it is definitely not. Especially since the varying nature of the field seems to be an important reason for its increased biological activity.

A plausible biophysical mechanism explaining how man-made EMFs can alter cell function by irregular gating of electrosensitive ion channels on the cell membranes is published (Panagopoulos et al. 2000a, 2002, 2015b) and verified by numerical test while previously suggested other mechanisms failed to pass the same test for realistic conditions within living cells (Halgamuge and Abbetrathne 2011). This mechanism known as "Ion Forced-Oscillation Mechanism" is based on the property of polarization of all man-made EMFs. Any externally applied polarized EMF will cause a parallel and in phase forced-oscillation of all charged particles - such as the mobile/free ions existing in large concentrations in all living cells - which can then exert constructive (additive) forces on any other charge such as the voltage-sensors of electrosensitive ion channels. These additive forces can be much more effective in gating (opening or closing) this type of ion channels than the chaotic forces in every possible direction exerted by the same ions due to their random thermal motion, or due to any random individual oscillations caused by any unpolarized (natural) EMF.

In the present chapter we review a series of experiments with real mobile phone exposures that show important effects on DNA in insect ovarian cells. Then we examine the opposition between the results of studies employing real telecommunication microwave emissions and the results of studies employing exposures by simulated microwave emissions, as we think this is a point of utmost importance. Finally, we analyze the differences in bioactivity between polarized and unpolarized EMFs emphasizing on mobile telephony EMFs, and we provide an explanation of the reported biological effects on a biophysical basis according to the "Ion Forced-Oscillation Mechanism" for man-made (polarized) EMFs.

1.2 Effect of GSM Mobile Phone EMFs on Insect Ovarian Cells

1.2.1 Exposure Device and EMFs Measurements

We were the first to use a commercially available mobile phone handset to expose biological samples to microwave radiation (Panagopoulos et al. 2000b, 2004; Panagopoulos and Margaritis 2002). The reason was (and still is) obvious: We wanted to test the effects of the real EMFs which expose daily billions of humans globally.

In each study, in spite of the fact that we used a commercially available and therefore approved for the market mobile phone handset, we measured the intensity of radiation in the RF/microwave region during a normal phone-conversation. Moreover, we measured the electric and magnetic field intensities in the ELF region, since it is known that GSM emissions except for their carrier microwave frequency around 900, 1800 (Europe), or 1900 (North America) MHz, employ ELF pulsing fields at 217 Hz, and other even lower ELFs such as 8.34 Hz (Tisal 1998; Hyland 2000). The measurements and the experiments were always carried out with the battery of the handset fully charged, and at the same bench within the lab with full signal reception, and same positions of all items around (Panagopoulos et al. 2004, 2007a, b, 2010).

From the first sets of measurements we found out that when the mobile phone user talked during a phone conversation ("talk signal" or "voice-modulated signal" or "GSM basic") the intensities of the emissions increased about tenfold than when the user listened and there was no sound in the room ("listening mode" or "non-modulated signal" or discontinuous transmission mode-DTX). The biological effect of the voice-modulated exposures on the fecundity of young adult Drosophila insects was similarly much more intense than the corresponding effect of the DTX exposures with the same handset. While the DTX 900 MHz GSM exposures

Table 1.1 GSM 900 and 1800 radiation and field intensities ± SD, in the microwave and ELF bands averaged over 6 min of voice-modulated emission, for different distances from a mobile phone antenna[a]

Distance from mobile phone Antenna (cm)	**GSM 900** Radiation Intensity at 900 MHz, (mW/cm^2)	GSM 900 ELF Electric Field Intensity, (V/m)	GSM 900 ELF Magnetic Field Intensity, (mG)	**GSM 1800** Radiation Intensity at 1800 MHz, (mW/cm^2)	GSM 1800 ELF Electric Field Intensity, (V/m)	GSM 1800 ELF Magnetic Field Intensity, (mG)
0	0.378 ±0.059	19 ±2.5	0.9 ±0.15	0.252 ±0.050	13 ±2.1	0.6 ±0.08
1	0.262 ±0.046	12 ±1.7	0.7 ±0.13	0.065 ±0.015	6 ±0.8	0.4 ±0.07
10	0.062 ±0.020	7 ±0.8	0.3 ±0.05	0.029 ±0.005	2.7 ±0.5	0.2 ±0.05
20	0.032 ±0.008	2.8±0.4	0.2 ±0.04	0.011 ±0.003	0.6 ±0.12	0.1±0.02
30	0.010 ±0.002	0.7 ±0.09	0.1 ±0.02	0.007 ±0.001	0.3 ±0.06	0.06 ±0.01
40	0.006 ±0.001	0.2 ±0.03	0.05 ±0.01	0.004 ±0.0007	0.1 ±0.04	–
50	0.004 ±0.0006	0.1 ±0.02	–	0.002 ±0.0003	–	–
60	0.002 ±0.0003	–	–	0.0016 ±0.0002	–	–
70	0.0017 ±0.0002	–	–	0.0013 ±0.0002	–	–
80	0.0012 ±0.0002	–	–	0.0011 ±0.0002	–	–
90	0.0010 ±0.0001	–	–	0.0005 ±0.0001	–	–
100	0.0004 ±0.0001	–	–	0.0002 ±0.0001	–	–

[a]For distances longer than 30–50 cm from the mobile phone antenna, the ELF electric and magnetic field components of both GSM 900 and 1800 radiations, fall within the background of the stray 50 Hz fields within the lab

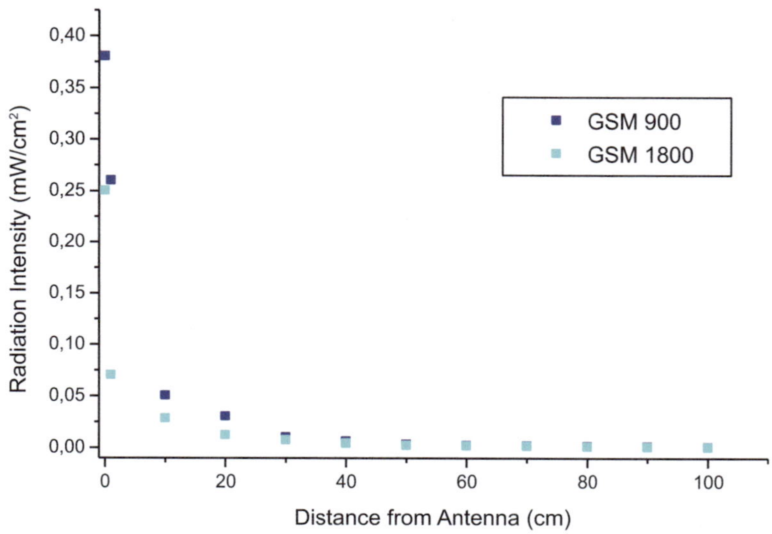

Fig. 1.1 Mobile phone radiation intensity values for GSM 900 and GSM 1800 MHz, according to the distance from the mobile phone antenna

of six min daily for 5 days reduced fecundity by approximately 18% compared to the unexposed insects, the voice-modulated ("talk") corresponding exposures reduced fecundity by approximately 53% respectively (Panagopoulos et al. 2004).

Representative intensity measurements of voice-modulated emissions averaged over 6 min, both in the microwave and ELF bands, and at different distances from the antenna of a mobile phone operating at 900 MHz and (the same handset with a different SIM - "subscriber identity module" card) at 1800 MHz, are shown in Table 1.1 (Panagopoulos et al. 2010). All mesured intensities are within the exposure criteria established by health authorities (ICNIRP 1998). The average radiation intensity levels of the microwave carrier (900 and 1800 MHz) according to the distance from the mobile phone antenna are graphically represented in Fig. 1.1. The standard deviations (SD) of the different intensity values given in Table 1.1 are not shown in Fig. 1.1. The ELF electric and magnetic field values also are given in Table 1.1 but not represented in the diagram as they are more or less proportional to the radiation intensity values.

1.2.2 Experimental Animals

Our experimental animals were *Drosophila melanogaster* flies, strain Oregon R wild-type, held in glass bottles with their food and kept in an incubator at 25 °C, with 12-h periods of light and darkness, and 70–75% relative humidity slightly varying from one experimental series to another but kept constant during each specific

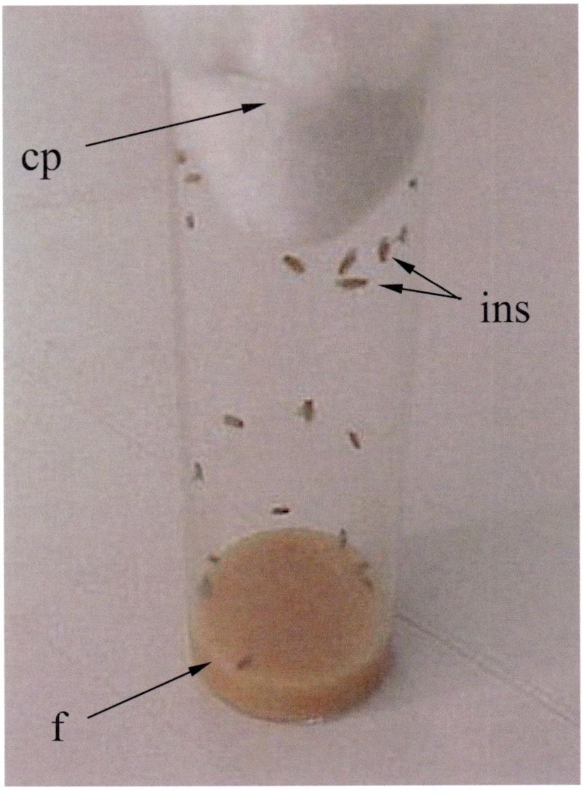

Fig. 1.2 A typical glass vial used in our experiments, containing a group of 20 insects (ins) and food (f), closed with cotton plug (cp)

experimental series. In each individual experiment we collected newly emerged adult flies which we separated into different groups of usually ten males and ten females (except when otherwise mentioned), and put within identical 50-ml cylindrical glass vials with 2.5 cm diameter and 10 cm height, with equal amounts of food of identical quality prepared at the same time, as previously described (Panagopoulos et al. 2004). Figure 1.2 shows one group of flies within a 50-ml glass vial with food.

The oogenesis of this insect is a model biological system, with a very good timing of developmental processes under controlled conditions (King 1970; Panagopoulos 2012b).

The reproductive capacity was assessed by the number of first filial generation (F_1) pupae derived from eggs laid during the first 3 days of the insect's maximum fecundity (oviposition). This number - under the conditions of our experiments - equals the number of fertilized laid eggs, since there is no statistically significant mortality of fertilized eggs, larvae or pupae derived from newly eclosed adult flies during the first days of their maximum oviposition (Panagopoulos et al. 2004, 2007a, b, 2010, 2013a; Panagopoulos 2012b).

1.2.3 Exposure Procedure

In each experiment, the collected newly emerged flies from the stock were anesthetised very lightly with diethyl ether and separated males from females. We put the collected flies in groups of usually ten males and ten females (twenty flies in each group) in the 50-ml vials with food, closed with cotton plugs.

In each experiment, the males and the females of each group were kept and exposed/sham-exposed in separate vials during the first 48 h (two separate vials for each exposed/sham-exposed group). Newly emerged adult *Drosophila* flies are not sexually mature immediately after eclosion. Male flies become sexually mature about 12 h after eclosion and females about 45 h after eclosion (King 1970; Panagopoulos 2012b). Keeping males separately from females for the first 48 h of each experiment ensures that when the two sexes are put together they are in complete sexual maturity and ready for immediate mating and laying of fertilized eggs. This in turn ensures that all laid eggs during the next 72 h are fertilized, and this minimizes variability in oviposition counts.

After the first 48 h, the males and females of each group (20 flies) were lightly anesthetized again and put together in a single vial with fresh food (one vial for each group) to continue being daily exposed/sham-exposed and allowed to mate and lay eggs for the next 72 h, during which the insect's oviposition is at its maximum (King 1970; Bos and Boerema 1981; Panagopoulos et al. 2004; Panagopoulos 2012b).

The exposure to the GSM fields started from the first day of the experiments, 1 h after the insects were fully awaken from the first anesthesia and - in most experiments - lasted for 6 min daily in a single exposure. In experiments testing the effect of exposure duration the daily exposures ranged from 1 to 21 min between the different groups (Panagopoulos and Margaritis 2010a). The exposures took place for the first 2–6 days of the insects' adult lives, depending on the experimental protocol. In experiments comparing the effect of the EMF on the reproductive capacity between the two sexes the exposures took place only for the first 2 days (while the two sexes in all groups were separated) (Panagopoulos et al. 2004). In experiments testing the effect on reproductive capacity of both sexes the EMF exposures took place for the first 5 days of the insects' adult lives (Panagopoulos et al. 2004, 2007b; Panagopoulos and Margaritis 2010a, b). In experiments testing the effect on DNA and proteins, there was an additional exposure in the sixth day before dissection and treatment of the ovaries (Panagopoulos et al. 2007a, 2010; Chavdoula et al. 2010).

In the different sets of experiments we separated the insects into: a) the Exposed group(s) and b) the Sham-Exposed group(s). The sham-exposed groups had identical treatment with the exposed groups, except that the mobile phone handset was turned off during the sham-exposures. We performed several replicate experiments for each different experimental protocol in the different studies.

The number of groups depended on the specific experimental protocol. For example, in the first experiments testing the effect of GSM radiation on reproduction (fecundity) we separated the insects into two groups (one exposed and one sham-exposed) (Panagopoulos et al. 2000b, 2004; Panagopoulos and Margaritis 2002). In the set of experiments testing the dependence of fecundity/DNA damage on the intensity of the radiation/fields, we exposed simultaneously 12 groups at 0, 1, 10, 20,. . ., 100 cm distance from the mobile phone, and thus we had 12 Exposed groups plus one sham-exposed (13 groups in each replicate experiment) (Panagopoulos et al. 2010). In the set of experiments testing the effect of exposure duration on reproductive capacity (Panagopoulos and Margaritis 2010a), we separated the insects into six groups: (a) a group exposed to the GSM EMFs for 1 min, (b) a group exposed for 6 min, (c) a group exposed for 11 min, (d) a group exposed for 16 min, (e) a group exposed for 21 min, and (f) the sham-exposed group (SE). [Details for each study can be found in the original publications].

The temperature in the laboratory during the experimental procedures was maintained at 24 ± 0.5 °C. No detectable temperature increase took place within the vials with the insects or within the mass of the food due to the exposures at all distances between the handset and the vials, and with all tested exposure durations. That was checked during preliminary exposures in each study by the use of a Hg-thermometer with 0.05 °C sensitivity.

1.2.4 Reproductive Capacity Assessment

When the male and female flies of each group had been together in the same vial for 3 days (72 h), that is after 5 days from the beginning of each experiment, the flies were removed from the glass vials. These vials containing the developing embryos were then kept in the culture room for at least six additional days without any further EMF exposure/sham-exposure.

After the six additional days that the vials without the parental flies were kept in the culture room without exposure, most F_1 embryos (deriving from the fertilized laid eggs) were at the stage of pupation, where they could be clearly seen macroscopically and easily counted on the walls of the glass vials. [At the last larvae stages the larvae go out of the food, crawl up on the walls of the glass vials and get immobilized there to become pupae (Panagopoulos 2012b)].

By counting the F_1 pupae 11 days after the beginning of each experiment (2 days separated males from females, plus 3 days mating and egg laying, plus 6 days for the F_1 embryos to reach the pupation stage), we had a representative estimate of each group's reproductive capacity. In this way, instead of counting eggs under a stereo microscope on the surface and within the food which is subject to large errors, we simply counted the number of pupae with bare eyes on the walls of the glass tubes with no error at all. That was also an important innovation in assessing the fecundity of this insect, in addition to keeping the males and females of each group in separate vials for the first 48 h of each experiment (Panagopoulos et al. 2004). All procedures and counts were performed blindly (the experimenter did not know the identity of the groups).

The blinded scoring was repeated for the next 2 days in order to ensure that no larvae had remained within the food and all F_1 embryos had reached the pupation stage and were counted. In this way, no error can occur in the scoring of F_1 pupae.

Comparing the numbers of F_1 pupae between exposed and sham-exposed groups, we get a credible estimate of the effect of EMF exposure on reproductive capacity.

1.2.5 Assessment of DNA Fragmentation in the Ovarian Cells

Oogenesis in *Drosophila* starts during the last stages of pupation. At eclosion, the ovaries of female flies contain already eggs at the first preyolk stages. The eggs develop through 14 distinct stages until they are ready to be fertilized and laid. This process for every egg lasts about 48 h at 25 °C (King 1970; Panagopoulos 2012b).

In order to investigate the ability of the EMFs to damage DNA during early and mid oogenesis (when programmed cell death does not occur) we applied the TUNEL (Terminal deoxynucleotide transferase dUTP Nick End Labeling) assay which is a marker for DNA fragmentation. In this assay fluorescein dUTP (a fluorescent substance) binds through the action of terminal transferase (an enzyme that catalyzes the specific biochemical reaction), onto broken chains (phosphodiester bonds) of genomic DNA which then become labelled by this fluorescent substance. The label incorporated at the damaged sites of DNA is visualized by fluorescence microscopy, as it emits characteristic visible radiation when it is irradiated by ultraviolet ("TUNEL-positive signal").

In the studies testing the effect of EMFs on ovarian DNA and proteins there was an additional exposure in the morning of the sixth day of each individual experiment as already reported. After 1 h from this additional exposure, the parental flies were removed from the glass vials, and they were anesthetized and sacrificed. The ovaries from all exposed and sham-exposed female flies were dissected in Ringer's solution and either were collected intact to measure the ovarian size (Panagopoulos 2012a), or separated into individual ovarioles and egg chambers from which the egg chambers of stages 11–14 were excluded (Panagopoulos et al. 2007a, 2010, 2013a; Chavdoula et al. 2010; Panagopoulos 2016). In egg chambers of stages 11–14 programmed cell death takes place physiologically in the nurse and follicle cells (McCall 2004; Nezis et al. 2000, 2002). Thereby we kept and treated ovarioles and individual egg chambers from germarium up to stage 10. A few representative samples were selected from both ovaries of all the exposed and sham-exposed females in each experiment. The selected samples of ovarioles and egg chambers were then fixed for the TUNEL assay or the rhodamine-conjugated phalloidin staining assay, to search for DNA damage or actin cytoskeleton damage (Panagopoulos et al. 2007a, 2010, 2013a; Chavdoula et al. 2010; Panagopoulos 2012a, 2016).

The application of the TUNEL assay is described in detail in Panagopoulos et al. (2007a). The samples were viewed under a Nikon EZ-C1 fluorescence microscope (Nikon Instruments, Japan). Samples from different groups were blindly observed under the fluorescence microscope (i.e. the observer did not know the origin of the sample) and the percentage of egg chambers with TUNEL-positive signal was scored in each sample.

1.2.6 Assessment of Actin Cytoskeleton Damage in the Ovarian Cells

In this set of experiments - a detailed description can be found in Chavdoula et al. (2010) - we examined whether the GSM exposure induces disorganization of the actin cytoskeleton in the reproductive cells of exposed female insects during early and mid oogenesis. The disorganization of the actin cytoskeleton is a known aspect of cellular death during both apoptosis and necrosis. For this we applied the rhodamine-conjugated phalloidin staining assay. Rhoramine is a fluorescent substance that gets attached to the actin cytoskeleton through the binding of phalloidin. When this is done, the morphology of the actin cytoskeleton is observed by fluorescence microscopy.

We also examined whether follicles with TUNEL-positive signal in their constituent cells had at the same time alterations in their actin cytoskeleton. For this we used double staining with rhodamine-conjugated phalloidin and TUNEL assay at the same samples. The simultaneous observation of DNA fragmentation and actin cytoskeleton damage was accomplished by double action of two different lasers on the samples and observation of the corresponding two types of fluorescence through a Nikon EZ-C1 Confocal Laser Scanning Microscope (Nikon Instruments, Japan) (Chavdoula et al. 2010).

1.2.7 Results

1.2.7.1 GSM EMFs Dramatically Decrease Reproduction

The results of the experiments investigating the effect on reproduction showed that the 6 min daily exposure to GSM EMFs in "talk" mode at close proximity (near-field) to the mobile phone antenna during the first 5 days of the insects' adult lives, induced an impressive average decrease by 53.01% in their reproductive capacity (fecundity). The corresponding average decrease in DTX mode was 18.24%. The GSM field was found to decrease the reproductive capacity in both male and female insects (Panagopoulos et al. 2004).

1.2.7.2 The Decrease in Reproduction Is Due to Elimination of Egg Chambers After DNA Fragmentation and Consequent Cell Death of Their Constituent Cells

The dramatic decrease in reproductive capacity was found to be due to destruction of significant numbers of egg chambers after severe DNA damage in the reproductive cells (gametes). This was found both for GSM 900 and GSM 1800 radiation types. The effect with GSM 900 was more intense than the corresponding effect with GSM 1800 mainly due to its higher intensity, and in a much lesser degree due to its lower carrier frequency (Panagopoulos et al. 2007a, b).

Figure 1.3 shows an ovariole of a sham-exposed female insect with no DNA fragmentation at all developmental stages. This TUNEL-negative picture was representative of the ovarioles of all sham-exposed insects. Figure 1.4 shows an ovariole of an exposed insect with DNA fragmentation (TUNEL-positive signal) only at the two most sensitive developmental stages/checkpoints (germarium and stage 7–8), and no DNA fragmentation (TUNEL-negtive signal) at all other developmental stages. Figure 1.5 shows ovarioles of exposed insects with DNA fragmentation (TUNEL-positive signal) at all developmental stages from germarium up to stage 8. Figure 1.6 shows a stage 10 egg chamber with DNA fragmentation (TUNEL-positive signal) in the nurse and follicle cells.

What was novel and impressive with these findings was that the GSM EMFs induced DNA fragmentation, (1) at all developmental stages of early and mid oogenesis while previously examined stress factors such as starvation or chemicals were found to induce DNA damage only at the two most sensitive developmental stages/checkpoints (germarium and stages 7–8). (2) The DNA fragmentation was observed in all three types of egg chamber cells (nurse cells, follicle cells, and the oocyte), while the above previously examined stress factors induced DNA damage only in the nurse and follicle cells, not in the oocyte (Nezis et al. 2000, 2002; Drummond-Barbosa and Spradling 2001; Panagopoulos et al. 2007a, 2010).

1.2.7.3 The DNA Damage in the Gametes and the Consequent Decrease in Reproduction Is Primarily Dependent on the Intensity of the GSM Fields

The effects on reproduction and on DNA damage decreased non-linearly with increasing distance from the mobile phone handset, or with decreasing EMFs intensities. The effects were maximum for radiation intensities higher than 250 μW/cm^2 (close proximity with the mobile phone antenna) and within a "window" around 10 μW/cm^2 (20–30 cm from the handset) (Panagopoulos et al. 2010). More specific experiments showed that the discovered "window" of maximum bioactivity was not dependent on any specific position but on the specific of EMF-intensity values (Panagopoulos and Margaritis 2010b).

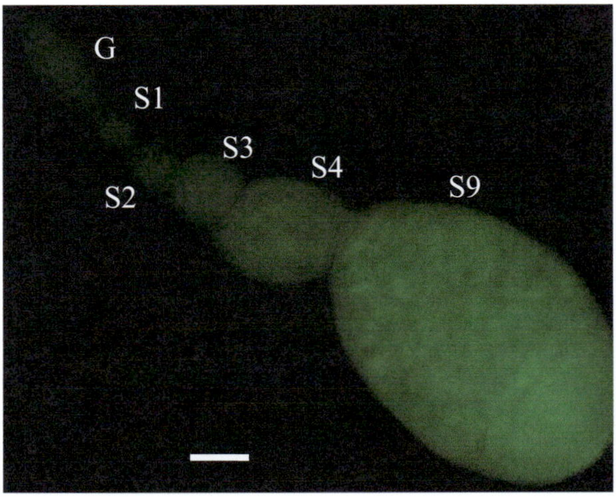

Fig. 1.3 Typical TUNEL-negative fluorescent picture of an ovariole of a sham-exposed female insect, containing egg chambers from germarium to stage 9. Bar: 10 μm

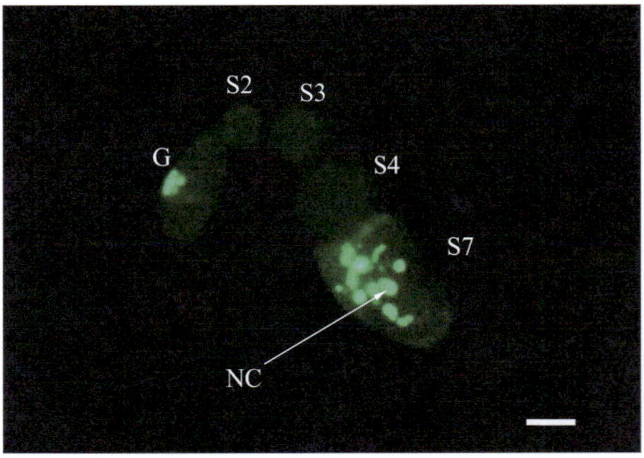

Fig. 1.4 Ovariole of an exposed female insect with TUNEL-positive signal in the nurse cells (NC) only at the two check points, germarium (G) and stage 7 egg chamber, and TUNEL-negative intermediate stages. Bar: 10 μm

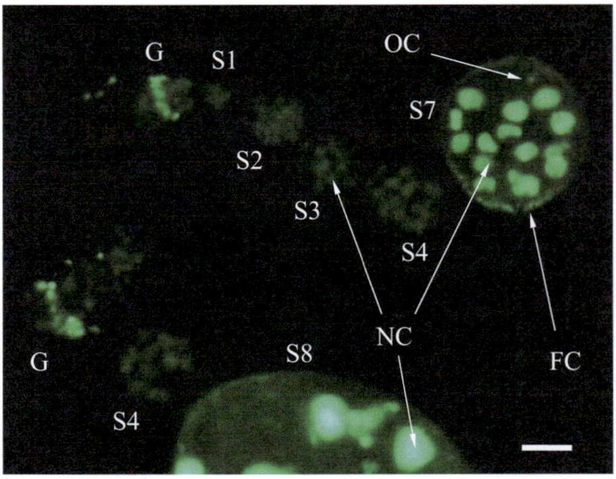

Fig. 1.5 Ovarioles of an exposed female insect with fragmented DNA at all developmental stages from germarium (G) to stage 8, and in all kinds of egg chamber cells, (NC: nurse cells, FC: follicle cells, OC: oocyte). Bar: 10 μm

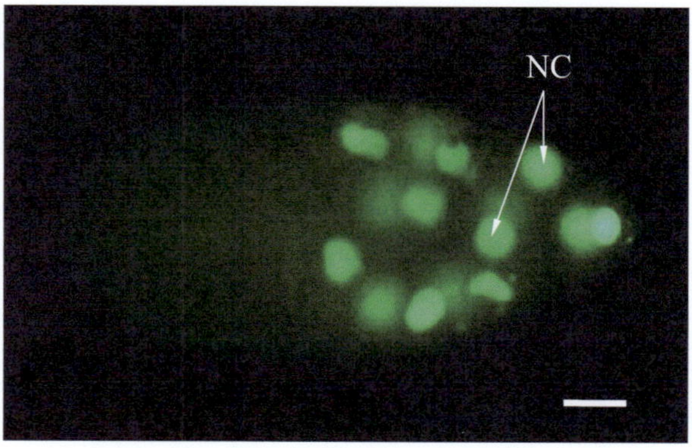

Fig. 1.6 A stage 10 egg chamber of an exposed female insect with fragmented DNA in the nurse cells, (NC). Bar: 10 μm

Figures 1.7 and 1.8, show the percentage of egg chambers with fragmented DNA and the reproductive capacity, in accordance to the distance of the vials from the mobile phone handset for GSM 900 and 1800 respectively. The effects manifested for intensities down to 1 $\mu W/cm^2$ corresponding to approximately 1 m distance from the handset (Panagopoulos et al. 2010).

The carrier frequency of the GSM radiation (900 or 1800 MHz) was found to have only a minimal effect on ovarian DNA fragmentation and the reproductive capacity, with 900 MHz being slightly more bioactive than 1800 MHz under the same EMFs intensities (Panagopoulos et al. 2007a, b, 2010). For example with radiation intensity approximately 285 $\mu W/cm^2$, GSM 900 decreased fecundity by 32.75% in relation to the sham-exposed group while GSM 1800 decreased fecundity by 31.08% (Panagopoulos et al. 2007b). [The "Ion Forced-Oscillation Mechanism" described below shows that the bioactivity is inversely proportional to the frequency of the EMF (Eq. 1.24)].

1.2.7.4 The Effect on Reproduction Increased with Increasing Daily Exposure Durations

In this set of experiments the insects were exposed at the far field of the antenna where previous experiments (Panagopoulos et al. 2010) showed that we have the largest effect, that is at 30 cm distance from the handset for GSM 900 or at 20 cm distance for GSM 1800 where the intensity of both radiation types was ~10 $\mu W/cm^2$. The groups were exposed simultaneously, placed along constant intensity sectors of an arc with a 30- or 20-cm radius - for GSM 900 or GSM 1800 respectively - at the center of which the handset was placed. Then, during each exposure session, the different groups were taken away from the exposure bench one by one, as soon as the exposure duration of each one was completed. (Panagopoulos and Margaritis 2010a).

The experiments showed that even 1 min of daily exposure during the first 5 days of the insects' adult lives is capable to induce a significant decrease in reproduction on the order of 35% in relation to the sham-exposed insects. Then, as the duration of

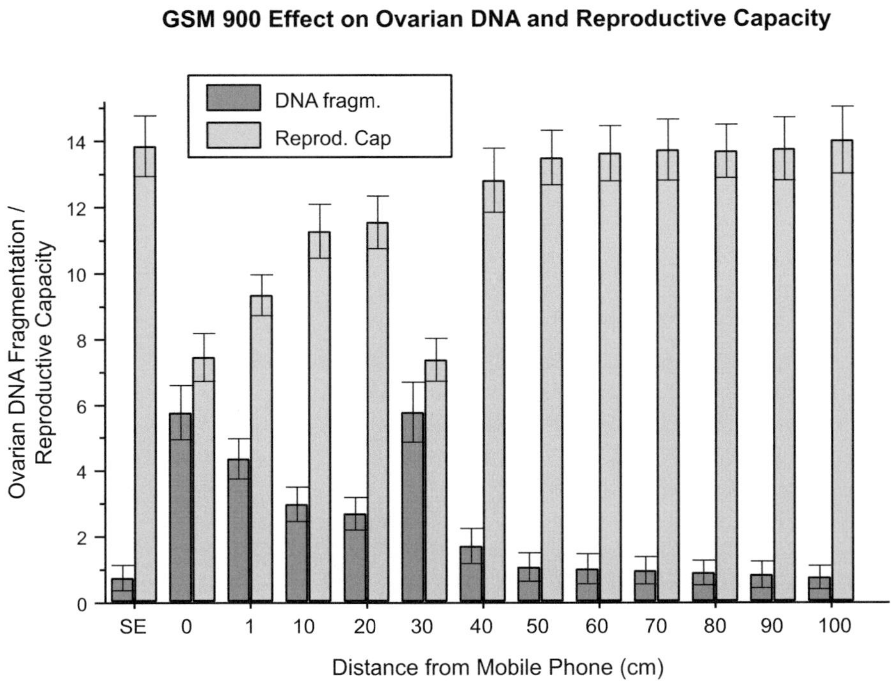

Fig. 1.7 [Mean ratio of ovarian DNA fragmentation (number of TUNEL- positive to total number of egg-chambers) ± SD]×10, and Mean number of F_1 pupae per maternal insect ± SD, versus Distance from mobile phone antenna (cm), for GSM 900

the single daily GSM exposure gradually increased from 1 to 21 min, fecundity was decreasing almost linearly. The effect was shown both with GSM 900 or GSM 1800 fields, but again, GSM 900 was found to be slightly more bioactive than GSM 1800 (Panagopoulos and Margaritis 2010a).

Figure 1.9 shows the decrease in reproduction in relation to the exposure duration for GSM 900, and 1800.

1.2.7.5 The DNA Damage in the Exposed Ovarian Cells Was Found to be Accompanied by Actin Cytoskeleton Damage

The same cells that suffered DNA fragmentation after exposure to the GSM fields, also suffered actin cytoskeleton disorganization as it was shown by double staining of the ovarian samples with both TUNEL and rhodamine-conjugated phalloidin, and observation by two different lasers under a Confocal Laser Scanning Microscope. The percentages of egg chambers with actin cytoskeleton damage in both exposed and sham-exposed samples were very close to corresponding percentages of DNA damage. The analysis by double-staining with TUNEL and rhodamine conjugated phalloidin revealed that DNA fragmentation and actin-cytoskeleton

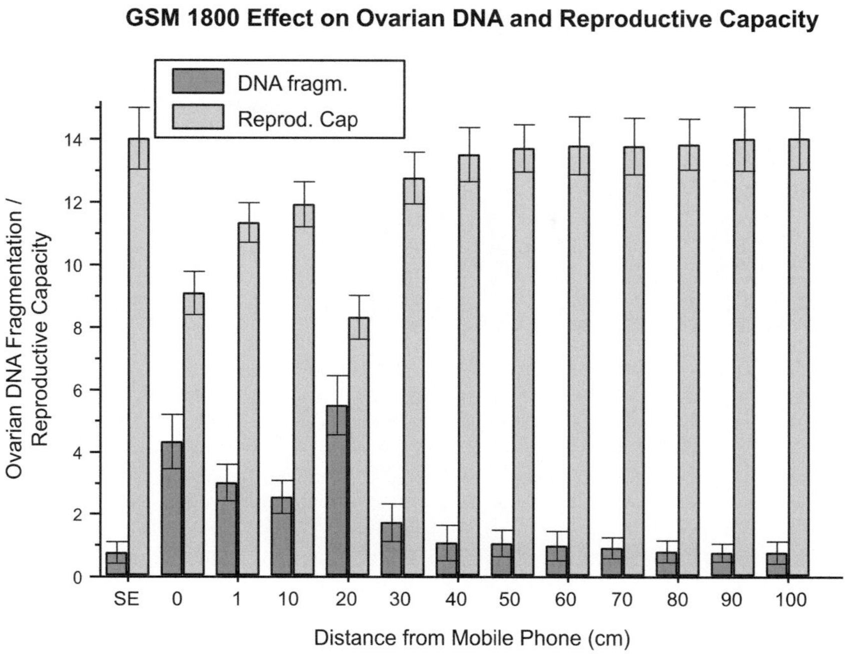

Fig. 1.8 [Mean ratio of ovarian DNA fragmentation (number of TUNEL- positive to total number of egg-chambers) ± SD]×10, and Mean number of F_1 pupae per maternal insect ± SD, versus Distance from mobile phone antenna (cm) for GSM 1800

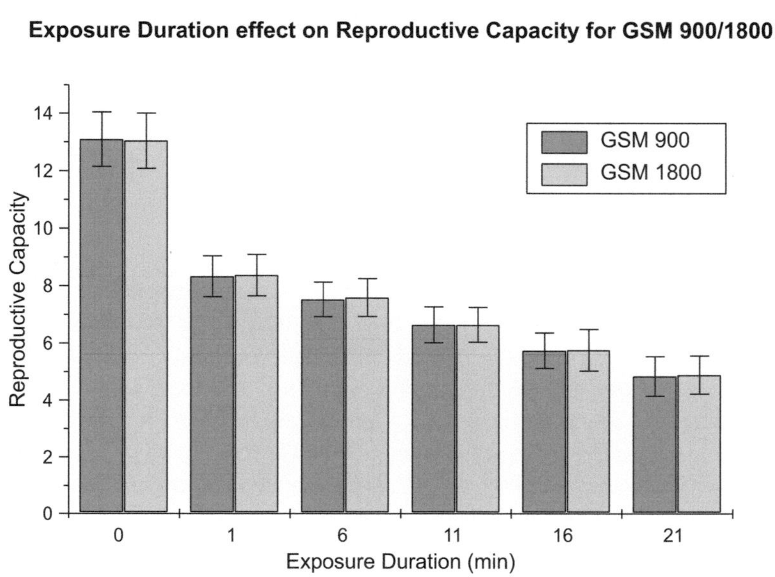

Fig. 1.9 Reproductive capacity ± SD of groups exposed to GSM 900 or 1800 fields for different daily exposure durations (1, 6, 11, 16, and 21 min) and of sham-exposed groups (0 min)

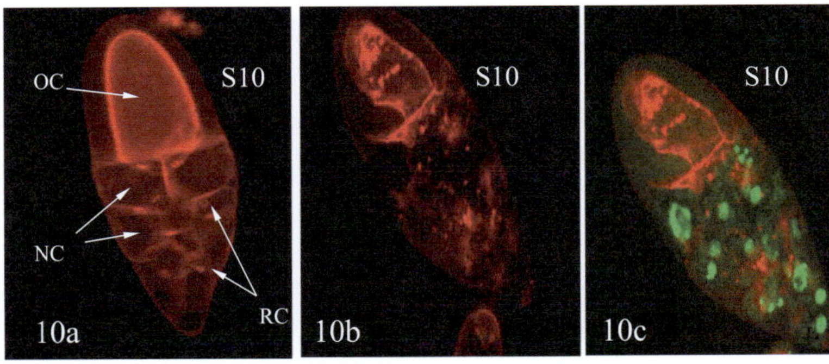

Fig. 1.10 (**a**) A stage 10 egg chamber from a sham exposed insect, treated with rhodamine-conjugated phalloidin assay, with normal cytoskeleton morphology. Characteristic features of the actin cytoskeleton like the ring channels (RC) can be observed. *NC* nurse cells, *OC* oocyte. (**b**) A stage 10 egg chamber of an exposed insect with disorganized actin cytoskeleton. (**c**) The same stage 10 egg chamber as in (**b**), treated with both TUNEL (*green* fluorescence) and rhodamine-conjugated phalloidin (*orange* fluorescence) assays, revealing that DNA fragmentation and actin cytoskeleton disorganization coexist in the damaged follicles of the exposed insects

disorganization coincide in the affected egg chambers. Since the actin cytoskeleton damage is a known sign of cell death, this result shows that the affected cells most usually do not survive, and are led to apoptosis (Chavdoula et al. 2010).

Figure 1.10a shows the actin cytoskeleton of a stage 10 egg chamber of a sham-exposed insect with normal morphology. Specific features such as the ring channels can be clearly seen. Figure 1.10b shows a stage 10 egg chamber of an exposed insect with damaged actin cytoskeleton. Figure 1.10c shows that the actin cytoskeleton damage accompanies the DNA damage in the same cells (Chavdoula et al. 2010).

1.2.7.6 The Ovarian Development in the Exposed Females Was Significantly Decreased

Due to the DNA and actin cytoskeleton damage induced in the egg chamber cells by the GSM EMFs and the consequent cell death in the affected cells, a large percentage of the developing egg chambers in the ovaries of exposed females were eliminated. As a result, the ovarian size of the exposed females was significantly decreased compared to the ovarian size of the unexposed insects. This we showed with just eclosed virgin female adult insects which we exposed for 6 min 3 h after eclosion, and subsequently every 10 h. Then the ovaries of exposed and sham-exposed insects were dissected and photographed under the same magnification at different developmental stages during the first 2 days (48 h) of their adult lives which is a little more than the time needed for the complete development of the first egg chambers in the ovaries under certain controlled laboratory conditions (~ 45 h). The average ovarian size was compared between exposed and sham-exposed insects according to the photographs. Figure 1.11a, b show ovaries of

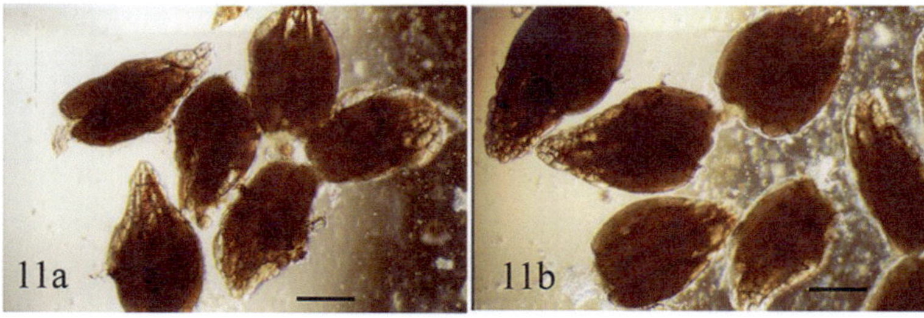

Fig. 1.11 Ovaries of exposed (**a**) and sham-exposed (**b**) female insects 45 h after eclosion. Ovaries of exposed insects are significantly smaller than those of sham-exposed, due to elimination of egg chambers after cell death of their constituent cells induced by the GSM field. Bars: 50 μm

exposed and sham-exposed insects respectively, 45 h after eclosion. The average "descriptive ovarian size" (DOS) of the exposed ovaries was decreased by 29.75% compared to the sham-exposed (Panagopoulos 2012a).

1.2.7.7 Microwave GSM Field Was Found to be Considerably More Bioactive than ELF Magnetic Field or Pulsed Electric Field. The Differential Bioactivity Between Different Types of EMFs Revealed a Differential Sensitivity Between Different Types of Cells, and Between the Two Checkpoints of Oogenesis

The exposure to the GSM field induced more DNA damage on the same biological system than exposure to 50 Hz alternating magnetic field with intensities 1–21 G, or 8 kHz electric field pulsed on 44.4 Hz with intensities 100–400 kV/m, even though the exposures to these fields were of significantly longer daily exposure durations, and of intensities exceeding the environmentally accounted ones (Panagopoulos et al. 2007a, 2013a; Panagopoulos 2016). While the GSM 6-min exposures for a few days induced DNA damage in the ovarian cells up to +55% compared to the sham-exposed cells, the 50 Hz alternating magnetic field induced DNA damage up to +7.52% with the strongest intensity (21 G), and the pulsed electric field up to +3.87% with the strongest intensity (400 kV/m).

This differential bioactivity of the different types of EMFs, revealed that there is a differential sensitivity between the three types of egg chamber cells (NC-nurse cells, FC-follicle cells, OC-oocyte). It was found that cellular sensitivity to EMFs decreases (or resistivity increases) from the NC to the FC, and from the FC to the OC. In other words it was found that the OC (the single cell in each egg chamber that will give the offspring after fertilization) is the most resistant cell type, and the NC the most vulnerable from the three cell types to EMF-exposure (Panagopoulos 2007a, 2013a; Panagopoulos 2016).

Moreover, it was found that the two checkpoints of oogenesis (germarium, and stage 7–8) present different sensitivities in regards to exposures to different types of

EMFs. With the GSM exposure which was found to be the most bioactive/stressful one, germarium was found to be more sensitive than stage 7–8 (the percentage of TUNEL-positive germaria was higher than the percentage of TUNEL-positive egg chambers of stages 7–8) (Panagopoulos et al. 2007a, 2010). On the contrary, with the 50 Hz magnetic field and with the 8 kHz pulsed electric field there was more DNA damage at the stage 7–8 than at the germaria (Panagopoulos et al. 2013a; Panagopoulos 2016).

1.3 The Key Role of Field-Variability in the Bioactivity of EMFs

1.3.1 Adaptation to EMFs

Living organisms have been constantly exposed throughout evolution to terrestrial static electric and magnetic fields of average intensities ~130 V/m and ~0.5 G respectively. While no adverse health effects are connected with normal exposure to these natural ambient fields, variations in their intensities on the order of 20% during "magnetic storms" or "geomagnetic pulsations" due to changes in solar activity with an average periodicity of about 11 years are connected with increased rates of animal/human health incidents, including nervous and psychic diseases, hypertensive crises, heart attacks, cerebral accidents, and mortality (Presman 1977; Dubrov 1978).

It is clear that living organisms perceive EMFs as environmental stressors (Presman 1977; Goodman et al. 1995; Weisbrot et al. 2003; Panagopoulos 2013). But since man-made EMFs constitute a very new stressor for living organisms within the billions of years of biological evolution, the cells have not developed defensive mechanisms, e.g. special genes to be activated for protection against electromagnetic stress of man-made EMFs. This is probably the reason why in response to man-made EMFs, cells are found to activate heat-shock genes and produce heat-shock proteins very rapidly (within minutes) and at a much higher rate than for heat itself (Weisbrot et al. 2003). It seems to be for the same reason that mobile phone radiation is found to induce DNA damage and cell death in insect reproductive cells at a higher degree than other types of external stressors examined before like food deprivation or chemicals, (Nezis et al. 2000, 2002; Drummond-Barbosa and Spradling 2001; Panagopoulos et al. 2007a). Thus it appears that cells are much more sensitive to man-made EMFs than to other types of stress previously experienced by living organisms in the course of evolution such as heat, cold, starvation, or chemicals.

One reason for the increased biological activity of man-made EMFs can be that cells/organisms adapt more easily to any external stressor - and to EMFs - when this stressor is not of significantly varying type, in other words when its parameters are kept constant or vary only slightly. Since living organisms do not have defense mechanisms against variations on the order of 20% of natural EMFs as reported, it is realistic to expect that they do not have innate defenses against unnatural (man-made) EMFs, which are mostly not static but varying (alternating, pulsed, modulated fields, including simultaneously several different frequencies, etc.) and totally polarized in contrast to natural EMFs.

Indeed, RF signals pulsed or modulated by ELF are found in numerous studies since the mid-seventies to be more bioactive than continuous RF signals of identical other parameters (intensity, frequency, duration, waveform, etc) (Bawin et al. 1975, 1978; Bawin and Adey 1976; Blackman et al. 1980; Lin-Liu and Adey 1982; Somosy et al. 1991; Veyret et al. 1991; Bolshakov and Alekseev 1992; Thuroczy et al. 1994; Penafiel et al. 1997; Höytö et al. 2008; Franzellitti et al. 2010; Campisi et al. 2010). Moreover, intermittent exposure to mobile phone radiation with short intermittence durations (which makes the field even more variable) is repeatedly found to be more bioactive than continuous exposure both with simulated and real fields (Diem et al. 2005; Chavdoula et al. 2010). This experimental evidence further supports the argument that the more complicated and variable the field/stressor is, the more difficult it is for a living organism to adapt to it.

1.3.2 Increased Variability of Mobile Telephony EMFs

All types of digital mobile telephony radiation, except for their RF carrier signal, employ ELF fields necessary for the modulation and for increasing the capacity of transmitted information by pulsing the signal. The combination of the RF carrier and the ELF pulsing and modulation frequencies has been found to be more bioactive than the RF carrier alone (Lin-Liu and Adey 1982; Veyret et al. 1991; Penafiel et al. 1997). Moreover, according to the "Ion Forced-Oscillation Mechanism" (Panagopoulos et al. 2000a, 2002, 2015b), (a) the ELF frequencies included in any pulsed or modulated RF signal, are those more responsible for the biological effects, (b) changes in field intensity play a major role, and (c) the pulsing of the signal makes it twice more bioactive.

A constant carrier RF wave modulated by a constant ELF field can certainly be simulated but this is not the case in real mobile telephony signals, in which both the carrier and the modulation are constantly and unpredictably varying in intensity, frequency, and waveform during a phone-conversation (Tisal 1998; Hillebrand 2002; Curwen and Whalley 2008; Holma and Toskala 2004; Panagopoulos 2011).

The intensity of radiation varies significantly each moment during a usual phone-conversation depending on signal reception, number of subscribers sharing the frequency band each moment, air conductivity, location within the wireless infrastructure, presence of objects and metallic surfaces, "speaking" versus "non-speaking" mode, etc. These variations are much larger than 20% of the average signal intensity (as opposed to the periodical variations in the terrestrial fields known to cause health effects). Moreover the phase of the carrier signal varies continuously during a phone-conversation, and the RF frequency constantly changes between different available frequency channels, especially in third generation (3G) radiation. The wave shape is also constantly changing depending on how the changing information transmitted each moment modulates the carrier wave. Thus, the parameters of this radiation change constantly and unpredictably each moment and large, sudden, unpredictable variations in the emitted EMFs/radiation take place constantly during a usual phone-conversation. The more the amount of carried information is increased (by adding text, speech, pictures, music, video, internet, etc) in more recent phone generations (G)/types (2G, 3G, 4G, etc), the more complicated and unpredictably varying the cell phone signals become (Tisal 1998; Hillebrand 2002; Curwen and Whalley 2008; Holma and Toskala 2004; Panagopoulos 2011).

Thus, real digital mobile phone - and other wireless communication devices - emissions change constantly and unpredictably. As a consequence, living organisms cannot adapt to such a highly varying type of stress. Moreover, due to the unpredictably varying type of the real emissions, it is impossible to simulate them by EMFs of fixed parameters.

1.3.3 Real Exposure Studies as Opposed to Studies with Simulated Exposures

A significant number of studies which employed commercially available mobile phones for exposure to a wide variety of animals (including humans)/biological samples have already been published. These include human volunteers *in vivo* (Ferreri et al. 2006; Vecchio et al. 2007, 2010, 2012; Yadav and Sharma 2008; D'Costa et al. 2003; Cam and Seyhan 2012; Luo et al. 2013; Mandala et al. 2014; Movvahedi et al. 2014), human sperm *in vitro* (Agarwal et al. 2009; Gorpinchenko et al. 2014), mice or rats or guinea-pigs or rabbits *in vivo* (Irmak et al. 2002; Dasdag et al. 2003; Ilhan et al. 2004; Ferreira et al. 2006; Elhag et al. 2007; Yan et al. 2007; Meral et al. 2007, 2014; Balci et al. 2007; Mailankot et al. 2009; Gul et al. 2009; Imge et al. 2010; Aldad et al. 2013; Al-Damegh 2012; Koca et al. 2013; Meo and Rubeaan 2013; Motawi et al. 2014), *Drosophila* (Weisbrot et al. 2003; Panagopoulos et al. 2004, 2007a, b, 2010; Panagopoulos and Margaritis 2010a, b; Panagopoulos 2012a; Margaritis et al. 2014; Chavdoula et al. 2010), bees (Sharma and Kumar 2010; Kumar et al. 2011; Favre 2011), ants (Cammaerts and Johansson 2014), chicken eggs (Batelier et al. 2008; Ingole and Ghosh 2012), quails (Tsybulin et al. 2013), mouse cells *in vitro* (Liu et al. 2013), human breast cancer cells *in vitro* (Cig and Naziroglu 2015), protozoa (Cammaerts et al. 2011), and even purified proteins *in vitro* (Barteri et al. 2005). An impressive percentage (96%) of these studies (48 out of 50 studies with real-life exposures) have recorded significant adverse biological or clinical effects, ranging from loss of orientation, kinetic changes, behavioral, or electroencephalographic (EEG) changes, to decrease in male and female reproductive capacity, reproductive declines, molecular changes,

changes in enzymatic activity, DNA damage and cell death, protein damage, and histopathological changes in the brain. From the remaining two studies, one reported no effect (Dasdag et al. 2003) and one reported an increase in short-term memory of children (Movvahedi et al. 2014) which we did not count as an adverse effect although it may be.

On the contrary, about 50% of the studies performed with simulated signals have reported no effects (IARC 2013; Health Protection Agency 2012; Verschaeve et al. 2010; Bourthoumieu et al. 2010; Waldmann et al. 2013; Shahin-Jafari et al. 2016), even though several recent review studies suggest an overall predominance of studies showing effects regardless of real or simulated exposures (Verschaeve 2009; Vignera et al. 2012; Cucurachi et al. 2013; Balmori 2014; Manna and Ghosh 2016). A meta-analysis of 88 studies published during 1990–2011 investigating genetic damage in human cells from RF radiation, 87 of which did not employ real telecommunication EMFs, reported no overall association with genotoxicity (Vijayalaxmi 2012).

Although we may have missed a few more studies with real mobile phone exposures, it becomes evident that there is a strong conflict between the overall results of studies performed with real mobile phone exposures and the overall results of studies with simulated exposures from generators and "test" phones. Moreover, while within the group of studies with simulated exposures there is also a conflict between studies that find effects and studies that do not, the group of studies with real exposures demonstrates an impressive consistency in showing effects almost at 100%. This impressive consistency is corroborated by increasing epidemiological evidence - especially during the last years - for an association between (real-life) mobile phone use and brain tumors (Kundi 2004; Khurana et al. 2009; Hardell et al. 2007, 2013), by statistical studies reporting symptoms of unwellness among people residing around mobile telephony base station antennas or among mobile phone users (Navarro et al. 2003; Salama and Abou El Naga 2004; Hutter et al. 2006; Blettner et al. 2009; Kundi and Hutter 2009; Viel et al. 2009; Singh et al. 2016), as well as by open field studies reporting declines in amphibian and bird populations around mobile telephony base station antennas (Balmori 2005, 2010; Balmori and Hallberg 2007; Everaert and Bauwens 2007; Bhattacharya and Roy 2014).

Although in most studies with real mobile phone exposures the biological samples were exposed in close proximity (within the near-field up to approximately 5 cm) with the mobile phone handset, in several studies the samples/animals were exposed at greater distances in the far-field up to 1 m (Ilhan et al. 2004; Ferreira et al. 2006; Yan et al. 2007; Balci et al. 2007; Batellier et al. 2008; Panagopoulos et al. 2010; Panagopoulos and Margaritis 2010a, b; Vecchio et al. 2010) where the intensity variations are much smaller and the dosimetry absolutely "reliable" as is generally accepted for far-field antenna measurements (Slater 1991). In one of these studies it was found that at 20–30 cm distance from the mobile phone the biological effect (DNA damage) was even more intense than at zero distance (Panagopoulos et al. 2010). [A mobile phone antenna's near-field extends to a distance of 5.2 or 2.6 cm, for 900 or 1800 MHz respectively (most commonly employed carrier frequencies in 2G mobile telephony radiation), according to the relation $r = \lambda/2\pi$, (r the distance of near-field far limit from the antenna when the length of the antenna is smaller than the wavelength λ of the emitted radiation (WHO 1993)].

In studies with real mobile phone exposures investigating the dependence of observed effects on dose (radiation intensity and/or exposure duration) the effects were found to be dose-dependent (Panagopoulos et al. 2004, 2007a, b, 2010; Panagopoulos and Margaritis 2010a, b; Batellier et al. 2008; Tsybulin et al. 2013; Aldad et al. 2013). The dependence on dose was in most cases non-linear, although in two studies the dependence of certain effects on exposure duration was approximating linearity (Panagopoulos and Margaritis 2010a; Aldad et al. 2013).

Certainly the results of experiments with real-life (variable) mobile phone EMFs cannot be identically reproducible, since between successive exposures at any specific location, the exact characteristics of the emitted signal are always different. But the average field values over a few minutes (or more) period, are close to each other, and thus the results of different replicate experiments with real emissions as the independent variable, although not identical quantitatively, are qualitatively similar. Statistical significance in the results can be increased by increasing the number of experimental replications while keeping rigorous control of all other parameters (such as animal/sample conditions, temperature, humidity, light, stray EMFs within the lab, etc). Then, as the number of replications increases, the variability of the field becomes less significant (Maber 1999). Moreover, in order to have a measure of the emitted EMFs variability, RF and ELF measurements of average intensity $\pm$ SD of the emitted real EMFs should be included in the studies, in addition to the Specific Absorption Rate (*SAR*) information supplied by the manufacturer (referring to a simulated human head (Gandhi et al. 2012)). With increasing number of measurements the SD decreases enough for the dosimetry to be judged as reliable (Maber 1999; Panagopoulos et al. 2015a).

1.4 The Key Role of Polarization in the Bioactivity of EMFs

We shall show now why man-made EMFs are actually very different and much more bioactive than natural EMFs, due to the fact that the first are always totally polarized, a property that only partially exists in natural EMFs.

1.4.1 Man-Made EMR Is More Bioactive Than Natural Non-ionizing EMR

Many studies during the past few decades have indicated a variety of adverse biological effects to be triggered by exposure to man-made EMFs, especially RF, and ELF. The recorded biological effects - many of which already described in the previous pages - range from alterations in the synthesis rates and intracellular concentrations of different biomolecules, to DNA and protein damage which may result in cell death, reproductive declines, or even cancer (Goodman et al. 1995; Phillips et al. 2009; Blackman 2009; Johansson 2009; Khurana et al. 2009; Panagopoulos 2011; Panagopoulos et al. 2013a). Under the weight of this evidence the International Agency for Research on Cancer (IARC) has classified both ELF magnetic fields and RF EMFs as possibly carcinogenic to humans (IARC 2002, 2013). The intensities of radiation and durations of exposure in the majority these studies were significantly smaller than those of corresponding exposures from natural EMFs in the terrestrial environment. Moreover, the field intensities applied in the studies were several orders of magnitude smaller than the physiological fields in cell membranes, or fields generated by nerve and muscle excitations (Alberts et al. 1994; Stryer 1996).

Solar EMR intensity incident upon a human body ranges normally between 8 and 24 mW/cm^2 (depending on season, atmospheric conditions, geographical location, etc) while corresponding intensity from a digital mobile phone handset upon a human head (even in contact with the ear) during "talk" emission is normally less than 0.2 mW/cm^2 (Roller and Goldman 1968; Parsons 1993; Panagopoulos et al. 2010). Similarly, terrestrial electric and magnetic fields, or infrared radiation from every human body at normal temperature, have significantly larger incident intensities and exposure durations on any human than most artificial EMF sources (Presman 1977; Dubrov 1978; Gulyaev et al. 1995). Why is then the first beneficial while the latter seem to be detrimental?

Below we shall explain theoretically that the increased adverse biological action of man-made EMFs is due to the fact that they are polarized in contrast to the natural ones. First we must provide some definitions and equations on polarization, field intensity, wave intensity, and superposition/interference of EMFs/EMR which will be necessary in order to explain our reasoning.

1.4.2 *Man-Made EMR Is Polarized, While Natural EMR Is Not*

A field/wave is called linearly polarized when it oscillates on a certain plane which is called the "polarization plane". A combination of linearly polarized fields/waves can give circularly or elliptically polarized fields/waves. Circularly and elliptically polarized 50–60 Hz electric and magnetic fields produced by 3-phase electric power transmission lines are accused for association with cancer (IARC 2002; Panagopoulos et al. 2013a).

Natural EMR/EMFs (cosmic microwaves, infrared, visible light, ultraviolet, gamma rays) and several forms of artificially triggered electromagnetic emissions (such as from light bulbs with thermal filaments, gas discharge lamps, x-rays, lasers, etc.) are not polarized. They are produced by large numbers of molecular, atomic, or nuclear transitions of random orientation and random phase difference between them (except for the lasers which are coherent). These are de-excitations of molecules, atoms, or atomic nuclei (Beiser 1987). Each photon they consist of oscillates on a distinct random plane, and therefore it has a different polarization. Moreover the different photons are not produced simultaneously but they have random phase differences among them.

In contrast, man-made electromagnetic waves are produced by electromagnetic oscillation circuits ("Thomson" circuits), forcing free electrons to oscillate back and forth along a metal wire (electric circuit). Thus, they are not produced by excitations/de-excitations of molecules, atoms, or nuclei, and because the electronic oscillations take place macroscopically in specific directions/orientations they are polarized (most usually linearly polarized). The plane of polarization is determined by the geometry of the circuit. [Lasers are coherent light emissions, not necessarily polarized, and condensed within a narrow beam with high intensity, but they may also be polarized].

Oscillating polarized EMFs/EMR (in contrast to unpolarized) have the ability to induce coherent forced-oscillations on charged/polar molecules within a medium. In case that the medium is biological tissue, the result is that all charged molecules will be forced to oscillate in phase with the field and on planes parallel to its polarization (Panagopoulos et al. 2000a, 2002, 2015b). Several oscillating electromagnetic fields of the same polarization - such as the fields from different antennas vertically oriented - may also produce constructive interference effects and thus, amplify at certain locations the local field intensity, and the amplitude of oscillation of any charged particle within the medium (and within living tissue) (Sangeetha et al. 2014). At such locations, living tissue becomes more susceptible to the initiation of biological effects.

Only coherent polarized fields/waves of the same polarization and frequency are able to produce standing interference effects (fringes of maximum and minimum intensity) (Arago and Fresnel 1819). When the polarization is fixed (e.g. vertically oriented antennas) but there are differences in coherence and/or frequency between the sources, the interference effects are not standing at fixed locations, but change with time creating transient peaks at changing locations.

Natural light from two or more different sources does not produce interference effects, except under the specific conditions of the Young experiment, where the light from a single source passes through two identical slits which - in turn - become two identical-coherent secondary sources (Pohl 1960; Alonso and Finn 1967).

Unpolarized electromagnetic radiation can become polarized when it passes through anisotropic media, as are certain crystals. In fluids (gases and liquids) the molecules are randomly oriented, and macroscopically are considered isotropic inducing no polarization in the electromagnetic waves transmitted through them. Unpolarized natural light can become partly polarized to a small average degree after diffraction on atmospheric molecules, or reflection on water, mirrors, metallic surfaces, etc. (Alonso and Finn 1967). Thus, living organisms exposed to natural radiation since the beginning of life on Earth, although have been exposed to partially polarized light at a small average degree under certain circumstances

(Chen and Rao 1968; Cronin et al. 2006), have never been exposed to totally polarized radiation as is EMR/EMFs of modern human technology.

1.4.3 Field Intensity and Wave Intensity of Electromagnetic Waves

A plane harmonic electromagnetic wave in the vacuum or the air has electric and magnetic field intensity components, given by the equations:

$$E = E_o \sin(k_w r - \omega t) \tag{1.1}$$

$$B = B_o \sin(k_w r - \omega t) \tag{1.2}$$

E_o, B_o are the amplitudes of electric and magnetic field intensities respectively, r is the distance from the source, t is the time, $\omega = 2\pi\nu = k_w \cdot c$ is the circular frequency of the wave (ν the frequency), k_w ($=2\pi/\lambda$) is the wave number (λ the wavelength), and c the velocity of the wave.

The velocity of the electromagnetic wave (and of any wave), is:

$$c = \lambda \cdot \nu \tag{1.3}$$

The wave intensity $\vec{J}$ (also called "Poynting vector"), is:

$$\vec{J} = \vec{c}\,\varepsilon_o E^2 = c^2\,\varepsilon_o\,\vec{E} \times \vec{B} \tag{1.4}$$

And the average value of its amplitude:

$$J_{ave} = \frac{1}{2} c \varepsilon_o E_o^2 \tag{1.5}$$

Thus, the wave intensity depends upon the square of the electric field intensity.

1.4.4 Superposition of Unpolarized EMR/EMFs

Consider two incoherent, unpolarized electromagnetic rays with electric components E_1, E_2, reaching a certain point P in space at a certain moment t in time. Each ray consists of innumerous elementary plane harmonic waves (e.g. photons) of random but discrete polarization. Let us pick for simplicity two elementary plane harmonic waves, one from each ray. The two vectors $\vec{E}_1$, $\vec{E}_2$ due to the different polarizations oscillate on different planes. Since the two rays are not polarized, the polarizations of their constituent plane harmonic elementary waves vary randomly at point P each moment. The total angle ϕ between the two vectors each moment at point P is determined by the different polarizations, plus the different phases, and varies randomly in time.

The resultant electric field $\vec{E}$ (electric component of the resultant electromagnetic wave) each moment at point P, is given by the equation:

$$E = \sqrt{{E_1}^2 + {E_2}^2 + 2E_1E_2\cos\phi} \tag{1.6}$$

E varies with time due to the temporal variations of E_1, E_2, $\cos\phi$. But the average value of $\cos\phi$ is zero: $\frac{1}{2\pi}\int_0^{2\pi}\cos\phi\, d\phi = 0$, and the averages of E^2, ${E_1}^2$, and ${E_2}^2$ are ${E_o}^2/2$, ${E_{o1}}^2/2$ and ${E_{o2}}^2/2$ respectively (E_o, E_{o1}, E_{o2} the amplitudes of E, E_1, E_2).The average resultant electric field is then:

$$E_{ave} = \sqrt{\frac{1}{2}\left(E_{o1}^2 + E_{o2}^2\right)} \text{ or } E_o^2 = E_{o1}^2 + E_{o2}^2 \quad (= \text{constant})$$

and (according to Eq. 1.5):

$$J_{ave} = J_{1,ave} + J_{2,ave}\,(= \text{constant}) \tag{1.7}$$

Even when the two component waves have the same frequency and phase, due to the randomly changing polarizations, the result is still the same.

Thus, the total time average wave intensity due to the superposition of two (or more) rays consisting of elementary plane harmonic waves of random polarizations (natural EMR/EMFs) is the sum of the two individual average intensities, and it is constant at every point and - macroscopically - there is no local variation in the resultant intensity, i.e. no interference effects.

1.4.5 Wave Intensity Versus Field Intensity of Unpolarized EMR

Although the sum average wave intensity due to superposition of natural unpolarized rays is the sum of individual average intensities each one depending on the square amplitude of individual electric field (Eq. 1.7), the sum electric field intensity from infinite number of individual elementary waves constituting each ray (as e.g. with natural light), is zero:

$$\lim_{n\to\infty}\sum_{i=1}^{n}\vec{E}_i = \vec{E}_1 + \vec{E}_2 + \vec{E}_3 + \ldots + \vec{E}_n = 0 \tag{1.8}$$

Let us explain this in more detail: Consider many photons of natural unpolarized light superposed on each other at a particular point in space. Let us assume for simplicity that these photons have equal amplitudes and are of the same frequency but have different polarizations meaning that their electric vectors have all possible orientations forming angles between each two of them from 0° to 360°. Since all possible orientations have equal probabilities, the superposition of a large number of such equal vectors applied on the same point in space will be the sum of vectors applied on the centre of a sphere with their ends equally distributed around the surface of the sphere. The sum of an infinite number of such vectors (all applied on the same point – centre of the sphere – and with their ends evenly distributed at all points of the sphere surface) tends to be zero.

In other words, at any given location, any moment, the sum electric field of a large number of incident photons of random polarization tends to be null, since the individual vectors are in all possible directions diminishing each other when superimposed (destructive interference of electric vectors). Similarly for the sum magnetic field: $\lim_{n\to\infty}\sum_{i=1}^{n}\vec{B}_i = 0$

Thus, the result of superposition of a large number of incident natural waves is increased wave intensity, but negligible electric and magnetic fields approaching zero with infinite number of individual waves/photons. Since the electric forces on charged particles depend directly not on the wave intensity$\vec{J}$, but on the electric and magnetic field intensities ($\vec{E}$, $\vec{B}$), unpolarized EMFs/EMR cannot induce any net forced-oscillations on any charged particles (e.g. biological molecules). They may only induce heat, i.e. random oscillations in all possible directions due to momentary non-zero field intensities, but this does not result to any net electric or magnetic field, or to any net forced-oscillation of charged molecules. This conclusion is very important for our whole reasoning.

1.4.6 Constructive and Destructive Interference of Polarized Waves/Fields

When two or more waves/fields of the same polarization and frequency are in addition coherent, in other words, when their phase difference at the location of superposition is:

$$\varphi = 2n\pi, (\text{with}\, n = 1, 2, 3, \ldots), \tag{1.9}$$

the result is constructive interference, meaning that the resultant wave has an amplitude (intensity) equal to the sum of amplitudes of the single waves that interfere at the particular location.

When two waves of same polarization have opposite phases at another location, in other words, when their phase difference is:

$$\varphi = (2n + 1)\pi, \tag{1.10}$$

then the result of their superposition is destructive interference, i.e. a wave of the same polarization but with diminished intensity.

The electrical components of two such waves (plane harmonic waves of the same polarization and frequency) reaching a certain location after having run different distances r_1, and r_2 from their two coherent sources, are given by the equations:

$$E_1 = E_{o1} \sin(k_w r_1 - \omega t) \tag{1.11}$$

$$E_2 = E_{o2} \sin(k_w r_2 - \omega t) \tag{1.12}$$

Again, the amplitude E_o of the resultant electric field $\vec{E}$ (electric component of the resultant electromagnetic wave), is:

$$E_o = \sqrt{E_{o1}{}^2 + E_{o2}{}^2 + 2E_{o1}E_{o2} \cos\varphi} \tag{1.13}$$

where $\varphi = \frac{2\pi}{\lambda}(r_1 - r_2)$ depending in this case only upon the difference in the distances run by the two waves, and not upon polarization.

At any location where: $\varphi = 2n\pi$, Eq. 1.13 gives:

$$E_o = \sqrt{E_{o1}^2 + E_{o2}^2 + 2E_{o1}E_{o2}} \quad (= |E_{o1} + E_{o2}|) \tag{1.14}$$

At these locations we have constructive interference.

At any location where: $\varphi = (2n{+}1)\pi$, Eq. 1.13 gives:

$$E_o = \sqrt{E_{o1}^2 + E_{o2}^2 - 2E_{o1}E_{o2}} \quad (= |E_{o1} - E_{o2}|) \tag{1.15}$$

At these locations we have destructive interference.

The intensity of the resultant wave at any location is:

$$\vec{J} = \vec{J}_1 + \vec{J}_2 \tag{1.16}$$

The amplitude of the resultant wave intensity will be, correspondingly:

$$J_o = c\varepsilon_o(E_{o1} + E_{o2})^2 \tag{1.17}$$

(at the locations of constructive interference), and

$$J_o = c\varepsilon_o(E_{o1} - E_{o2})^2 \tag{1.18}$$

(at the locations of destructive interference).

Thus, at the locations of constructive interference, the electric field vectors of the two waves/fields are parallel and in the same direction, and both the resultant field and the resultant wave intensity are maximum (Eqs. 1.14 and 1.17).

For two identical sources $(E_{o1} = E_{o2}) : E_o = 2E_{o1}$ and $J_o = 4c\varepsilon_o E_{o1}^2 = 4J_{o1}$

$$\text{For } N \text{ identical sources} : E_o = NE_{o1} \tag{1.19}$$

$$\text{and } J_o = N^2 J_{o1} \tag{1.20}$$

This is why series of parallel RF/microwave antennas are often used to produce high-intensity beams in certain directions (Alonso and Finn 1967).

At the locations of destructive interference the electric field vectors of the two waves are anti-parallel, and thus, both the resultant field and the resultant wave intensity are minimum (Eqs. 1.15 and 1.18). For identical sources ($E_{o1} = E_{o2}$): $E = 0, J = 0$.

Thus, for N number of polarized coherent electromagnetic sources of the same polarization, frequency, and different intensities, with electric components E_1, E_2,. . ., E_N, it comes that at the locations of constructive interference, the resultant electric field is the sum electric field from all the individual sources (e.g. antennas):

$$E = E_1 + E_2 + E_3 + \ldots + E_N \tag{1.21}$$

The bigger the number of coherent superimposed waves/fields (from the same or different sources), the higher and narrower the peaks (Alonso and Finn 1967). That situation can create very sharp peaks of wave and field intensities at certain locations, not easily detectable by field meters, where any living organism may be exposed to peak electric and magnetic field intensities.

Thus, the difference between superposition of unpolarized and polarized electromagnetic waves/fields, is that while in the first case we have increased average wave intensity but zeroed net fields at any location, in the second case we have increased both wave intensity and fields at certain locations where constructive interference occurs. This difference is of crucial importance for understanding the differences in biological activity between natural and man-made EMFs/non-ionizing EMR.

1.5 Theoretical Explanation of the Effects: "Ion Forced-Oscillation Mechanism" for Polarized EMFs. Why ELF and Pulsed EMFs Are More Bioactive

A review of the whole EMF-bioeffects literature reveals that the most bioactive EMFs are the lower frequency ones, especially the ELF fields (Goodman et al. 1995). Moreover it is shown that pulsed EMFs are more bioactive than continuous fields of same rest characteristics (Goodman et al. 1995; Veyret et al. 1991; Penafiel et al. 1997). The pulse repetition frequency is always a low frequency, most usually ELF. We argue that the reason of the intense bioactivity of modern low-intensity microwave fields is most likely the ELF pulsing and modulation frequencies that they include and not the RF carrier wave itself. Below we shall explain this.

All critical biomolecules are either electrically charged or polar (Alberts et al. 1994). While natural unpolarised EMF/EMR at any intensity cannot induce any specific/coherent oscillation on these molecules, polarized man-made EMFs/EMR will induce a coherent and parallel forced-oscillation on every charged/polar molecule within biological tissue. This is fundamental to our understanding of the biological phenomena. This oscillation will be most evident on the free (mobile) ions which carry a net electric charge and exist in large concentrations in all types of cells or extracellular tissue determining practically all cellular/biological functions (Alberts et al. 1994). Although all molecules oscillate randomly with much higher velocities due to thermal motion, this has no biological effect other than increase in tissue temperature. But a coherent polarized oscillation of even millions of times smaller energy than average thermal molecular energy (Panagopoulos et al. 2013b) can initiate biological effects.

A forced-oscillation of mobile ions, induced by an external polarized EMF, can result in irregular gating of electrosensitive ion channels on the cell membranes. That was described in detail in Panagopoulos et al. (2000a, 2002). According to this theory - the plausibility of which in actual biological conditions was verified by numerical test (Halgamuge and Abeyrathne 2011) - the forced-oscillation of ions in the vicinity of the voltage-sensors of voltage-gated ion channels can exert forces on these sensors equal to or greater than the forces known to physiologically gate these channels. Irregular gating of these channels can potentially disrupt any cell's electrochemical balance and function (Alberts et al. 1994), leading to a variety of biological/health effects including the most detrimental ones, such as DNA damage, cell death, or cancer (Pall 2013, 2015).

Most cation channels (Ca^{+2}, K^{+}, Na^{+}, etc) on the membranes of all animal cells, are voltage-gated, or as they are usually called, "electrosensitive" (Alberts et al. 1994). They interconvert between open and closed state, when the electrostatic force on the electric charges of their voltage sensors due to transmembrane voltage changes, transcends some critical value. The voltage sensors of these channels are four symmetrically arranged, transmembrane, positively charged helical domains, each one designated S4. Changes in the transmembrane potential on the order of 30 mV, are normally required to gate electrosensitive channels (Noda et al. 1986;

Liman et al. 1991). Several ions may interact simultaneously each moment with an S4 domain from a distance on the order of 1 nm, since - except for the single ion that may be passing through the channel pore when the channel is opened - a few more ions are bound close to the pore of the channel at specific ion-binding sites (e.g. three in potassium channels) (Miller 2000).

Consider e.g. four potassium ions at distances on the order of 1 nm from the channel-sensors (S4), and an externally applied oscillating EMF/EMR. The electric (and the magnetic) force on each ion due to any unpolarized field is zero (Eq. 1.8). On the contrary, the force due to a polarized field with an electrical component E, is $F = Ezq_e$. It is shown that for a sinusoidal alternating field $E = E_o \ sin\omega \ t$, the movement equation of a free ion of mass m_i, is (Panagopoulos et al. 2000a, 2002):

$$m_i \frac{d^2r}{dt^2} + \lambda \frac{dr}{dt} + m_i \ \omega_o^2 \ r = E_o \ z \ q_e \sin \omega \ t \tag{1.22}$$

r is the ion displacement due to the forced-oscillation, z is the ion's valence ($z = 1$ for potassium ions), $q_e = 1.6 \times 10^{-19}$C the elementary charge, λ the damping coefficient for the ion's displacement (calculated to have a value within a channel $\lambda \cong 6.4 \times 10^{-12}$Kg/s), $\omega_o = 2\pi\nu_o$ (ν_o the ion's oscillation self-frequency taken equal to the ion's recorded spontaneous intracellular oscillation frequency on the order of 0.1 Hz), $\omega = 2\pi\nu$ (ν the frequency of the field/radiation), and E_o the amplitude of the applied electric field.

The general solution of Eq. 1.22, is:

$$r = \frac{E_o z q_e}{\lambda\omega} \cos \omega t + \frac{E_o z q_e}{\lambda\omega} \tag{1.23}$$

The term $\frac{E_o z q_e}{\lambda\omega}$ in the solution, represents a constant displacement, but has no effect on the oscillating term $\frac{E_o z q_e}{\lambda\omega} \ cos\omega \ t$. This constant displacement doubles the amplitude $\frac{E_o z q_e}{\lambda\omega}$ of the forced-oscillation at the moment when the field is applied or interrupted, or during its first and last periods, and the ion's displacement will be twice the amplitude of the forced-oscillation. For pulsed fields (such as most fields of modern digital telecommunications) this will be taking place constantly with every repeated pulse. Thus, pulsed fields are - theoretically - twice more drastic than continuous/non-interrupted fields of the same other parameters, in agreement with several experimental data (Goodman et al. 1995; Veyret et al. 1991; Penafiel et al. 1997).

The amplitude of the forced-oscillation (ignoring the constant term in Eq. 1.23), is:

$$A = \frac{E_o z q_e}{\lambda\omega} \tag{1.24}$$

The force acting on the effective charge q of an S4 domain, via an oscillating single-valence free cation, is: $F = \frac{1}{4\pi\varepsilon\varepsilon_o} \cdot \frac{q \cdot q_e}{r^2}$, ($r$ is the distance of the free ion from the effective charge of S4). Each oscillating cation displaced by dr, induces a force on each S4 sensor:

$$dF = -\frac{q \cdot q_e}{2\pi\varepsilon\varepsilon_o r^3} dr \tag{1.25}$$

While in the case of a non-polarized applied field $\sum d\vec{r} = 0$, and $\sum d\vec{F} = 0$, in the case of a polarized applied field, the sum force on the channel sensor from all four cations, is: $4dF = -2\frac{q \cdot q_e}{\pi\varepsilon\varepsilon_o r^3} dr$.

This is an even more crucial difference between polarized and unpolarized EMFs in regard to biological activity than the ability of constructive interference.

The effective charge of each S4 domain is found to be: $q = 1.7\, q_e$ (Liman et al. 1991). The minimum force on this charge required normally to gate the channel – equal to the force generated by a change of 30 mV in the membrane potential (Liman et al. 1991) – is calculated (Panagopoulos et al. 2000a) to be: $dF = 8.16 \times 10^{-13}$ N.

The displacement of one single-valence cation within the channel, necessary to exert this minimum force is calculated from Eq. 1.25 to be:

$$dr = 4 \times 10^{-12}\text{m}$$

For 4 cations oscillating in phase and on parallel planes due to an external polarized field/radiation, the minimum displacement is decreased to: $dr = 10^{-12}$ m.

Therefore, any external polarized oscillating EMF able to force free ions to oscillate with amplitude $\frac{E_o z q_e}{\lambda\omega} \geq 10^{-12}$m, is able to irregularly gate cation channels on cell membranes. For $z = 1$ (potassium ions), and substituting the values for q_e, λ on the last condition, we get:

$$E_o \geq 0.25\nu \times 10^{-3} \tag{1.26}$$

(ν in Hz, E_o in V/m)

For double-valence cations ($z = 2$) (e.g. Ca^{+2}) the condition becomes,

$$E_o \geq \nu \times 10^{-4} \tag{1.27}$$

(ν in Hz, E_o in V/m)

For pulsed fields the second part of Condition 27 is divided by 2, and becomes:

$$E_o \geq 0.5\nu \times 10^{-4} \tag{1.28}$$

(ν in Hz, E_o in V/m)

For digital mobile telephony fields/radiation emitting ELF pulses with a pulse repetition frequency $\nu = 217$ Hz (among other ELF frequencies they transmit) (Tisal 1998), Condition 28 becomes:

$$E_o \geq 0.01\,\text{V/m} \tag{1.29}$$

For the pulse repetition frequency of $\nu = 8.34$ Hz (also included in mobile telephony signals) (Tisal 1998; Hyland 2000; Tuor et al. 2005), Condition 28 becomes:

$$E_o \geq 0.0004\,\text{V/m} \tag{1.30}$$

Thus, ELF electric fields emitted by mobile phones and base stations stronger than 0.0004 V/m are also potentially able to disrupt the function of any living cell. This ELF intensity value is emitted by regular cell phones at distances up to a few meters and base stations at distances up to a few hundred meters (Tuor et al. 2005; Panagopoulos 2011). For *N* number of mobile telephony antennas vertically oriented, the last value is divided by *N* (according to Eq. 1.19) at locations of constructive interference.

We do not distinguish between externally applied EMFs and internally induced ones within living tissue, especially in the case of ELF for the following reasons: (1) Living tissue is not metal to shield from electric fields and certainly is not ferromagnetic metal (Fe, Co, Ni) to shield from magnetic fields. Moreover, it is known that especially ELF fields cannot be easily shielded even by Faraday cages and in order to significantly minimize them it is recommended to totally enclose them in closed metal boxes (Panagopoulos 2011). Thus, ELF electric fields penetrate living tissue with certain degree of attenuation, and magnetic fields penetrate with zero attenuation. (2) Even in case that the ELF fields are significantly attenuated in the inner tissues of a living body, the eyes, the brain, the skin cells, or the myriads of nerve fiber terminals that end up on the outer epidermis, are directly exposed to the field intensities measured externally on the surface of the living tissue.

It has been shown that tissue preparations (such as bovine fibroblasts or chicken tendons) respond to externally applied pulsed or sinusoidal ELF electric fields (by changes in DNA or protein synthesis rates, proliferation rates, alignment with respect to the field direction, etc), at very low thresholds $\sim 10^{-3}$ V/m (Goodman et al. 1995; McLeod et al. 1987; Cleary et al. 1988; Lee et al. 1993). These thresholds are very close to those predicted above by the described mechanism.

We note that the external field does not gate the channel by forces exerted directly on the channel sensors. It would take a field on the order of the transmembrane field (10^6–10^7 V/m) for that. It is the mediation of the oscillating ions in close proximity to the S4 channel sensors that allows such weak fields to be able to exert the necessary forces to gate the channel.

Except for direct electric field exposure by an external field, there can be an electric field within tissues induced by an externally applied oscillating magnetic one, which - as explained - penetrates living tissue with zero attenuation. Tuor et al. (2005) measured ELF magnetic fields from mobile phones on the order of 1 G (= 10^{-4} T) at 217 Hz. This can induce electric fields on the order of ~0.1 V/m within the human body, as can be shown by application of Maxwell's law of electromagnetic induction:

$$\oint_l \vec{E}_{ind} \cdot d\vec{l} = -\frac{d}{dt} \int_S \vec{B} \cdot \vec{u}_N dS \quad (1.31)$$

($\vec{B}$, $\vec{E}_{ind}$, the magnetic and the induced electric field intensities respectively, $d\vec{l}$ an incremental length along a closed path l of induced electric field circulation. $\vec{u}_N$ is the unit vector vertical to the surface S).

Assuming $\vec{E}_{ind}$ parallel to and independent of l, $\vec{B}$ vertical to and independent of S, and l a circular path of radius α including the surface S, Eq. 1.31 becomes:

$E_{ind} \oint_l dl = -\frac{dB}{dt} \int_S dS$

which gives:

$$E_{ind} = 0.5\alpha\frac{dB}{dt} \quad (1.32)$$

(E_{ind} in V/m, B in T, α in m.).

By replacing in the last equation $\alpha = 0.20$ m (a reasonably large radius for a circumference within an adult human body), and $\frac{dB}{dt} = 1\,\mathrm{T/s}$, [according to Tuor et al. (2005)], we get $E_{ind} \sim 0.1$ V/m. This is the induced electric field intensity within a human body by the 217 Hz pulses of mobile telephony, and it is about ten times larger than the minimum estimated value able to initiate biological effects at this frequency according to Condition 29.

Thus, we have shown that - according to the presented "Ion Forced-Oscillation Mechanism" - the EMFs emitted by mobile telephony antennas possess enough energy to disrupt any living cell's electrochemical balance and function, and consequently induce adverse biological/health effects.

1.6 Discussion

Our experiments showed that GSM mobile telephony EMFs induced DNA damage and actin cytoskeleton damage in a high degree on female insect ovarian cells. This in combination with a corresponding effect on the male gametes (sperm cells) is what causes the impressive decrease in reproduction recorded since our first experiments (Panagopoulos et al. 2000b, 2004). The affected cells are most usually led to apoptosis, as we showed that the induced DNA damage is accompanied by actin cytoskeleton disorganization which is a known aspect of cellular death. But in cases that this will not occur, cancer induction (in the case of somatic cells), or inherited mutations transferred to the next generations (in the case of the oocyte or the sperm cells), may take place. Therefore our studies have shown that this type of microwave radiation is highly bioactive and dangerous for health, and a cautious use of mobile phones should necessarily be adopted, as well as a strict governmental policy regarding the positioning of the base station antennas. Moreover, our experimental results explain biologically the increased cancer rates reported by recent epidemiological studies among long-term mobile phone users (Khurana et al. 2009; Hardell et al. 2007, 2013).

The effects found in our experiments (DNA damage, actin cytoskeleton damage, reproductive decrease, decrease in ovarian development) are also found by other investigators in a variety of other animal (including human) cells. For example the decrease in ovarian development was also found for female rats exposed to GSM radiation (Gul et al. 2009). The large decrease in reproduction was also found for bees (Sharma and Kumar 2010), and this explains the disappearing of bees reported during the past ~15 years especially in Europe and North America (Hamzelou 2007; Cucurachi et al. 2013). Corresponding effects were found for birds (Balmori 2005), amphibia (Balmori 2010), and decrease in human male fertility (Agarwal et al. 2008, 2009). This unique similarity of reproductive/fertility declines in different organisms found by different research groups is explained by the observed cell death induction in reproductive cells due to DNA damage found in our experiments (Panagopoulos et al. 2007b, 2010; Panagopoulos 2012a), and reported also for human and rat sperm cells (De Iuliis et al. 2009; Yan et al. 2007; Mailankot et al. 2009). It is evident that such a similarity of effects cannot be due to randomness.

Moreover, this similarity of effects in different organisms is not surprising, since cells are essentially the same in all animals (including humans). Both insect and mammalian (including human) cells (as well as cells from all other animals) have identically constructed membranes and membrane channels, identical intracellular organelles (nuclei, mitochondria, ribosomes, endoplasmic reticulum, Golgi apparatus, etc), identical free/mobile ions initiating practically every cell function, identically constructed biomolecules such as DNA, RNA, proteins, etc. (Alberts et al. 1994; Stryer 1996).

In the present chapter we have shown that polarized EMFs/EMR, such as every type of man-made EMF, have the ability to create interference effects and amplify their field intensities at specific locations where constructive interference occurs, and that this phenomenon cannot occur with natural EMFs/EMR which are not polarized. Locations at equal distances from identical sources (antennas), in other words locations along the midperpendicular to the distance d between the two sources, are locations of constructive interference and increased field and wave intensities. As the number of sources (e.g. antennas) increases, the amplification of the resultant field intensities (E, B) at certain locations increases as well (Eq. 1.19), and for a large number of sources, field intensities may become very sharp. This explains the detected "hot spots" from mobile telephony base stations in urban environments (Sangeetha et al. 2014). The result of field superposition at those locations are standing waves (i.e. they do not change with time) when the two or more sources of the same polarization are in addition coherent (i.e. same frequency, same phase difference). Within biological tissue, at those locations of constructive interference we can have increased biological activity. The most usual case is when the multiple incident fields/waves are of the same polarization but not coherent (i.e. different frequencies and/or varying phase differences), as e.g. the waves from all different radio, television, and mobile telephony antennas vertically oriented. Then, the resultant fields/waves are not standing but timely varying, creating momentary constructive interference at unpredictably different locations each moment. This fact may represent an extraordinary ability of man-made/polarized EMFs to trigger biological effects.

By use of the Ion Forced-Oscillation Mechanism (Panagopoulos et al. 2000a, 2002, 2015b) we showed that the resultant force exerted on the S4 sensors of electrosensitive ion channels on cell membranes by several ions forced to oscillate on parallel planes and in phase by an applied polarized EMF (and even more by constructively superimposed fields from several polarized EMF-sources), is able to irregularly gate these channels. The result can then be the disruption of the cell's electrochemical balance, leading to a variety of biological/health effects (Pall 2013, 2015). This is in contrast to the null force exerted by any number of ions oscillating on non-parallel random planes and with different phases from each other due to any number of non-polarized applied EMFs, and in contrast to the null force exerted by the random thermal movement of the same ions (Panagopoulos et al. 2002, 2013b).

Thus, we showed that polarized EMFs/EMR can have increased biological activity, due to: (1) Ability to produce constructive interference effects and amplify their intensities at many locations. (2) Ability to force all charged/polar molecules and especially free ions within and around all living cells to oscillate on parallel planes and in phase with the applied polarized field. These features render man-made EMFs/EMR considerably more bioactive than natural non-ionizing EMFs/EMR. This explains the increasing number of biological effects discovered during the past few decades to be induced by man-made EMFs, in contrast to the absence of effects attributed to natural EMFs at normal intensities in the terrestrial environment which have always been present throughout evolution, although human exposure to the latter ones is normally of significantly higher intensities/energy and longer durations.

This is the reason why polarized microwave radiation of maximum power 1–2 W emitted by a mobile phone can damage DNA and cause adverse health effects (Philips et al. 2009; Blackman 2009; Khurana et al. 2009; Panagopoulos 2011), while non-polarized infrared, visible, and ultraviolet radiation from a 100 W light bulb, or ~400 W infrared and visible EMR from a human body (Gulyaev et al. 1995), cannot. Similarly with solar EMR the intensity of which incident upon a human body (~ 8–24 mW/cm^2) is hundreds of times higher than radiation intensity incident from e.g. a mobile phone on a user's head/body during a usual phone-conversation with the handset in touch with the head (less than 0.2 mW/cm^2), or incident intensities from other RF, ELF sources of human technology (Roller and Goldman 1968; Parsons 1993; Panagopoulos et al. 2010). The total daily duration of human exposure to the sunlight is also much longer than the total daily duration of mobile phone exposure during conversations. Yet, there are no adverse biological effects due to normal/non-excessive exposure to sunlight. On the contrary, it is beneficial and vital/necessary for human/animal health, in contrast to mobile phone radiation. Similarly, there are no adverse biological effects due to exposure (mainly in the infrared and visible bands) from one human body to another (with an incident intensity ~ 20 mW/cm^2) (Gulyaev et al. 1995).

Although all animals on Earth have adapted throughout evolution to exposures to EMFs from the sun and the earth, these fields are non-polarized (even though natural light may become partially polarized in a small average degree due to atmospheric scattering or reflections). Moreover, terrestrial electric and magnetic fields are mainly static, emitting very weak non-polarized ELF radiation due to slight variations in their intensities. However, larger variations on the order of 20% of their normal intensities due to solar activity with a periodicity of about 11 years result in increase of human/animal health incidents (Dubrov 1978). Therefore, living organisms on Earth are adapted to natural (non-polarized or even partially polarized) EMFs since the beginning of life, but not to variations in their normal intensities on the order of 20%, and thus we would not expect them to adapt to unnatural (man-made) and totally polarized EMFs/EMR.

The role of polarization in the ability of EMFs/non-ionizing EMR to induce biological effects, is - up to today - largely underestimated in the EMF-bioeffects literature. Exposure to polarized EMFs may even be beneficial in certain cases of applied static electric or magnetic fields of specified orientation and intensities that enhance the action of endogenous physiological fields within living cells/organisms e.g. during development, wound healing, bone fracture healing etc. (Lee et al. 1993; Panagopoulos 2013).

We should emphasize that the increased bioactivity of man-made EMFs does not necessarily result in observable biological/health effects, since there are adaptive mechanisms operating at cellular-tissue-organism levels in response to ever occurring changes. However, these mechanisms may not always be totally effective, especially when the organism is under additional stress or increased metabolic needs (e.g. sickness, childhood/development, old age, etc.). Then exposure to polarized (man-made) and highly variable EMFs may considerably increase the probability for the initiation of adverse health effects.

In the present chapter we also underscored the need for real-life exposure assessment in experimental studies investigating the biological/health effects of EMFs. We showed that the percentages of positive results differ significantly between studies employing real mobile phone exposures and studies employing simulated exposures, regardless of biological samples or other procedure details. The basic difference between real and simulated mobile telephony EMFs is the inherent significant variability of the first which we believe is the reason for their increased bioactivity.

Any variability in the field and correspondingly in the dosimetry, does not change the fact that people are actually exposed daily for increasing periods to this "highly variable" field in contact with their heads/bodies and at different distances. For this reason it is unrealistic for experimental studies to use simulated mobile phone signals with fixed parameters to expose biological samples. Using non-realistic simulations, especially when real conditions are easily accessible to be studied in the lab with well-controlled other parameters, is in our opinion a serious scientific flaw. Simulated signals with fixed parameters bear little - if any - resemblance to what mobile phone users actually experience, even when they employ combinations of simulated signals (Kuster and Schoenborn 2000; Ndoumbè Mbonjo Mbonjo et al. 2004; Czyz et al. 2004). To investigate the biological/health effects from a widely accessible device exposing daily billions of humans we should not try to simulate its emissions, but simply use the device itself. Especially, we should not try to simulate its real varying emissions with totally unrealistic invariant ones. This is a serious scientific flaw that may lead to totally devious results with enormous adverse consequences for public health. In order for the biological/clinical studies testing the bioactivity of mobile telephony radiation to account for real conditions, exposures should be performed by real EMFs as these are emitted by commercially available mobile phones. The same holds for experiments with other types of EMFs employed in modern telecommunication systems such as DECT phones, Wi-Fi, etc. Simulated emissions may be used complementarily to real-life ones, in order to study e.g. the effects of separate parameters of the real EMFs, but in no way should simulated emissions substitute the real ones.

In Panagopoulos et al. (2015a) we analyzed the need for real-life exposures assessment in experimental studies in detail. One of the co-authors of the study, Dr. Olle Johansson wrote about the study:

> *Any paper moving mankind closer to the final mechanistic understanding of the association between electromagnetic fields and health effects is very important and valuable. I count this paper as one of the very best and most important in my career. To work together with Dr. Panagopoulos and Dr. Carlo is a genuine blessing. [Professor Dr. Olle Johansson, Department of Neuroscience, Karolinska Institute, Stockholm, Sweden]*

In the present chapter we also provided a detailed theoretical explanation for the increased bioactivity of man-made EMFs (emphasizing on mobile telephony EMFs) - based on the irregular gating of membrane ion channels in all living cells - due to their totally polarized nature. One of the main arguments of those who support that low-intensity EMFs of human technology do not justify any biological/health effects has always been the lack of a plausible biophysical mechanism to theoretically explain the reported effects (Adair 1991). We did show that the Ion Forced-Oscillation Mechanism is indeed a plausible biophysical mechanism that explains the reported biological effects, and therefore, "lack of mechanism" is not the case anymore. Dr. Martin Pall wrote about this issue:

> *The argument that has been made by the advocates of the current safety standards is that the low intensity, non-thermal EMFs produce only very weak forces on charged groups, weaker than those due to thermal motions at body temperature. They argue, therefore that any effect would be no more than effects produced all the time spontaneously in the body. That physics argument has been disproven by Panagopoulos and his colleagues when they published two biophysical studies in 2000 and 2002. The problem with that can be seen when one looks at the voltage-gated calcium channels (VGCCs). The VGCCs have a specific number of charged amino acid residues each of which has a role in the opening and closing of the channels. Each of these are pushed by weak forces, acting in the same direction when a change in the electrical potential across the plasma membrane opens the channel. In much the same way, weak forces ACTING IN THE SAME DIRECTION*

produced by the EMFs should be able to open the channel as well. Whereas thermal motions act randomly in three dimensions and will therefore only extremely rarely be all acting in the same direction, the forces produced by these fields like the forces produced by changes in plasma membrane electrical potential, do act coordinately in the same direction and can, therefore open these channels. This was the insight that led Panagopoulos to formulate his "Ion Forced-Oscillation" theory and it is, in my view, a brilliant insight! The whole basis of the heating/thermal/SAR paradigm of action of these fields is entirely based on the claim that "there is no biophysically viable mechanism for the action of these weak non-thermal or micro-thermal fields and that claim was shown by Panagopoulos to be wrong and the empirical evidence shows that Panagopoulos is right. In addition to that there are literally thousands of studies that falsify the heating/thermal/SAR paradigm. This is THE best example I have seen of a clearly strongly supported paradigm shift within the last 50 years." [Dr. Martin L. Pall, Professor Emeritus, Washington State University, Portland, USA]

References

Adair RK (1991) Biological effects on the cellular level of electric field pulses. Health Phys 61 (3):395–399

Agarwal A, Deepinder F, Sharma RK, Ranga G, Li J (2008) Effect of cell phone usage on semen analysis in men attending infertility clinic: an observational study. Fertil Steril 89(1):124–128

Agarwal A, Desai NR, Makker K et al (2009) Effects of radiofrequency electromagnetic waves (RF-EMW) from cellular phones on human ejaculated semen: an in vitro pilot study. Fertil Steril 92(4):1318–1325

Alberts B et al (1994) Molecular biology of the cell. Garland Publishing, Inc., New York

Aldad TS, Gan G, Gao XB, Taylor HS (2013) Fetal radiofrequency radiation exposure from 800-1900 mhz-rated cellular telephones affects neurodevelopment and behavior in mice. Sci Rep 2:312 Erratum in: Sci Rep. 2013; 3:1320

Al-Damegh MA (2012) Rat testicular impairment induced by electromagnetic radiation from a conventional cellular telephone and the protective effects of the antioxidants vitamins C and E. Clinics (Sao Paulo) 67(7):785–792

Alonso M, Finn EJ (1967) Fundamental university physics, vol. 2: fields and waves. Addison-Wesley, Reading

Arago DFJ, Fresnel AJ (1819) On the action of rays of polarized light upon each other. Ann Chim Phys 2:288–304

Balci M, Devrim E, Durak I (2007) Effects of mobile phones on oxidant/antioxidant balance in cornea and lens of rats. Curr Eye Res 32:21–25

Balmori A (2005) Possible effects of electromagnetic fields from phone masts on a population of white stork (*Ciconia ciconia*). Electromagn Biol Med 24:109–119

Balmori A (2010) Mobile phone mast effects on common frog (*Rana temporaria*) tadpoles: the city turned into a laboratory. Electromagn Biol Med 29:31–35

Balmori A (2014) Electrosmog and species conservation. Sci Total Environ 496:314–316 Review

Balmori A, Hallberg O (2007) The urban decline of the house sparrow (*Passer domesticus*): a possible link with electromagnetic radiation. Electromagn Biol Med 26:141–151

Barteri M, Pala A, Rotella S (2005) Structural and kinetic effects of mobile phone microwaves on acetylcholinesterase activity. Biophys Chem 113(3):245–253

Batellier F, Couty I, Picard D, Brillard JP (2008) Effects of exposing chicken eggs to a cell phone in "call" position over the entire incubation period. Theriogenology 69(6):737–745

Bawin SM, Adey WR (1976) Sensitivity of calcium binding in cerebral tissue to weak environmental electric fields oscillating at low frequency. Proc Natl Acad Sci U S A 73:1999–2003

Bawin SM, Kaczmarek LK, Adey WR (1975) Effects of modulated VMF fields, on the central nervous system. Ann N Y Acad Sci 247:74–81

Bawin SM, Adey WR, Sabbot IM (1978) Ionic factors in release of $^{45}Ca^{2+}$ from chick cerebral tissue by electromagnetic fields. Proc Natl Acad Sci U S A 75:6314–6318

Beiser A (1987) Concepts of modern physics. McGraw-Hill, Inc, New York

Bhattacharya R, Roy R (2014) Impact of electromagnetic pollution from mobile phone towers on local birds. Int J Innov Res Sci Eng Technol 3:32–36

Blackman C (2009) Cell phone radiation: evidence from ELF and RF studies supporting more inclusive risk identification and assessment. Pathophysiology 16:205–216

Blackman CF, Benane SG, Elder JA, House D.E, Lampe JA, Faulk JM (1980) Induction of calcium – ion efflux from brain tissue by radiofrequency radiation: effect of sample number and modulation frequency on the power – density window. Bioelectromagnetics 1:35–43

Blettner M, Schlehofer B, Breckenkamp J, Kowall B, Schmiedel S, Reis U, Potthoff P, Schuz J, Berg-Beckhoff G (2009) Mobile phone base stations and adverse health effects: phase 1 of a population-based, cross-sectional study in Germany. Occup Environ Med 66:118–123

Bolshakov MA, Alekseev SI (1992) Bursting responses of Lymnea neurons to microwave radiation. Bioelectromagnetics 13(2):119–129

Bos M, Boerema A (1981) Phenetic distances in the *Drosophila melanogaster* – subgroup species and oviposition-site preference for food components. Genetica 56:175–183

Bourthoumieu S, Joubert V, Marin B, Collin A et al (2010) Cytogenetic studies in human cells exposed in vitro to GSM-900 MHz radiofrequency radiation using R-banded karyotyping. Radiat Res 174(6):712–718

Cam ST, Seyhan N (2012) Single-strand DNA breaks in human hair root cells exposed to mobile phone radiation. Int J Radiat Biol 88(5):420–424

Cammaerts MC, Johansson O (2014) Ants can be used as bio-indicators to reveal biological effects of electromagnetic waves from some wireless apparatus. Electromagn Biol Med 33 (4):282–288

Cammaerts MC, Debeir O, Cammaerts R (2011) Changes in *Paramecium caudatum* (Protozoa) near a switched-on GSM telephone. Electromagn Biol Med 30:57–66

Campisi A, Gulino M, Acquaviva R, Bellia P, Raciti G, Grasso R, Musumeci F, Vanella A, Triglia A (2010) Reactive oxygen species levels and DNA fragmentation on astrocytes in primary culture after acute exposure to low intensity microwave electromagnetic field. Neurosci Lett 473(1):52–55

Chavdoula ED, Panagopoulos DJ, Margaritis LH (2010) Comparison of biological effects between continuous and intermittent exposure to GSM-900 MHz mobile phone radiation. Detection of apoptotic cell death features. Mutat Res 700:51–61

Chen HS, Rao CRN (1968) Polarization of light on reflection by some natural surfaces. Br J Appl Phys 1:1191–1200

Çiğ B, Nazıroğlu M (2015) Investigation of the effects of distance from sources on apoptosis, oxidative stress and cytosolic calcium accumulation via TRPV1 channels induced by mobile phones and Wi-Fi in breast cancer cells. Biochim Biophys Acta 1848(10 Pt B):2756–2765

Cleary SF, Liu LM, Graham R, Diegelmann RF (1988) Modulation of tendon fibroplasia by exogenous electric currents. Bioelectromagnetics 9:183–194

Cronin TW, Warrant EJ, Greiner B (2006) Celestial polarization patterns during twilight. Appl Opt 22:5582–5589

Cucurachi S, Tamis WL, Vijver MG, Peijnenburg WJ, Bolte JF, de Snoo GR (2013) A review of the ecological effects of radiofrequency electromagnetic fields (RF-EMF). Environ Int 51:116–140

Curwen P, Whalley J (2008) Mobile communications in the 21st century. In: Harper AC, Buress RV (eds) Mobile telephones: networks, applications and performance. Nova Science Publishers, New York, pp 29–75

Czyz J, Guan K, Zeng Q, Nikolova T, Meister A, Schönborn F, Schuderer J, Kuster N, Wobus AM (2004) High frequency electromagnetic fields (GSM signals) affect gene expression levels in tumor suppressor p53-deficient embryonic stem cells. Bioelectromagnetics 25(4):296–307

Dasdag S, Zulkuf Akdag M, Aksen F et al (2003) Whole body exposure of rats to microwaves emitted from a cell phone does not affect the testes. Bioelectromagnetics 24:182–188

D'Costa H, Trueman G, Tang L, Abdel-rahman U, Abdel-rahman W, Ong K, Cosic I (2003) Human brain wave activity during exposure to radiofrequency field emissions from mobile phones. Australas Phys Eng Sci Med 26(4):162–167

De Iuliis GN, Newey RJ, King BV, Aitken RJ (2009) Mobile phone radiation induces reactive oxygen species production and DNA damage in human spermatozoa in vitro. PLoS One 4(7): e6446

Diem E, Schwarz C, Adlkofer F, Jahn O, Rudiger H (2005) Non-thermal DNA breakage by mobile-phone radiation (1800 MHz) in human fibroblasts and in transformed GFSH-R17 rat granulosa cells in vitro. Mutat Res 583(2):178–183

Drummond-Barbosa D, Spradling AC (2001) Stem cells and their progeny respond to nutritional changes during *Drosophila oogenesis*. Dev Biol 231:265–278

Dubrov AP (1978) The geomagnetic field and life. Plenum Press, New York

Elhag MA, Nabil GM, AMM A (2007) Effects of electromagnetic field produced by mobile phones on the oxidant and antioxidant status of rats. Pak J Biol Sci 10:4271–4274

Everaert J, Bauwens D (2007) A possible effect of electromagnetic radiation from mobile phone base stations on the number of breeding house sparrows (*Passer domesticus*). Electromagn Biol Med 26:63–72

Favre D (2011) Mobile phone-induced honeybee worker piping. Apidologie 42:270–279

Ferreira AR, Knakievicz T, de Bittencourt Pasquali MA et al (2006) Ultra high frequency-electromagnetic field irradiation during pregnancy leads to an increase in erythrocytes micronuclei incidence in rat offspring. Life Sci 80:8043–8050

Ferreri F, Curcio G, Pasqualetti P, De Gennaro L, Fini R, Rossini PM (2006) Mobile phone emissions and human brain excitability. Ann Neurol 60(2):188–196

Franzellitti S, Valbonesi P, Ciancaglini N, Biondi C, Contin A, Bersani F, Fabbri E (2010) Transient DNA damage induced by high-frequency electromagnetic fields (GSM 1.8 GHz) in the human trophoblast HTR-8/SVneo cell line evaluated with the alkaline comet assay. Mutat Res 683(1–2):35–42

Gandhi OP, Morgan LL, de Salles AA, Han Y-Y, Herberman RB, Davis DL (2012) Exposure limits: the underestimation of absorbed cell phone radiation, especially in children. Electromagn Biol Med 31(1):34–51

Goodman EM, Greenebaum B, Marron MT (1995) Effects of electro- magnetic fields on mollecules and cells. Int Rev Cytol 158:279–338

Gorpinchenko I, Nikitin O, Banyra O, Shulyak A (2014) The influence of direct mobile phone radiation on sperm quality. Cent Eur J Urol 67(1):65–71

Gul A, Celebi H, Uğraş S (2009) The effects of microwave emitted by cellular phones on ovarian follicles in rats. Arch Gynecol Obstet 280(5):729–733

Gulyaev YV, Markov AG, Koreneva LG, Zakharov PV (1995) Dynamical infrared thermography in humans. Eng Med Biol Mag, IEEE 14:766–771

Halgamuge MN, Abeyrathne CD (2011) A study of charged particle's behavior in a biological cell exposed to AC-DC electromagnetic fields. Environ Eng Sci 28:1–10

Hamzelou J (2007) Where have all the bees gone? Lancet 370:639

Hardell L, Carlberg M, Söderqvist F, Mild KH, Morgan LL (2007) Long-term use of cellular phones and brain tumours: increased risk associated with use for > or =10 years. Occup Environ Med 64(9):626–632 Review

Hardell L, Carlberg M, Söderqvist F, Mild KH (2013) Case-control study of the association between malignant brain tumours diagnosed between 2007 and 2009 and mobile and cordless phone use. Int J Oncol 43(6):1833–1845

Health Protection Agency (2012) Health effects from radiofrequency electromagnetic fields. ISBN 978–0–85951-714-0
Hillebrand F (2002) GMS and UMTS. The creation of global mobile communication. Wiley, Chichester, p 580
Holma H, Toskala A (2004) WCDMA for UMTS radio access for third generation mobile communications. Wiley, Hoboken
Höytö A, Luukkonen J, Juutilainen J, Naarala J (2008) Proliferation, oxidative stress and cell death in cells exposed to 872 MHz radiofrequency radiation and oxidants. Radiat Res 170 (2):235–243
Hutter H-P, Moshammer H, Wallner P, Kundi M (2006) Subjective symptoms, sleeping problems, and cognitive performance in subjects living near mobile phone base stations. Occup Environ Med 63:307–313
Hyland GJ (2000) Physics and biology of mobile telephony. Lancet 356:1833–1836
IARC (2002) Non-ionizing radiation, part 1: static and extremely low-frequency (ELF) electric and magnetic fields, vol 80. World Health Organization, Geneva
IARC (2013) Non-ionizing radiation, part 2: radiofrequency electromagnetic fields, vol 102. World Health Organization, Geneva
ICNIRP (1998) Guidelines for limiting exposure to time-varying electric, magnetic and electromagnetic fields (up to 300GHz). Health Phys 74:494–522
Ilhan A, Gurel A, Armutcu F et al (2004) *Ginkgo biloba* prevents mobile phone-induced oxidative stress in rat brain. Clin Chim Acta 340:153–162
Imge EB, Kilicoğlu B, Devrim E et al (2010) Effects of mobile phone use on brain tissue from the rat and a possible protective role of vitamin C – a preliminary study. Int J Radiat Biol 86:1044–1049
Ingole IV, Ghosh SK (2012) Effect of exposure to radio frequency radiation emitted by cell phone on the developing dorsal root ganglion of chick embryo: a light microscopic study. Nepal Med Coll J 14(4):337–341
Irmak MK, Fadillioğlu E, Gulec M et al (2002) Effects of electromagnetic radiation from a cellular telephone on the oxidant and antioxidant levels in rabbits. Cell Biochem Funct 20:279–283
Johansson O (2009) Disturbance of the immune system by electromagnetic fields-a potentially underlying cause for cellular damage and tissue repair reduction which could lead to disease and impairment. Pathophysiology 16:157–177
Karaca E, Durmaz B, Aktug H, Yildiz T, Guducu C, Irgi M, Koksal MG, Ozkinay F, Gunduz C, Cogulu O (2012) The genotoxic effect of radiofrequency waves on mouse brain. J Neuro-Oncol 106(1):53–58
Khurana VG, Teo C, Kundi M, Hardell L, Carlberg M (2009) Cell phones and brain tumors: a review including the long-term epidemiologic data. Surg Neurol 72(3):205–214
King RC (1970) Ovarian development in *Drosophila melanogaster*. Academic Press, New York
Koca O, Gökçe AM, Öztürk MI, Ercan F, Yurdakul N, Karaman MI (2013) Effects of intensive cell phone (Philips Genic 900) use on the rat kidney tissue. Urol J 10(2):886–891
Kumar NR, Sangwan S, Badotra P (2011) Exposure to cell phone radiations produces biochemical changes in worker honey bees. Toxicol Int 18:70–72
Kundi M (2004) Mobile phone use and cancer. Occup Environ Med 61:560–570
Kundi M, Hutter HP (2009) Mobile phone base stations – effects on wellbeing and health. Pathophysiology 16:123–135
Kuster N, Schönborn F (2000) Recommended minimal requirements and development guidelines for exposure setups of bio-experiments addressing the health risk concern of wireless communications. Bioelectromagnetics 21(7):508–514
Lee RC, Canaday DJ, Doong H (1993) A review of the biophysical basis for the clinical application of electric fields in soft-tissue repair. J Burn Care Rehabil 14:319–335
Lerchl A, Klose M, Grote K, Wilhelm AF, Spathmann O, Fiedler T, Streckert J, Hansen V, Clemens M (2015) Tumor promotion by exposure to radiofrequency electromagnetic fields below exposure limits for humans. Biochem Biophys Res Commun 459(4):585–590

Liman ER, Hess P, Weaver F, Koren G (1991) Voltage-sensing residues in the S4 region of a mammalian K^+ channel. Nature 353:752–756
Lin-Liu S, Adey WR (1982) Low frequency amplitude modulated microwave fields change calcium efflux rates from synaptosomes. Bioelectromagnetics 3(3):309–322
Lioliousis C (1979) Microwaves, Athens, (Λιολιούσης Κ, μικροκύματα, Αθήνα 1979)
Liu C, Gao P, Xu SC et al (2013) Mobile phone radiation induces mode-dependent DNA damage in a mouse spermatocyte-derived cell line: a protective role of melatonin. Int J Radiat Biol 89 (11):993–1001
Luo Q, Jiang Y, Jin M, Xu J, Huang HF (2013) Proteomic analysis on the alteration of protein expression in the early-stage placental villous tissue of electromagnetic fields associated with cell phone exposure. Reprod Sci 20(9):1055–1061
Maber J (1999) Data analysis for biomolecular sciences. Longman, Harlow
Mailankot M, Kunnath AP, Jayalekshmi H, Koduru B, Valsalan R (2009) Radio frequency electromagnetic radiation (RF-EMR) from GSM (0.9/1.8GHz) mobile phones induces oxidative stress and reduces sperm motility in rats. Clinics (Sao Paulo) 64(6):561–565
Mandalà M, Colletti V, Sacchetto L et al (2014) Effect of bluetooth headset and mobile phone electromagnetic fields on the human auditory nerve. Laryngoscope 124(1):255–259
Manna D, Ghosh R (2016) Effect of radiofrequency radiation in cultured mammalian cells: a review. Electromagn Biol Med 6:0 [Epub ahead of print]
Margaritis LH, Manta AK, Kokkaliaris KD et al (2014) Drosophila oogenesis as a bio-marker responding to EMF sources. Electromagn Biol Med 33(3):165–189
McCall K (2004) Eggs over easy: cell death in the Drosophila ovary. Dev Biol 274(1):3–14
McLeod KJ, Lee RC, Ehrlich HP (1987) Frequency dependence of electric field modulation of fibroblast protein synthesis. Science 236:1465–1469
Meo SA, Al Rubeaan K (2013) Effects of exposure to electro-magnetic field radiation (EMFR) generated by activated mobile phones on fasting blood glucose. Int J Occup Med Environ Health 26(2):235–241
Meral I, Mert H, Mert N et al (2007) Effects of 900 MHz electromagnetic field emitted from cellular phone on brain oxidative stress and some vitamin levels ofguinea pigs. Brain Res 1169:120–124
Meral I, Tekintangac Y, Demir H (2014) Effects of 900 MHz electromagnetic field emitted by cellular phones on electrocardiograms of guinea pigs. Hum Exp Toxicol 33(2):164–169
Miller C (2000) An overview of the potassium channel family. Genome Biol 1(4)
Morgan LL, Miller AB, Sasco A, Davis DL (2015) Mobile phone radiation causes brain tumors and should be classified as a probable human carcinogen (2A) (review). Int J Oncol 46 (5):1865–1871
Motawi TK, Darwish HA, Moustafa YM, Labib MM (2014) Biochemical modifications and neuronal damage in brain of young and adult rats after long-term exposure to mobile phone radiations. Cell Biochem Biophys 70(2):845–855
Movvahedi MM, Tavakkoli-Golpayegani A, Mortazavi SA, Haghani M, Razi Z, Shojaie-Fard MB, Zare M, Mina E, Mansourabadi L, Nazari-Jahromi SA, Shokrpour N, Mortazavi SM (2014) Does exposure to GSM 900 MHz mobile phone radiation affect short-term memory of elementary school students? J Pediatr Neurosci 9(2):121–124
Navarro A. Enrique, J. Segura, M. Portolés, Claudio Gómez- Perretta de Mateo, (2003): The microwave syndrome: a preliminary study in Spain, Electromagn Biol Med, 22 (2–3), 161–169.
Ndoumbè Mbonjo Mbonjo H, Streckert J, Bitz A, Hansen V, Glasmachers A, Gencol S, Rozic D, German Research Foundation for Radio Waves (2004) Generic UMTS test signal for RF bioelectromagnetic studies. Bioelectromagnetics 25(6):415–425
Nezis IP, Stravopodis DJ, Papassideri I, Robert-Nicoud M, Margaritis LH (2000) Stage-specific apoptotic patterns during Drosophila oogenesis. Eur J Cell Biol 79:610–620

Nezis IP, Stravopodis DJ, Papassideri I, Robert-Nicoud M, Margaritis LH (2002) The dynamics of apoptosis in the ovarian follicle cells during the late stages of *Drosophila* oogenesis. Cell Tissue Res 307:401–409

Noda M, Ikeda T, Kayano T, Suzuki H, Takeshima H, Kurasaki M, Takahashi H, Numa S (1986) Existence of distinct sodium channel messenger RNAs in rat brain. Nature 320:188–192

Pall ML (2013) Electromagnetic fields act via activation of voltage-gated calcium channels to produce beneficial or adverse effects. J Cell Mol Med 17:958–965

Pall ML (2015) Microwave frequency electromagnetic fields (EMFs) produce widespread neuropsychiatric effects including depression. J Chem Neuroanat 2015, pii: S0891–0618(15) 00059–9. doi: 10.1016/j.jchemneu. 2015.08.001

Panagopoulos DJ (2011) Analyzing the health impacts of modern telecommu-nications microwaves. In: Berhardt LV (ed) Advances in medicine and biology, vol 17. Nova Science Publishers, Inc., New York

Panagopoulos DJ (2012a) Effect of microwave exposure on the ovarian development of *Drosophila melanogaster*. Cell Biochem Biophys 63:121–132

Panagopoulos DJ (2012b) Gametogenesis, embryonic and post-embryonic development of *Drosophila melanogaster*, as a model system for the assessment of radiation and environmental genotoxicity. In: Spindler-Barth M (ed) *Drosophila melanogaster*: life cycle, genetics and development. Nova Science Publishers, New York

Panagopoulos DJ (2013) Electromagnetic interaction between environmental fields and living systems determines health and well-being. In: Electromagnetic fields: principles, engineering applications and biophysical effects. Nova Science Publishers, New York

Panagopoulos DJ (2016) Pulsed electric field increases reproduction in *Drosophila melanogaster*. Int J Radiat Biol 92(2):94–106

Panagopoulos DJ, Margaritis LH (2002) Effects of different kinds of EMFs on the offspring production of insects. In: Kostarakis P (ed) 2nd international workshop, biological effects of electromagnetic fields, Proceedings, Rhodes, Greece, 7–11 October 2002, 438–452, ISBN: 960–86733-3-X

Panagopoulos DJ, Margaritis LH (2010a) The effect of exposure duration on the biological activity of mobile telephony radiation. Mutat Res 699(1/2):17–22

Panagopoulos DJ, Margaritis LH (2010b) The identification of an intensity "Window" on the bioeffects of mobile telephony radiation. Int J Radiat Biol 86(5):358–366

Panagopoulos DJ, Messini N, Karabarbounis A, Filippetis AL, Margaritis LH (2000a) A mechanism for action of oscillating electric fields on cells. Biochem Biophys Res Commun 272 (3):634–640

Panagopoulos DJ, Messini N, Karabarbounis A, Philippetis AL, Margaritis LH (2000b) Radio frequency electromagnetic radiation within "safety levels" alters the physiological function of insects. In: Kostarakis P, Stavroulakis P (eds) Millennium international workshop on biological effects of electromagnetic fields, Proceedings, Heraklion, Crete, Greece, 17–20 October, 2000, 169–175, ISBN: 960–86733–0-5

Panagopoulos DJ, Karabarbounis A, Margaritis LH (2002) Mechanism for action of electromagnetic fields on cells. Biochem Biophys Res Commun 298(1):95–102

Panagopoulos DJ, Karabarbounis A, Margaritis LH (2004) Effect of GSM 900-MHz mobile phone radiation on the reproductive capacity of *Drosophila melanogaster*. Electromagn Biol Med 23 (1):29–43

Panagopoulos DJ, Chavdoula ED, Nezis IP, Margaritis LH (2007a) Cell death induced by GSM 900 MHz and DCS 1800 MHz mobile telephony radiation. Mutat Res 626:69–78

Panagopoulos DJ, Chavdoula ED, Karabarbounis A, Margaritis LH (2007b) Comparison of bioactivity between GSM 900 MHz and DCS 1800 MHz mobile telephony radiation. Electromagn Biol Med 26(1):33–44

Panagopoulos DJ, Chavdoula ED, Margaritis LH (2010) Bioeffects of mobile telephony radiation in relation to its intensity or distance from the antenna. Int J Radiat Biol 86(5):345–357

Panagopoulos DJ, Karabarbounis A, Liolioussis C (2013a) ELF alternating magnetic field decreases reproduction by DNA damage induction. Cell Biochem Biophys 67:703–716
Panagopoulos DJ, Johansson O, Carlo GL (2013b) Evaluation of specific absorption rate as a dosimetric quantity for electromagnetic fields bioeffects. PLoS One 8:e62663. doi:10.1371/journal.pone.0062663
Panagopoulos DJ, Johansson O, Carlo GL (2015a) Real versus simulated mobile phone exposures in experimental studies. Biomed Res Int 2015:607053
Panagopoulos DJ, Johansson O, Carlo GL (2015b) Polarization: a key difference between man-made and natural electromagnetic fields, in regard to biological activity. Sci Rep 5:14914. doi:10.1038/srep14914
Parsons KC (1993) Human thermal environments. Taylor and Francis, London
Paul B, Saha I, Kumar S, Samim Ferdows SK, Ghose G (2015) Mobile phones: time to rethink and limit usage. Indian J Public Health 59:37–41
Penafiel LM, Litovitz T, Krause D, Desta A, Mullins MJ (1997) Role of modulation on the effect of microwaves on ornithine decarboxylase activity in L929 cells. Bioelectromagnetics 18 (2):132–141
Phillips JL, Singh NP, Lai H (2009) Electromagnetic fields and DNA damage. Pathophysiology 16:79–88
Pohl R (1960) Discovery of interference by Thomas Young. Am J Phys 28:530
Presman AS (1977) Electromagnetic fields and life. Plenum Press, New York
Roggeveen S, van Os J, Viechtbauer W, Lousberg R (2015) EEG changes due to experimentally induced 3G mobile phone radiation. PLoS One 10(6):e0129496. doi:10.1371/journal.pone.0129496
Roller WL, Goldman RF (1968) Prediction of solar heat load on man. J Appl Physiol 25:717–721
Ros-Llor I, Sanchez-Siles M, Camacho-Alonso F, Lopez-Jornet P (2012) Effect of mobile phones on micronucleus frequency in human exfoliated oral mucosal cells. Oral Dis 18(8):786–792
Salama OE, Abou El Naga RM (2004) Cellular phones: are they detrimental? J Egypt Public Health Assoc 79(3–4):197–223
Sangeetha M, Purushothaman BM, Suresh Babu S (2014) Estimating cell phone signal intensity and identifying radiation hotspot area for Tirunel Veli Taluk using RS and GIS. Int J Res Eng Technol 3:412–418
Shahin-Jafari A, Bayat M, Shahhosseiny MH, Tajik P, Roudbar-Mohammadi S (2016) Effect of long-term exposure to mobile phone radiation on alpha-Int1 gene sequence of *Candida albicans*. Saudi J Biol Sci 23(3):426–433
Sharma VP, Kumar NR (2010) Changes in honeybee behaviour and biology under the influence of cellphone radiations. Curr Sci 98:1376–1378
Singh K, Nagaraj A, Yousuf A, Ganta S, Pareek S, Vishnani P (2016) Effect of electromagnetic radiations from mobile phone base stations on general health and salivary function. J Int Soc Prev Community Dent 6(1):54–59
Slater D (1991) Near-field antenna measurements. Artech House, Inc, Boston
Somosy Z, Thuroczy G, Kubasova T, Kovacs J, Szabo LD (1991) Effects of modulated and continuous microwave irradiation on the morphology and cell surface negative charge of 3 T3 fibroblasts. Scanning Microsc 5(4):1145–1155
Stryer L (1996) Biochemistry, 4th edn. W.H. Freeman and Co, New York
Thuroczy G, Kubinyi G, Bodo M, Bakos J, Szabo LD (1994) Simultaneous response of brain electrical activity (EEG) and cerebral circulation (REG) to microwave exposure in rats. Rev Environ Health 10(2):135–148
Tisal J (1998) GSM cellular radio telephony. Wiley, West Sussex
Tsybulin O, Sidorik E, Brieieva O et al (2013) GSM 900 MHz cellular phone radiation can either stimulate or depress early embryogenesis in Japanese quails depending on the duration of exposure. Int J Radiat Biol 89(9):756–763

Tuor M, Ebert S, Schuderer J, Kuster N (2005) Assessment of ELF exposure from GSM handsets and development of an optimized RF/ELF exposure setup for studies of human volunteers, BAG Reg. No. 2.23.02.-18/02.001778, IT'IS Foundation

Vecchio F, Babiloni C, Ferreri F, Curcio G, Fini R, Del Percio C, Rossini PM (2007) Mobile phone emission modulates interhemispheric functional coupling of EEG alpha rhythms. Eur J Neurosci 25(6):1908–1913

Vecchio F, Babiloni C, Ferreri F, Buffo P, Cibelli G, Curcio G, van Dijkman S, Melgari JM, Giambattistelli F, Rossini PM (2010) Mobile phone emission modulates inter-hemispheric functional coupling of EEG alpha rhythms in elderly compared to young subjects. Clin Neurophysiol 121(2):163–171

Vecchio F, Tombini M, Buffo P, Assenza G, Pellegrino G, Benvenga A, Babiloni C, Rossini PM (2012) Mobile phone emission increases inter-hemispheric functional coupling of electroencephalographic α rhythms in epileptic patients. Int J Psychophysiol 84(2):164–171

Verschaeve L (2009) Genetic damage in subjects exposed to radiofrequency radiation. Mutat Res 681(2–3):259–270 Review

Verschaeve L, Juutilainen J, Lagroye I, Miyakoshi J, Saunders R, de Seze R, Tenforde T, van Rongen E, Veyret B, Xu Z (2010) In vitro and in vivo genotoxicity of radiofrequency fields. Mutat Res 705(3):252–268

Veyret B, Bouthet C, Deschaux P, de Seze R, Geffard M, Joussot-Dubien J, le Diraison M, Moreau JM, Caristan A (1991) Antibody responses of mice exposed to low-power microwaves under combined, pulse-and-amplitude modulation. Bioelectromagnetics 12(1):47–56

Viel JF, Clerc S, Barrera C, Rymzhanova R, Moissonnier M, Hours M, Cardis E (2009) Residential exposure to radiofrequency fields from mobile phone base stations, and broadcast transmitters: a population-based survey with personal meter. Occup Environ Med 66:550–556

Vignera SL, Condorelli RA, Vicardi E, D'Agata R, Calogero AE (2012) Effects of the exposure to mobile phones on male reproduction: a review of the literature. J Androl 33:350–356

Vijayalaxmi PTJ (2012) Genetic damage in human cells exposed to non-ionizing radiofrequency fields: a meta-analysis of the data from 88 publications (1990–2011). Mutat Res 749(1–2):1–16

Waldmann P, Bohnenberger S, Greinert R, Hermann-Then B, Heselich A et al (2013) Influence of GSM signals on human peripheral lymphocytes: study of genotoxicity. Radiat Res 179:243–253

Weisbrot D, Lin H, Ye L, Blank M, Goodman R (2003) Effects of mobile phone radiation on reproduction and development in *Drosophila melanogaster*. J Cell Biochem 89(1):48–55

WHO (1993) Environmental health criteria 137. Electromagnetic fields 300 Hz to 300GHz. World Health Organization, Geneva

Yadav AS, Sharma MK (2008) Increased frequency of micronucleated exfoliated cells among humans exposed in vivo to mobile telephone radiations. Mutat Res 650(2):175–180

Yan JG, Agresti M, Bruce T, Yan YH, Granlund A, Matloub HS (2007) Effects of cellular phone emissions on sperm motility in rats. Fertil Steril 88(4):957–964

Chapter 2
Effect of Microwaves on DNA and Proteins

Takeo Yoshimura, Jun-ichi Sugiyama, Shigeru Mineki, and Shokichi Ohuchi

Abstract The Rolling Circle Amplification (RCA) is an enzymatic synthesis method of deoxyribonucleic acid (DNA) strands with repeated sequence of a circulate template-DNA. Microwave heating technology is applied to organic and inorganic chemistry to produce useful effects such as rapid heating, selective heating, decreased reaction times, and improved product yield. However, the mechanisms underlying the thermal or non-thermal effects, i.e., reaction-promoting effects, of MW chemistry are still unclear. We focus on the molecular perspective to investigate the effect of MWs via a DNA amplification reaction that utilizes enzymes. The microwave reaction temperature was strictly controlled using a microwave applicator optimized for enzymatic-scale reactions. We show that microwave heating facilitated the synthesis of repetitive DNA through RCA using the four DNA polymerases. Analysis of the temperature profiles of each RCA component subjected to microwave heating revealed the selectivity heating of buffer components compared with primers, template DNA, dNTP, and RNase-free water. Further, we showed the relation between the microwave power loss and RCA components via dielectric measurements, cavity resonator feature measurement, and electromagnetic simulation. The buffer containing ions of the RCA components was selectively heated via microwave irradiation in the TM_{010} cavity resonator.

T. Yoshimura (✉)
Department of Chemical Science and Engineering, School of Materials and Chemical Technology, Tokyo Institute of Technology, 2-12-1 Ookayama, Meguro, Tokyo, Japan
e-mail: takeo.yoshimura@gmail.com

J.-i. Sugiyama
National Institute of Advanced Industrial Science and Technology (AIST), Tsukuba Central 5, 1-1-1 Higashi, Tsukuba, Ibaraki 305-8565, Japan

S. Mineki
Department of Applied Biological Science, Tokyo University of Science, 2641 Yamazaki, Noda-shi, Chiba 278-8510, Japan

S. Ohuchi
Graduate School of Life Science and System Engineering, Kyushu Institute of Technology, 680-4 Kawazu, Iizuka, Fukuoka 820-8502, Japan

C.D. Geddes (ed.), *Microwave Effects on DNA and Proteins*,
DOI 10.1007/978-3-319-50289-2_2

Keywords DNA polymerization • Rolling circle amplification • Dielectric-loss • Conductive-loss • Resonator

2.1 Introduction

Do microwaves (MWs) have any special influence on DNA and proteins? MWs represent a form of electromagnetic radiation with wavelengths between 1 m and 1 cm (30 GHz–300 MHz). MWs are used for multiple purposes, ranging from communication (e.g., smartphones) to heating food (e.g., MW ovens). The question of whether mobile phones have an influence on the brain has been a subject of extensive research. In this chapter, rather than focusing on the macro perspective of the effect of MWs on DNA and proteins with respect to the human body and cells, we focus on the molecular perspective to investigate the effect of MWs via a DNA amplification reaction that utilizes enzymes. If MW heating could be used to promote or induce new reactions via mechanisms other than those used for conventional heating, this would be useful as a new heating approach for proteins, enzymes, and DNA and could be expected to find practical applications in the fields of engineering and medicine and in cell and organism stimulation.

The first report of the promotion of a chemical reaction using MW heating was published in 1986 (Giguere et al. 1986; Gedye et al. 1986). Since then, there have been multiple reports on the utility of MW heating in both inorganic and organic chemistry, including the ability to increase yield, promote reactions, and generate products different from those generated by conventional heating (Chen et al. 2014; Bilecka and Niederberger 2010; Lidström et al. 2001; Kappe and Dallinger 2009). The thermal energy of MW heating is the sum of a molecule's dielectric loss, magnetic loss heating, and conduction loss heating(de la Hoz and Loupy 2013; Grant and Halstead 1998). In comparison with conventional heating, MW heating is characteristically rapid and selective and facilitates the superheating of certain materials. Nearly everyone has experienced the rapid heating utility of MW ovens. However, the mechanisms underlying the thermal or non-thermal effects, i.e., reaction-promoting effects, of MW chemistry are still unclear.

With respect to this dispute, Kappe revealed the problem with temperature measurements in MW environments (Obermayer and Kappe 2010). Traditional alcohol thermometers and thermocouples cannot be used under MW exposure because they are directly influenced by MWs, and consideration for the location of measurements is essential even for radiation and fiber-optic thermometers because of the presence of hot spots. In other works, Kappe suggested that the results that are considered to be a consequence of reaction promotion by MWs were actually inadequate owing to MW-heating problems (Hayden et al. 2013). Moreover, thermometers are not the only issue with MW chemistry. In MW-heating experiments, attention toward the container material, stirring methods, penetration depth, and reaction scale is necessary.

In addition to the aforementioned important points, the user requires an understanding of the MW devices. The devices used in MW chemistry are a combination of two types of cavities (multimode or single mode) and oscillators (magnetron or semiconductor oscillator). Users must understand the characteristics of both and use them in accordance with their purpose. Often when researchers first engage in MW heating, they believe that turning on the MW device causes oscillations at the frequency indicated in the device manual and the displayed electrical power is imparted to the heated object. If possible, one should measure the oscillated frequency and the output power of the MW device using a network analyzer and a power meter. With the exception of simply using MWs as a heat source for decomposition reactions, it is possible that replication of the result would be impossible if a different device is used.

Precisely conducting MW-heating experiments is particularly important when handling enzymatic reactions because enzymes are proteins that have an irreversible deactivation temperature at which they lose their function as well as an optimum temperature that is most suited to their reaction. With the exception of experiments that aim to decompose biomass by heating, experiments conducted at the solvent boiling point, such as those in organic chemistry, are not viable for MW-heating experiments involving DNA and proteins. Thus, there is a considerable demand for MW-heating devices that could be accurately controlled for use with enzymatic reactions.

The use of enzymes require limited reaction conditions, such as optimal temperatures and pH, to demonstrate their catalytic functions. The optimal conditions for exhibiting this activity vary by enzyme, and they cannot sufficiently exert their catalytic functions if not under their optimal conditions. The temperature must be regulated when comparing MW and conventional heating for an accurate comparison of results. Further, it should be noted that enzymes cannot form three-dimensional structures under high-temperature conditions wherein they lose their catalytic functions. This deactivation of enzymes is irreversible, and they can no longer reform their original structures. Therefore, it is essential to accurately set the appropriate reaction conditions, including the temperature, in MW-assisted enzymatic reactions. To clarify the influence of MW heating on DNA and proteins from the perspective of molecular reactions, we used MW devices that can be precisely controlled for accurate heating during DNA polymerization.

2.2 The Research of Proteins Under MW Frequencies

In 2007, Collins and Leadbeater hypothesized that MW-effects on enzymatic digestion could be due to the increasing dipole moments of the α-helices of proteins (Collins and Leadbeater 2007). They illustrated the interaction between the dipole moment of α-helix and the three-dimensional structure of protein. They speculated that if the MW energy interacts with the dipole of α-helix, the perturbation of the protein structure could facilitate the digestion of previously enclosed areas of the protein.

In 2009, Campbell et al. examined the effect of ionic liquid properties on lipase stabilization under MW irradiation in order to understand enzyme stabilization and activity in ionic liquids (Zhao et al. 2009). Trans-esterification of ethyl butyrate and 1-butanol was performed using Novozyme 435 immobilized on acrylic resin in ionic liquids under MW heating. The Discover Lab-Mate equipped with both an infrared temperature probe and a pressure-monitoring sensor was used for the MW reactor. Consequently, no superheating of the enzyme's surface and no enhanced rate were observed under MW irradiation. They suggested a model of water-induced superheating on the free or immobilized enzyme in dry media or a small amount of dispersed water and concluded that the effect of MW irradiation on enzymes can be explained by the superheating of a water layer near the surface of the enzymes.

The MW technology has been used for enzymatic protein digestions in proteomics. In 2012, Kappe et al. revaluated the tryptic digests of three proteins (bovine serum albumin (BSA), cytochrome c, and β-casein) at 37 °C and 50 °C under both MW and conventional heating in order to reveal non-thermal MW effects in proteomics (Damm et al. 2012). Furthermore, they simulated the tertiary structure of trypsin and BSA in an electromagnetic field. An MW 300 single-mode MW reactor (Anton Paar) equipped with a fiber-optic thermometer was used. The results of their combined experimental and computational studies revealed that neither the enzyme stability itself nor the tryptic digest could be directly affected by an electromagnetic field. They concluded that the incorrect temperature measurement led to erroneous conclusions about the non-thermal effects in MW-assisted proteomics experiments.

Yadav is a leading researcher of between microwave irradiation and lipase. In 2014, Yadav and coworkers reported that the synthesis of ethyl 2-(4-aminophenyl) acetate using Novozym 435 (*Candida antarctica* lipase B immobilized on polyacrylic resin) in toluene as a solvent under microwave irradiation (modele: CEM-SP 1245) and conventional heating (B Sontakke and D Yadav 2014). Reaction parameters such as speed of agitation, temperature (40, 50, 60, and 70 °C), catalyst (Lipozyme TLIM, Lipozyme RMIM, and Novozyme 435), enzyme loading, and mole ratio of the reactants were optimized. As a result, a synergistic effect of microwave irradiation and lipases on reaction rate was observed in comparison with conventional heating. The authors concluded that the reaction followed a ping-pong bi–bi mechanism.

The change of dielectric constant of solution containing proteins and ions under wide range of MW frequencies were attracted attention. In 1996, Makoto Suzuki et al. described a method to evaluate protein hydration in an aqueous solution using MW dielectric measurements. Measurements were taken with a precision MW network analyzer and a thermostated glass cell at 20.0 °C ± 0.01 °C with an open-end flat-surface coaxial probe. The studied proteins were cytochrome *c*, myoglobin, ovalbumin, BSA, and hemoglobin. MW frequency dependence on the relative complex permittivity of a BSA was measured by a range from 0.2 to

20 GHz. A plot of the relative complex permittivity changed with a change in the frequency (Suzuki et al. 1996). As an influential study, Gabriel and coworkers reviewed the basic theory underlying MW dielectric heating and collates the dielectric data for a wide range of organic solvents which are commonly used in MW synthesis (Grant and Halstead 1998). Microwave heating significantly depends on the loss tangents of irradiated materials. The loss tangents of organic solvents depend on the relaxation time of the molecules with various functional groups. In 2015, Shuntaro Tsubaki et al. measured the dielectric spectra of sodium alginate L and M, λ- and κ-carrageenans, corn starch, and *Citrus* pectin in water (2.0 wt%) at MW frequencies between 100 MHz and 20 GHz and at temperatures between 27 °C and 80 °C to evaluate the effects of acidic functional groups (Tsubaki et al. 2015). They showed the temperature dependency of the relative permittivity and the loss factors of the six solutions containing proteins.

The frequency effect of MW irradiation of 5.8-GHz was reported below. In 2014, Hiroki Shimizu and coworkers presented the efficiency of the 2.45- and 5.80-GHz MW irradiation (MWS-1000 (EYELA) reactor for the 2.45-GHz MWs and an ATMW500B-5.8 G (Amil) reactor for the 5.80-GHz MWs) for a hydrolysis reaction by thermostable β-Glucosidase HT145(Nagashima et al. 2014). 4-Mehtoxyphenyl glucopyranoside was used as a substrate and β-Glucosidase HT1 was employed. Three types of heating (conventional heating, 2.45 GHz, and 5.8 GHz) were compared for 60 min at four temperatures. Consequently, the 2.45-GHz MWs were more effective than conventional heating, lowering the optimum temperature and accelerating the reaction. On the other hand, 5.80-GHz MWs did not improve this reaction. The authors suggested a heating mechanism where 5.8-GHz MWs affected only the water molecules in a buffer solution, whereas 2.45-GHz MWs affected both the water molecules and buffering ions. Horikoshi and coworkers studied the enzymatic proteolysis of peptide of 13-amino acids under precise temperature control with 5.8-GHz microwave irradiation (Horikoshi et al. 2015). The 5.8-GHz MW system was consisted of a single-mode TEM_{103} cavity, a short plunger, an iris, a four-stub tuner, a power monitor, an isolator, and MW generator (Fuji Electronic Industrial Co. Ltd). Proteolysis yields using the Asp-N metallo-endoproteinase at 30–42 °C under MW *E*-field and *H*-field heating and under conventional heating revealed that enhancement of the proteolysis of the peptide at 37 °C by E-field (42.6%) and H-field (63.5%) compared with conventional heating (27.5%). They suggested that in addition to a thermal factor, MW non-thermal factors also had a significant influence as MW considerable enhanced the proteolysis.

2.3 The Research of DNA Under MW Frequencies

DNA/DNA hybridization under controlled MW power and temperature conditions was studied (Edwards et al. 2009). A CEM discover system (CEM Corporation) with a jacketed reaction vessel enabling temperature control through the continuous flow of coolant while precisely measuring the temperature using a fiber optic probe was used to differentiate between MW effects and thermal effects on the melting of dsDNA. Edwards and coworkers demonstrated that MW led to the melting of double-stranded DNA well below their thermal melting temperature and independent of the length of the DNAs. They suggested that a specific interaction of MW with DNA under MW heating.

Geddes and coworkers reported that a series of papers on Microwave-Accelerated Metal-Enhanced Fluorescence (MEF) which is a method of protein detection. In the first report, low power microwave assisted-MEF was employed to kinetically accelerate assays an to dramatically increase the quantum yield and photo-stability of weakly fluorescing species (Aslan and Geddes 2005). Recently, they showed that low-power MW irradiation accelerates both the formation of the anchor-DNA scaffold on the silver island films-surface and anchor/probe DNA hybridization (Dragan and Geddes 2014).

The published studies on PCR using MW heating (MW-assisted PCR, MW-PCR) illustrate that amplicon synthesis is inefficient, large reaction volumes are required, and temperature is difficult to control or measure. We reasoned that these problems may be attributed to the use of three temperatures (for annealing, elongation, and denaturation) in the MW-PRC method, which are not precisely controlled by MW heating. The temperature measurement under MW irradiation is a matter of importance. For strict temperature control in MW irradiation, a gene amplification study using MWs was reported in milliliter-scale PCR (Orrling et al. 2004; Fermér et al. 2003). An MW device with several frequencies for PCR was developed to accurately control temperature (Shaw et al. 2010).

2.4 MW Rolling Circle Amplification

Rolling circle amplification (RCA), which is based on the mechanism of the replication of viral genomes, is an isothermal nucleic acid amplification technique that uses two primers, a circularized template, and a DNA polymerase with strand displacement activity (Branch and Robertson 1984; Fire and Xu 1995). RCA efficiently synthesizes multiple copies of repeated sequences from the circular DNA and RNA templates. RCA does not require complex temperature control compared with conventional PCR methods, and isothermal conditions are appropriate for controlling MW heating. In our previous study, we reported the MW-assisted RCA for the first time using a multimode MW generator, but reproducibility was difficult (Yoshimura et al. 2006). Therefore, to conduct in-depth research of the effect of MW heating on RCA, MW-RCA was performed using four thermostable DNA polymerases that had strand displacement activity and using a MW single mode cavity resonator.

MW-assisted heating experiments in a cold room (4 °C) were performed using a MW applicator in the TM_{010} mode (SAIDA FDS), equipped with a fiber-optic probe (Neoptix) for internal online temperature control (Fig. 2.1). This resonant

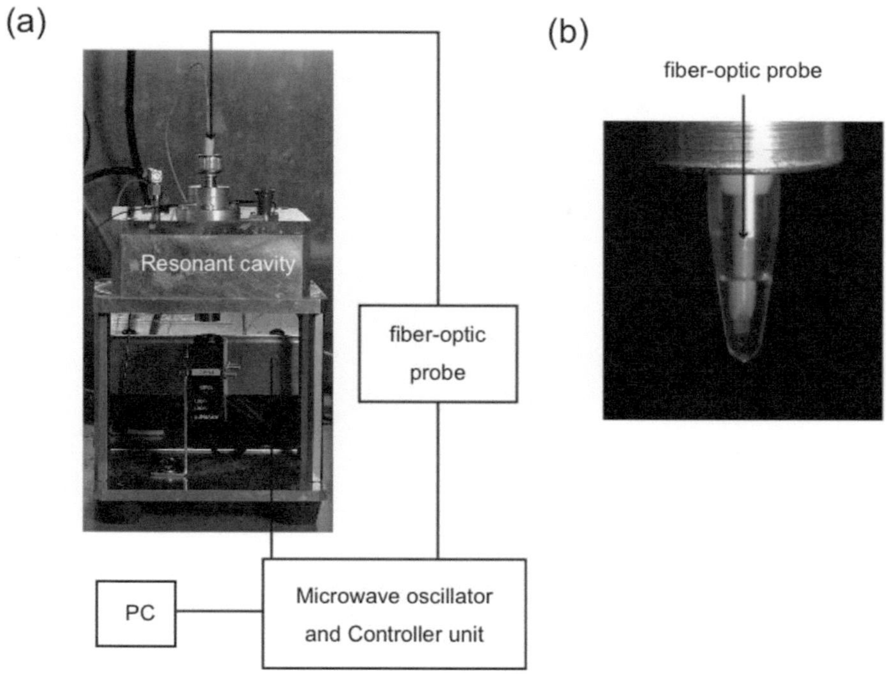

Fig. 2.1 Resonant cavity-type microwave system (**a**) Resonant cavity of TM_{010} mode (**b**) polymerase chain reaction tube with a fiber-optic probe

cavity-type MW applicator more precisely controls the output compared to the magnetron power generated by a solid-state oscillator (maximum input power: 10 W). This MW device can tune a resonant frequency using a voltage conversion system to maximize the inner electric power using a side loop antenna. The MW-heating device with a cylindrical cavity resonator mechanism was designed for resonance with the TM_{010} mode at 2.45 GHz (Fig. 2.2). The power for irradiation was supplied by a coaxial line through port 1, coupled with the monopole antenna of the cavity resonator. During MW irradiation, the resonant frequency is regularly revised by scanning for the highest power detected at the port 2-loop antenna. A monopole antenna was placed at port 3 at a symmetrical position to port 1 for the external measurement of the irradiation frequency. The cavity resonator having an irradiation target with a protective dielectric tube was equipped with a thick wall to prevent the diffusion of the sample. This wall plays a role in adjusting the resonant frequency. The surface-temperature distribution was determined using a radiation thermometer from a slit in the electric current direction of the resonator cavity.

The S parameter values determined at an unloaded condition were confirmed using a network analyzer (Fig. 2.3). The resonant frequencies of 2.44, 3.96, 5.51, and 5.70 GHz, obtained by observing S31 of the cavity resonator, were attributed to

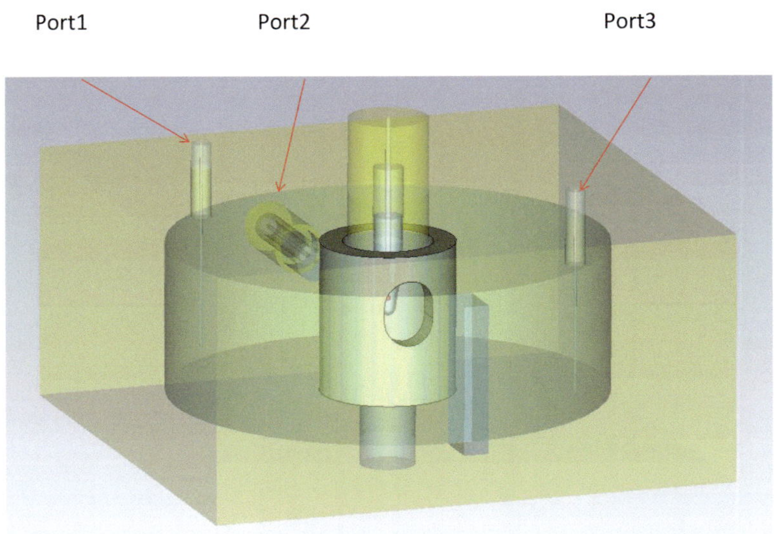

Fig. 2.2 Illustration of cavity resonator with three antennas. (copyright(c)2016 IEICE)

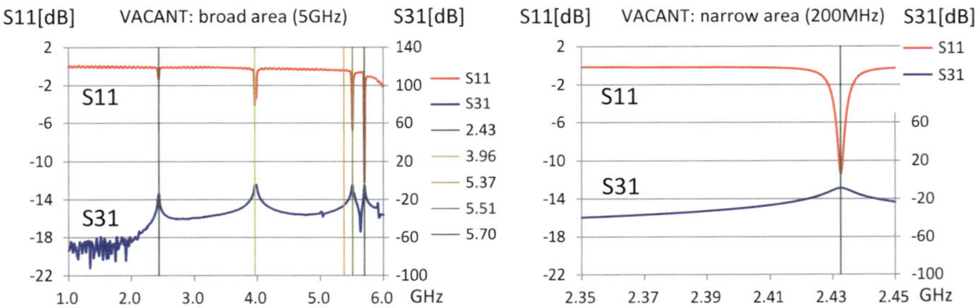

Fig. 2.3 Simulation of resonant frequency of microwave cavity resonator. *Left*: 1–5 GHz area, *Right*: 2.35–2.45 GHz area. (copyright(c)2016 IEICE)

the TM010, TM110, TM210, and TM020 modes, respectively. Thus, the assumed resonant frequency of the TM010 mode was shown to be approximately 2.45 GHz.

The temperature, power, and frequency profiles of MW-RCA using *Bst*-LF at 60 °C for 30 min is shown in Fig. 2.4. The power profile of MWs is the value obtained by subtracting reflected power from incident power. The temperature profile data proved that MW-RCA was performed at precise temperatures. The frequency profile in Fig. 2.4c showed that the frequency tuning system was achieved by this MW device with an auto-tuning system. The dielectric properties of the reaction samples are changed in accordance with the progress of reaction; thus, the resonance frequency of reaction samples changed over several MHz or kHz. It is important that MW-RCA samples in the cavity resonator were optimally tuned by the auto-frequency tuning system. Therefore, the temperature of MW-RCA samples was maintained constant by MW resonance heating.

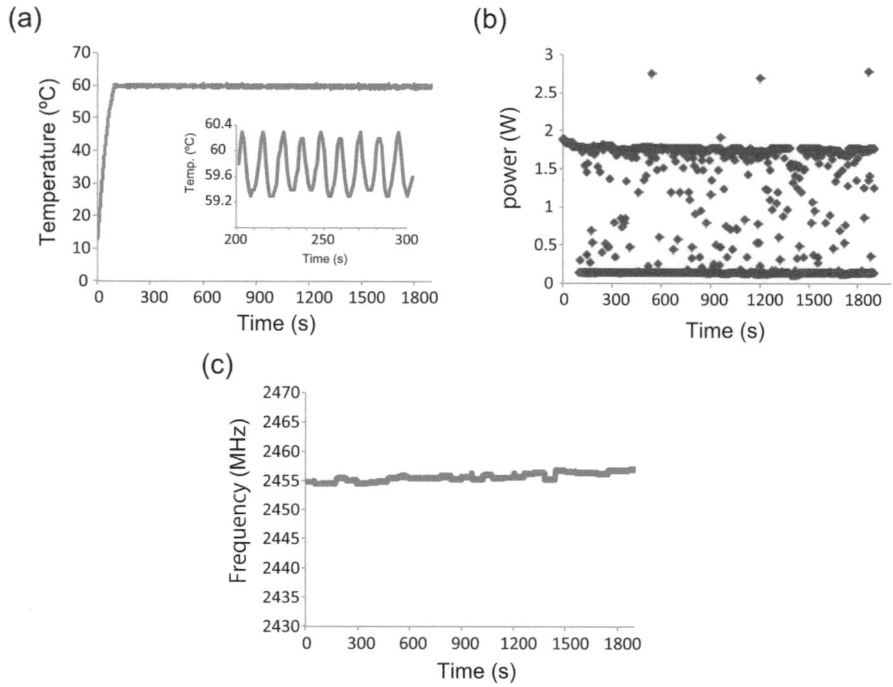

Fig. 2.4 Temperature profile (**a**), electric power profile (**b**), and frequency profile (**c**) of MW-RCA using *Bst* DNA polymerase-LF at 60 °C

Conventional RCA and MW-RCA reactions were sampled at intervals from 10 to 60 min and 10 to 30 min, respectively, and analyzed using agarose gel electrophoresis and fluorescence emission (Fig. 2.5). A comparison of MW-RCA with conventional RCA using *Bst*-LF revealed that the reaction product of MW-RCA increased by a factor of four compared to conventional RCA (Fig. 2.5a). Fluorescence intensity showed a fourfold increase compared to conventional RCA in 30 min. We showed that the MW-assisted RCA reactions catalyzed by either of the four thermostable DNA polymerases were accelerated over fourfold compared to conventional RCA.

Why was MW-RCA accelerated by MW heating even though the overall reaction temperatures were the same? The acceleration of MW-assisted RCA due to the application of a different heating method cannot be explained by conventional chemical thermodynamics and requires the inclusion of the analysis of electromagnetic fields. We focused on the role of buffer components under MW irradiation. To elucidate the mechanism of the MW-heating effect in RCA, we compared the temperature increase of the five components (circularized template with primers, dNTPs, ThermoPol Buffer, *Bst*-LF, and RNase-free water) of RCA under conventional or MW heating for 10 min from 13 °C to 60 °C (Fig. 2.6). All six samples (RCA and its five components) achieved 60 °C using a 60 °C heating block (Fig. 2.6a). In contrast, in the MW heating of the RCA mixture, only the

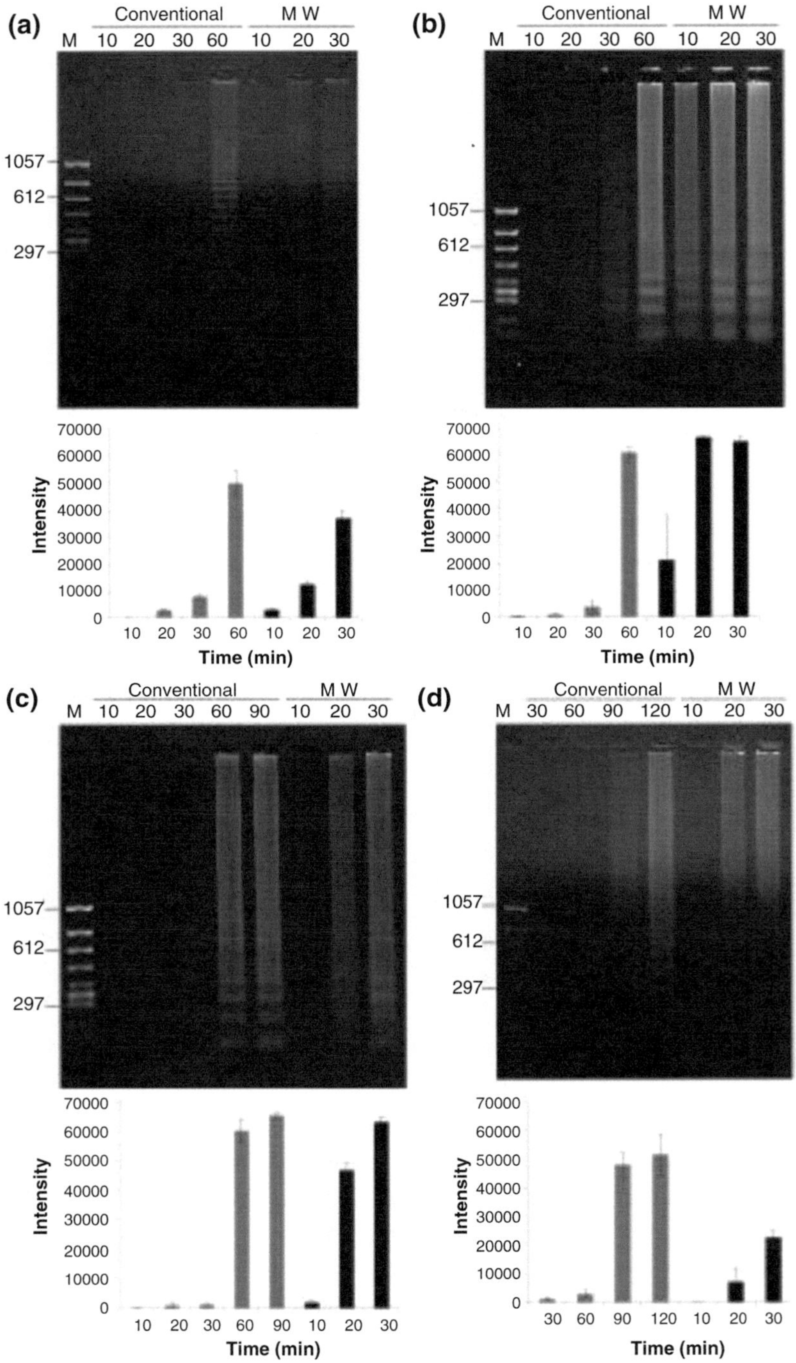

Fig. 2.5 Electrophoresis and fluorescence intensity of DNA synthesized by four DNA polymerases under conventional and microwave heating. MW = microwave. (**a**) *Bst* DNA polymerase-LF, 10–60-min rolling circle amplification (RCA) and10–30-min microwave-assisted (MW)-RCA. (**b**) *Bst* DNA polymerase, 10–60-min RCA and 10–30-min MW-RCA (**c**) *Csa* DNA polymerase, 10–90-min RCA and 10–30-min MW-RCA. (**d**) 96–7 DNA polymerase, 10–120-min RCA and 10–30 min MW-RCA

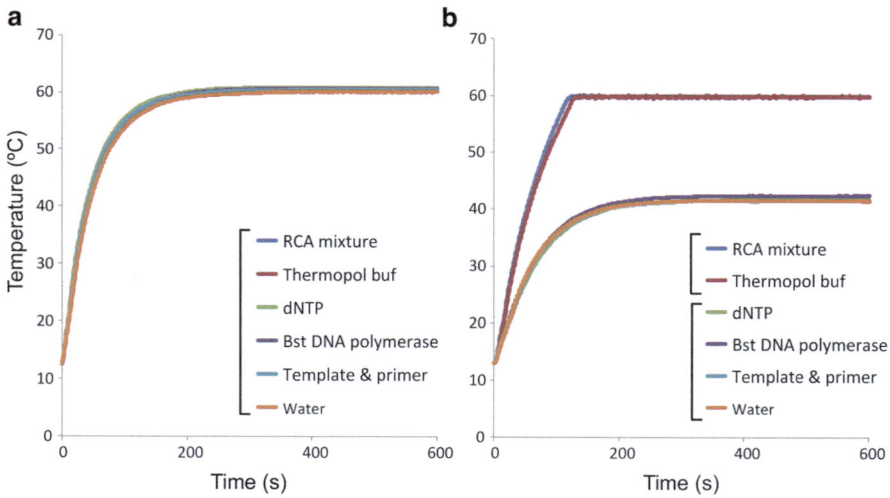

Fig. 2.6 Comparison of temperatures of rolling circle amplification components from 13 °C to 60 °C using conventional (**a**) and microwave heating (**b**). Conventional heating was conducted in a 60 °C heating block

ThermoPol Buffer reached 60 °C. Moreover, four samples (dNTPs with 46 μL of RNase-free water, circularized template with primers in 50 μL of RNase-free water, 8 U of the *Bst* DNA polymerase in 50 μL of RNase-free water, and 50 μL of RNase-free water) reached a maximum temperature of 42 °C by MW heating in 10 min (Fig. 2.6b). These data indicate that the ThermoPol Buffer in MW-RCA was selectively heated by MW irradiation.

To determine the permittivity of RCA components, a network analyzer (Rohde&Schwarz ZVB14, 10 MHz–14 GHz) and a spectrum analyzer (Agilent Technologies N9914A) were used. The complex relative permittivity was determined by the coaxial probe method using the εr and tan δ measurement system (Keycom Corp.).

The εr* value, which is the value obtained by dividing the dielectric constant ε* [F/m] by the dielectric constant of a vacuum (ε0[F/m]), of the RCA substrate should be determined at the reaction temperature. In our measurements, the head of the coaxial probe (3–5 mm) was soaked in water that was heated to 60 °C by an aluminum thermostatic bath. Subsequently, the calibration was performed using the standard εr* value. The results are presented in Table 2.1 (Yoshimura et al. 2015). At 2.45 GHz, the εr′, εr″, and tan δ values of pure water decrease with increasing temperature, resulting from the relaxation frequency of water (>17 GHz) that is higher than 2.45 GHz. In contrast, the εr′ and εr″ values of RCA at 2.45 GHz increased with increasing temperature. Furthermore, the tan δ value of RCA remained nearly unchanged. The Nyquist diagram of RCA exhibited a boundary between the dielectric and conductive losses at 3–4 GHz (Fig. 2.7). Thus, the unusual observed temperature dependence is caused by the dielectric loss due to water and the conductive loss for frequencies lower than 2.45 GHz. We

Table 2.1 εr* and tan δ values of RCA components

Target	RCA	Buffer	Bst	dNTP	DNA	Water
ε_r^*(20° C)	57.46-j 9.11	55.75-j 9.00	58.80-j 4.38	57.46-j 4.20	56.18-j 3.67	77.41-j 10.04
tanδ (20° C)	0.159	0.161	0.075	0.073	0.065	0.130
ε_r^*(60° C)	67.24-j 10.08	67.40-j 10.08	66.61-j 3.90	67.50-j 4.16	67.11-j 3.75	66.45-j 3.42
tanδ (60° C)	0.150	0.150	0.059	0.062	0.056	0.051

The εr* is a complex relative permittivity, and it means εr*=εr′ – jεr′′

Coaxial probe reflection method@2.45GHz

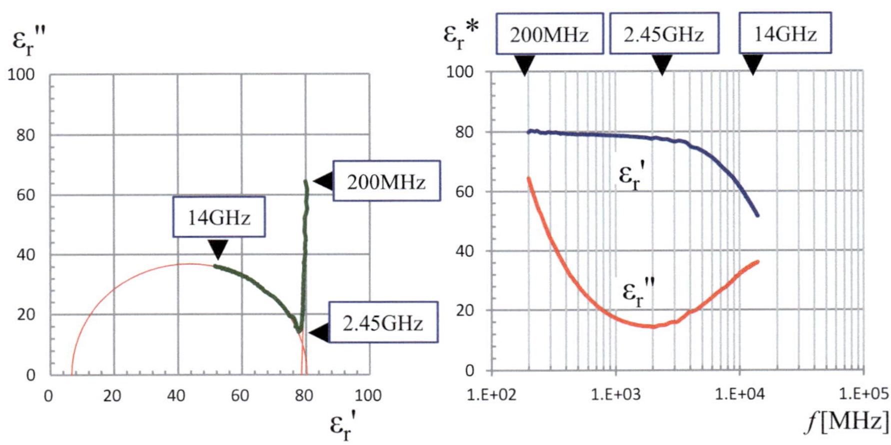

Fig. 2.7 ε_r^* of RCA at 60 °C from 200 MHz to 14 GHz. (*Left*) Nyquist diagram and (*Right*) Bode diagram. Table 2.1 ε_r^* and tan δ values of RCA components. (copyright(c)2016 IEICE)

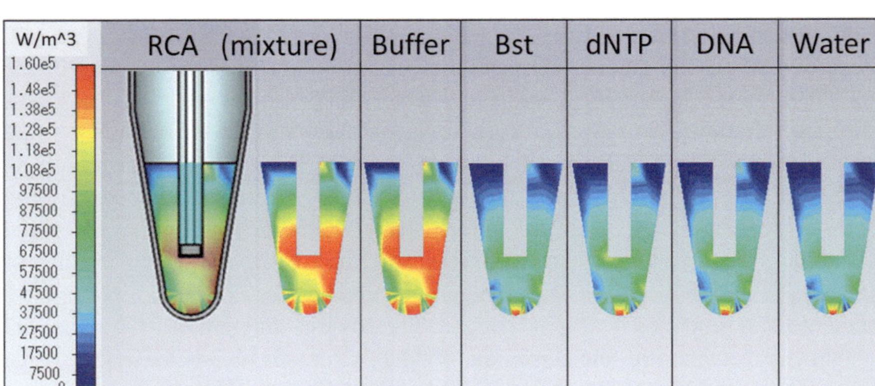

Fig. 2.8 Evaluation of power loss density of RCA (mixture) and RCA components

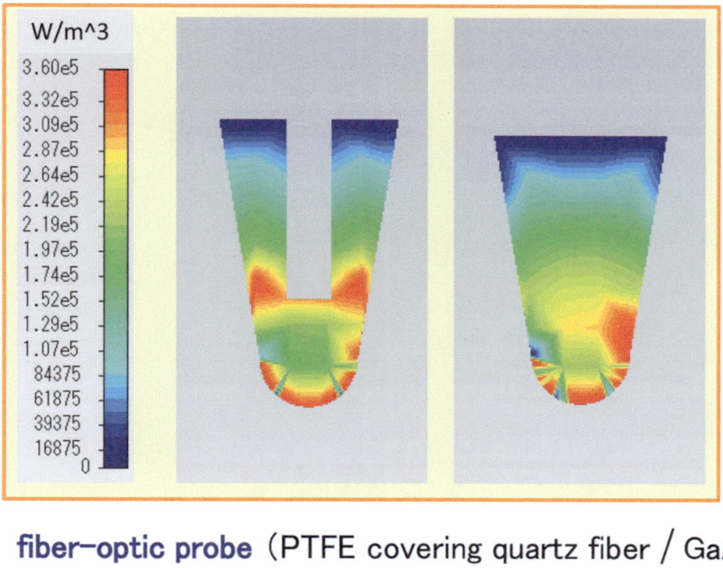

Fig. 2.9 Effect of fiber-optic probe on temperature measurement. (copyright(c)2016 IEICE)

concluded that the conductive loss of the total RCA solution is because of the contribution from the ion-based buffer.

Electromagnetic field distribution was simulated via the FDTD method using MW Studio (CST). The heating distribution was obtained by a simulation of the power loss density using the measured values of εr^* listed in Table 2.1. The power loss density of the buffer with high a tan δ value was higher than that of the RCA components and similar to that of RCA (Fig. 2.8). The simulated power loss density showed a correlation with the temperature profiles of the RCA components. During the irradiation experiments, a fiber-optic probe for measuring the temperature was inserted in the reaction solution; however, heating with and without this probe was similar to heating by convection (Fig. 2.9). The results shown in Figs. 2.6 and 2.8 suggest that the ThermoPol Buffer was a primary factor leading to an increase in temperature under MW irradiation.

We hypothesized that selective MW heating could be observed by heating only one of each buffer component. We accordingly prepared four samples of one-fold ThermoPol Buffer components [20 mM Tris–HCl, 10 mM $(NH_4)_2SO_4$, 10 mM KCl, and 2 mM $MgSO_4$] and four samples of ThermoPol Buffer components in fourfold excess concentrations [80 mM Tris–HCl, 40 mM $(NH_4)_2SO_4$, 40 mM KCl, and 8 mM $MgSO_4$] and measured their temperatures under MW heating (Fig. 2.10). Temperature measurements of the ThermoPol Buffer components at fourfold excess concentrations showed an increase in temperature compared to the one-fold concentrations. The MW heating method correlated with the increase in concentrations of the buffer components.

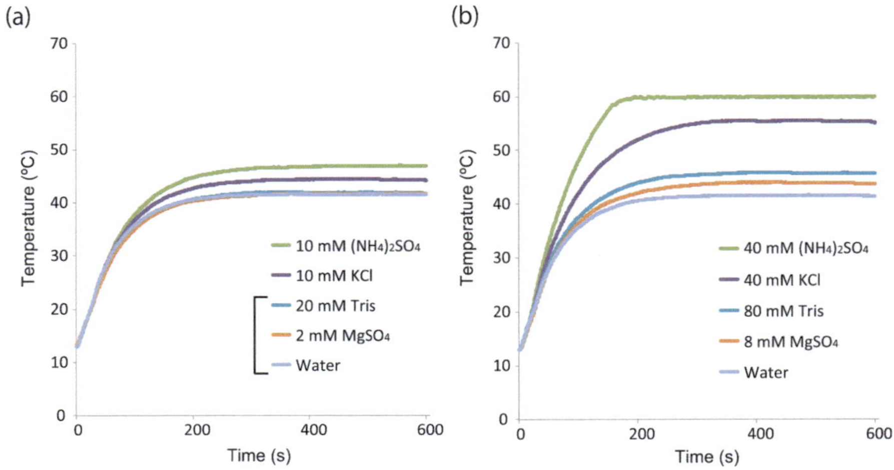

Fig. 2.10 The temperatures of ThermoPol Buffer individual components at (**a**) 1-fold (**b**) and 4-fold higher concentrations heated from 13 °C to 60 °C by microwave heating. (copyright(c)2016 IEICE)

We performed MW-RCA reactions containing a fourfold higher concentration of each RCA component [dNTP, template–primers, *Bst* DNA polymerase-LF, Tris–HCl, KCl, $(NH_4)_2SO_4$, and $MgSO_4$] to identify the link between MW selective heating and DNA amplification. All results for MW heating were accelerated compared to conventional heating (Fig. 2.11) with two notable outcomes. MW-RCA heating with 40 mM of $(NH_4)_2SO_4$ was accelerated relative to the control, whereas RCA with 40 mM of $(NH_4)_2SO_4$ was amplified similar to the control of conventional RCA. $(NH4)_2SO_4$ plays a role in buffering the solution and is unimportant for DNA polymerization. In contrast, Mg^{2+} is necessary for the function of DNA polymerase. The conventional RCA reaction mixture containing $MgSO_4$ (8 mM) had a slightly amplified concentration of DNA. These findings suggest a link between the amplification of MW-RCA and the temperature increase of $(NH_4)_2SO_4$ and $MgSO_4$ under MW irradiation.

2.5 Conclusions

In our study, we report the development of MW heating for enzymatic reactions and demonstrate the effect of MW heating on the RCA of DNA. The temperature profiles of RCA components by MW heating suggest that the ionic components of the ThermoPol Buffer are selectively heated and that the rate of temperature increase induced by MW heating depends on the components of the molecule and the concentrations of the buffer components. Further, the relation between MW power loss and RCA components was studied via dielectric measurements, cavity

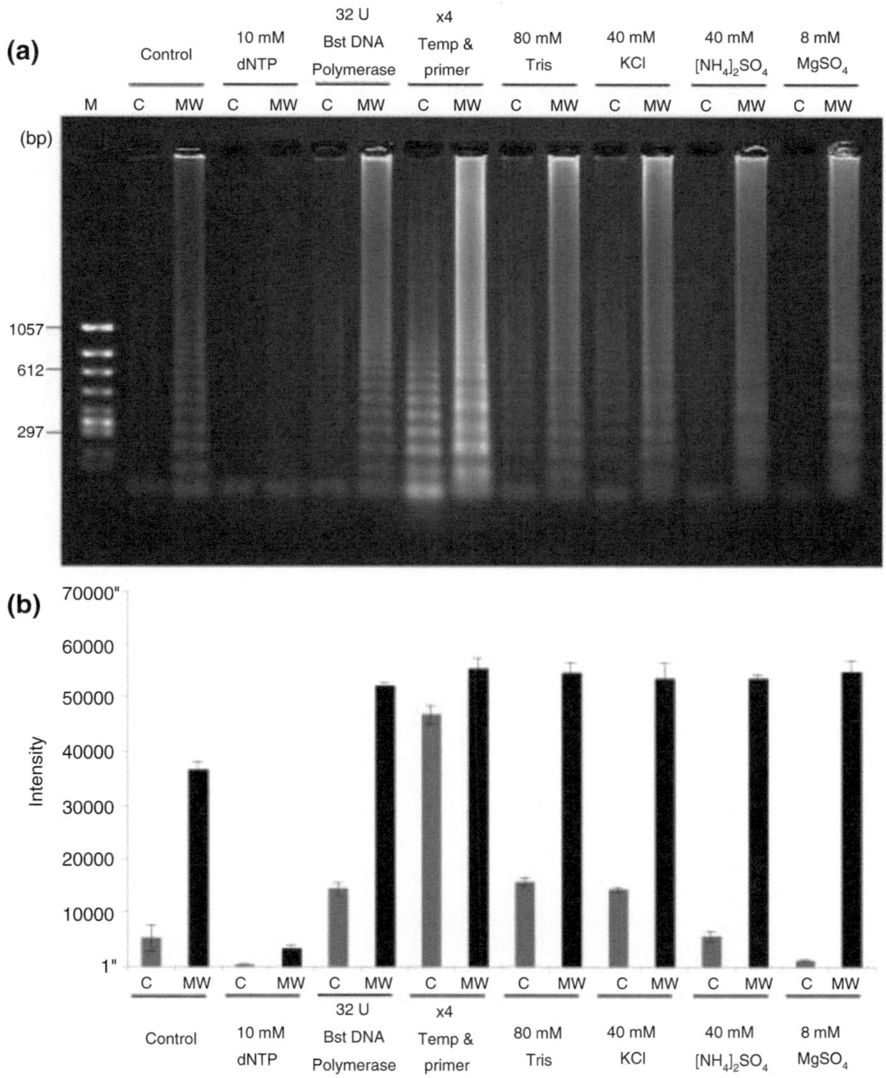

Fig. 2.11 Microwave-assisted rolling circle amplification reaction mixtures containing 4-fold increased concentrations of one component. (**a**) Agarose gel electrophoresis results. (**b**) Fluorescence intensity of SYBR *Green* I. *C* conventional, *MW* microwave

resonator feature measurement, and electromagnetic simulation. The εr* and tan δ values of all RCA components were determined at 60 °C. Additionally, the simulation indicated that the Thermopol Buffer contained ions that are selectively heated by MW irradiation. Controlled MWs can accelerate RCA-type gene amplification; however, the effect of MW irradiation on DNA and proteins is unclear. We speculate that the selectively heated ionic molecules under MW irradiation affect

DNA polymerization using DNA polymerase. Our research displayed the possibility of controlling gene amplification (RCA and PCR) by the concentration of ions under resonated MW irradiation.

References

Aslan K, Geddes CD (2005) Microwave-accelerated metal-enhanced fluorescence: platform technology for ultrafast and ultrabright assays. Anal Chem 77(24):8057–8067

Bilecka I, Niederberger M (2010) Microwave chemistry for inorganic nanomaterials synthesis. Nanoscale 2(8):1358. doi:10.1039/b9nr00377k

Branch AD, Robertson HD (1984) A replication cycle for viroids and other small infectious RNA's. Science 223(4635):450–455

Chen P-K, Rosana MR, Dudley GB, Stiegman A (2014) Parameters affecting the microwave-specific acceleration of a chemical reaction. J Org Chem 79(16):7425–7436

Collins JM, Leadbeater NE (2007) Microwave energy: a versatile tool for the biosciences. Org Biomol Chem 5(8):1141–1150. doi:10.1039/b617084f

Damm M, Nusshold C, Cantillo D, Rechberger GN, Gruber K, Sattler W, Kappe CO (2012) Can electromagnetic fields influence the structure and enzymatic digest of proteins? A critical evaluation of microwave-assisted proteomics protocols. J Proteome 75(18):5533–5543. doi:10.1016/j.jprot.2012.07.043

de la Hoz A, Loupy A (2013) Microwaves in organic synthesis, 2 volume set. Wiley, Weinheim

Dragan A, Geddes CD (2014) 5-color multiplexed microwave-accelerated metal-enhanced fluorescence: detection and analysis of multiple DNA sequences from within one sample well within a few seconds. J Fluoresc 24(6):1715–1722

Edwards WF, Young DD, Deiters A (2009) The effect of microwave irradiation on DNA hybridization. Org Biomol Chem 7(12):2506–2508. doi:10.1039/b903609a

Fermér C, Nilsson P, Larhed M (2003) Microwave-assisted high-speed PCR. Eur J Pharm Sci 18(2):129–132

Fire A, Xu S-Q (1995) Rolling replication of short DNA circles. Proc Natl Acad Sci 92(10):4641–4645

Gedye R, Smith F, Westaway K, Ali H, Baldisera L, Laberge L, Rousell J (1986) The use of microwave ovens for rapid organic synthesis. Tetrahedron Lett 27(3):279–282

Giguere RJ, Bray TL, Duncan SM, Majetich G (1986) Application of commercial microwave ovens to organic synthesis. Tetrahedron Lett 27(41):4945–4948

Grant E, Halstead BJ (1998) Dielectric parameters relevant to microwave dielectric heating. Chem Soc Rev 27(3):213–224

Hayden S, Damm M, Kappe CO (2013) On the importance of accurate internal temperature measurements in the microwave dielectric heating of viscous systems and polymer synthesis. Macromol Chem Phys 214(4):423–434

Horikoshi S, Nakamura T, Kawaguchi M, Serpone N (2015) Enzymatic proteolysis of peptide bonds by a metallo-endoproteinase under precise temperature control with 5.8-GHz microwave radiation. J Mol Catal B Enzym 116:52–59

Kappe CO, Dallinger D (2009) Controlled microwave heating in modern organic synthesis: highlights from the 2004–2008 literature. Mol Divers 13(2):71–193. doi:10.1007/s11030-009-9138-8

Lidström P, Tierney J, Wathey B, Westman J (2001) Microwave assisted organic synthesis—a review. Tetrahedron 57(45):9225–9283

Nagashima I, Sugiyama J, Sakuta T, Sasaki M, Shimizu H (2014) Efficiency of 2.45 and 5.80 GHz microwave irradiation for a hydrolysis reaction by thermostable beta-Glucosidase HT1. Biosci Biotechnol Biochem 78(5):758–760. doi:10.1080/09168451.2014.891931

Obermayer D, Kappe CO (2010) On the importance of simultaneous infrared/fiber-optic temperature monitoring in the microwave-assisted synthesis of ionic liquids. Org Biomol Chem 8(1):114–121

Orrling K, Nilsson P, Gullberg M, Larhed M (2004) An efficient method to perform milliliter-scale PCR utilizing highly controlled microwave thermocycling. Chem Commun (Camb) 7:790–791. doi:10.1039/b317049g

Shaw KJ, Docker PT, Yelland JV, Dyer CE, Greenman J, Greenway GM, Haswell SJ (2010) Rapid PCR amplification using a microfluidic device with integrated microwave heating and air impingement cooling. Lab Chip 10(13):1725–1728. doi:10.1039/c000357n

Sontakke JB, Yadav GD (2014) Microwave assisted synthesis of ethyl 2-(4-aminophenyl) acetate using novozyme 435. Curr Catal 3(1):27–34

Suzuki M, Shigematsu J, Kodama T (1996) Hydration study of proteins in solution by microwave dielectric analysis. J Phys Chem 100(17):7279–7282

Tsubaki S, Hiraoka M, Hadano S, Okamura K, Ueda T, Nishimura H, Kashimura K, Mitani T (2015) Effects of acidic functional groups on dielectric properties of sodium alginates and carrageenans in water. Carbohydr Polym 115:78–87

Yoshimura T, Nishida K, Uchibayashi K, Ohuchi S (2006) Microwave assisted rolling circle amplification. Nucleic Acids Symp Ser (Oxf) 50:305–306. doi:10.1093/nass/nrl152

Yoshimura T, Hanai T, Mineki S, Sugiyama J, Sato C, Ohneda N, Okamoto T, Odajima H (2015) Electromagnetic field analysis of the resonator with a microfuge tube to heat in enzymatic DNA replication reaction. IEICE Tech Rep 115(3):71–76

Zhao H, Baker GA, Song Z, Olubajo O, Zanders L, Campbell SM (2009) Effect of ionic liquid properties on lipase stabilization under microwave irradiation. J Mol Catal B Enzym 57(1–4):149–157. doi:10.1016/j.molcatb.2008.08.006

Chapter 3
Microwave Effects on DNA

Jitendra Behari and Tanu Jindal

Abstract The effects of nonionizing electromagnetic radiations on biological system is a vexed issue. Despite the fact that a large number of publications have appeared on the subject, the issue is far from settled. In the present work the data showing a distinct possibility in the mechanism of EMF biointeraction leading to DNA strand break is presented. This may have effects on physiology of humans, besides causing genetic damage. A confirmation to this lies in adopting safety criteria by various countries including India. However a quantitative picture still remains to be drawn. In support of this experimental data are summarized to take a holistic view.

Keywords Non ionizing electromagnetic radiations • DNA • Double strand break • Micronuclei • Reactive oxygen species

3.1 Introduction

During the last four decades or so there has been a growing concern about the effects of nonionizing electromagnetic radiations on biological systems. This is because of the fact that electronic devices are increasing in geometric proportion and emitting these radiations. For communication, the air surveillance system, industry, and diagnostic and therapeutic purposes in medicine, all are emitting microwaves/radiofrequency (RF) waves. In view of this, the importance of the electromagnetic radiations pervading the environment is now increasingly realized (Moulder et al. 1999; Brusick et al. 1998) and their adverse effect on human health is a matter of concern. This emitter has led to the addition of another pollutant in the environment (electropollution) after air, water, soil and noise. The effects are broadly classified into two: thermal and nonthermal, though the line of demarcation between the two is not very clear (Fig. 3.1). While the former leads to an increase in body temperature (SAR>2), the latter does not. The nonthermal effects are

J. Behari (✉) • T. Jindal
Amity Institute of Environmental Toxicology Safety and Management, Amity Univeristy, Sector-125, Noida, India
e-mail: jbehari@hotmail.com

C.D. Geddes (ed.), *Microwave Effects on DNA and Proteins*,
DOI 10.1007/978-3-319-50289-2_3

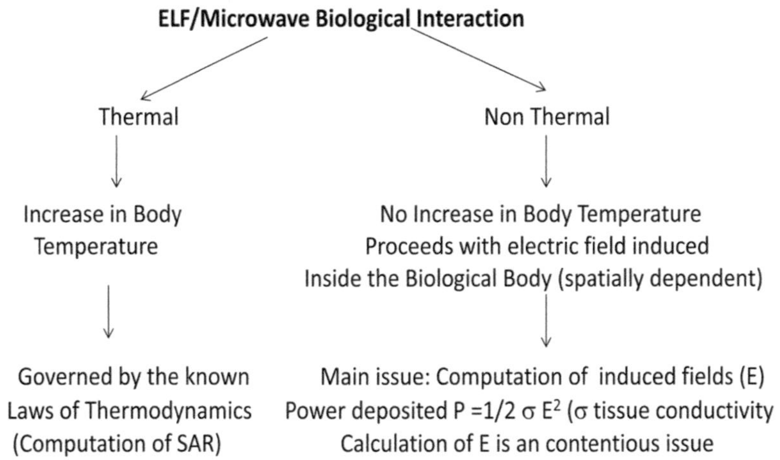

Fig. 3.1 The boundary line between thermal and non-thermal effects is thin and yet not well defined

attributed to the induced electromagnetic fields inside the biological body. The controversy over non-thermal effects of RF-EMFs continues further for two main reasons: Firstly, it is difficult to replicate a demonstrated effect, since many environmental variables impact the outcome than previously assumed. Secondly, the mechanism of non-thermal effects is not yet well understood, that has everything to do with its complexity but nothing with its biological non-existence. In fact differences in DNA repair capacity have become the accepted explanation for this range of intrinsic radio sensitivities, since it is generally believed that a subset of the DNA breaks is lethal if unrepaired (Iliakis 1991).

Scenario of the type as shown in Fig. 3.2 emerges when acting on the whole body, mediates its path through the central nervous system, targeting the whole body and may also interact with the reproductive system (Fig. 3.2).

While the issue concerning electro pollution is still being debated, devices emitting these radiations have found their use in everyday life. These principally include mobile phones and microwave ovens, to name a few. Microwaves are emitted by mobile phones of the GSM type (global system for mobile telecommunication). It is an extremely low frequency (ELF) modulated pulsed microwave carrier. It can be argued that digital mobile phones transmit information in bursts of microwaves and this raises a question regarding its negative effect on human health. It has been observed that the two low-frequency sine waves (8.3 and 217 Hz, respectively) act on the microwave (continuous wave CW) to make it a composite pulsed GSM signal. The issue has become all the more relevant for its possible interaction with tissues and DNA. Before we consider the mode of EMF interaction with DNA, it is pertinent to examine its (DNA) structure in a broad manner.

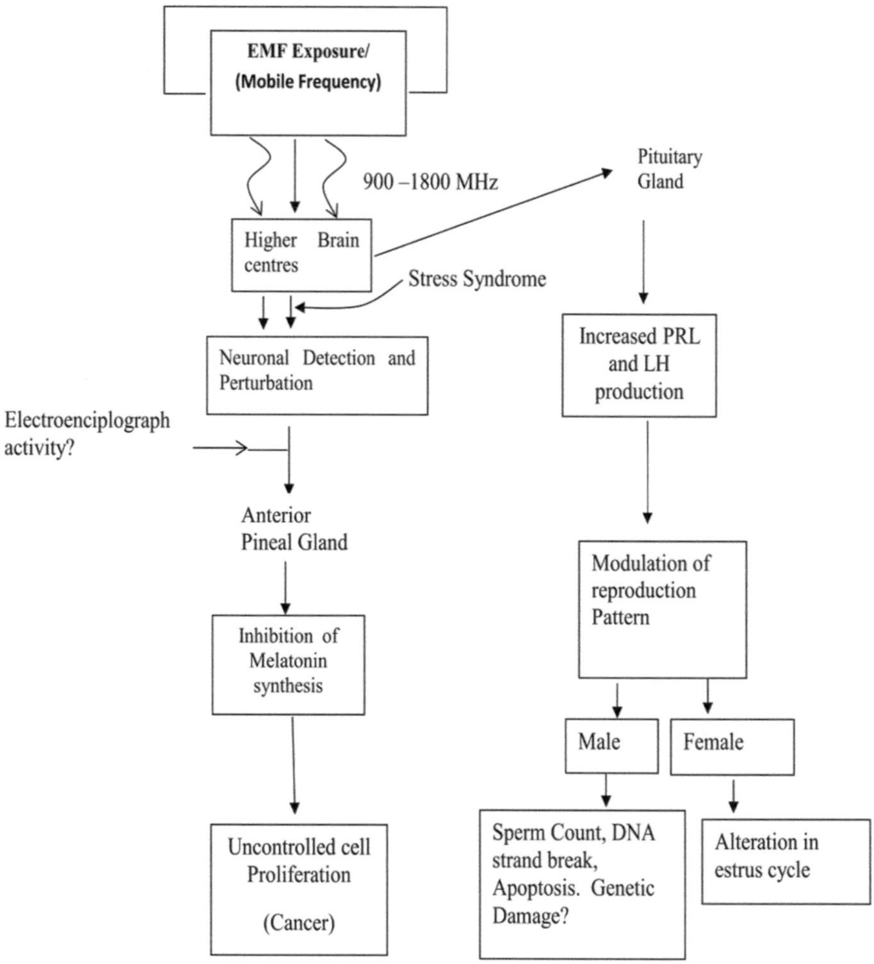

Fig. 3.2 Proposed model of EMF exposure, influence brain cells promotes the growth of initiated tumor cells and reproduction pattern

3.2 Structure of DNA

DNA is essentially a molecule that carries the genetic instructions used in the development, functioning and reproduction of all known living organisms and many viruses. This is composed of three major macromolecules essential for all known forms of life. DNA is a nucleic acid; alongside proteins and carbohydrates. Most DNA molecules consist of two biopolymer strands coiled around each other to form a double helix. The two DNA strands are known as polynucleotides since they are composed of simpler units called nucleotides. Each nucleotide is composed of a

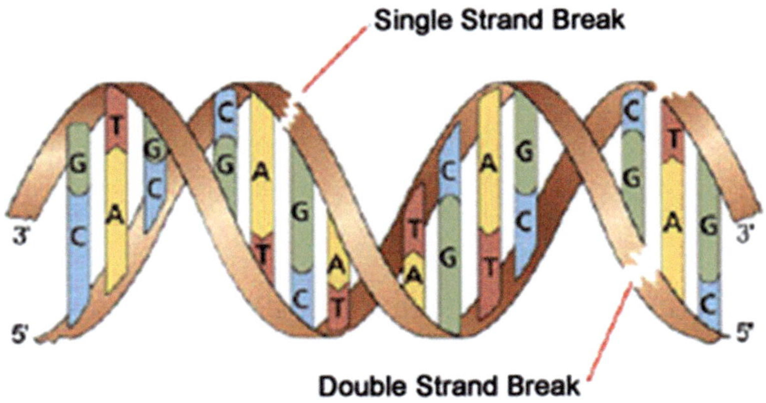

Fig. 3.3 Pictorial demonstration of single and double strand break

nitrogen-containing nucleobase—either cytosine (C), guanine (G), adenine (A), or thymine (T)—as well as a mono saccharide sugar called deoxyribose and a phos phate group. Adenine pairs with thymine and guanine pairs with cytosine. It is represented by A-T base pairs and G-C base pairs. The nucleotides are joined to one another in a chain by covalent bonds between the sugar of one nucleotide and the phosphate of the next, resulting in an alternating sugar-phosphate backbone. According to base pairing rules (A with T, and C with G), hydrogen bonds bind the nitrogenous bases of the two separate polynucleotide strands to make double-stranded DNA (Fig. 3.3).

DNA is a long polymer made from repeating units called nucleotides (Saenger 1984; Alberts et al. 2002). DNA was first identified and isolated by Friedrich Miescher in (1869), a substance he called nuclein. The double helix structure of DNA was first discovered by Watson and Crick (1953). The structure of DNA is non-static (Irobalieva et al. 2015), all species comprises two helical chains each coiled round the same axis, and each with a pitch of 34 Å and a radius of 10 Å. According to another study, when measured in a particular solution, the DNA chain measured 22–26 Å wide and one nucleotide unit measured 3.3 Å long (Mandelkern et al. 1981). Although each individual repeating unit is very small, DNA polymers can be very large molecules containing millions of nucleotides. For instance, the DNA in the largest human chromosome, chromosome number 1, consists of approximately 220 million base (Gregory et al. 2006) and would be 85 mm long if straightened. DNA damage is referred as any modification of DNA that changes its coding properties or normal function in transcription or replication (Lindahl 1993; Rao 1993). DNA lesions can occur in many different forms, including apurinic.apyrimidinic (AP) sites (a basic sites), adducts, single-strand breaks (SSBs), DSBs, DNA-protein cross-links, and insertion/deletion mismatches (Rao 1993).

3.2.1 Nitrogenous Bases

The bases that are responsible for the bonding of two DNA strands are known as nitrogenous bases. Each nucleotide has one nitrogenous base but there are a total of four different nitrogenous bases to choose from. The nitrogenous base molecules are bonded to a deoxyribose sugar. Differences in the arrangement of these four bases along a strand of DNA accounts for the formation of different genes. A specific sequence of nitrogenous bases provides the information for a cell to produce a specific protein.

In living organisms DNA does not usually exist as a single molecule, but instead as a pair of molecules that are held tightly together (Watson and Crick 1953; Berg et al. 2002). The nucleotide repeats contain both the segment of the backbone of the molecule, which holds the chain together, and a nucleobase, which interacts with the other DNA strand in the helix. A nucleobase linked to a sugar is called a nucleoside and a base linked to a sugar and one or more phosphate groups is called a nucleotide. A polymer comprising multiple linked nucleotides (as in DNA) is called a polynucleotide (IUPAC-IUB 2006). The two strands of DNA run in opposite directions to each other and are therefore anti-parallel. Under the genetic code, RNA strands are translated to specify the sequence of amino acids within proteins. These RNA strands are initially created using DNA strands as a template in a process called transcription.

Importance of DNA molecule is that it stores biological information. Attached to each sugar is one of four types of nucleobases (informally, bases). It is the sequence of these four nucleobases along the backbone that encodes biological information. The DNA backbone is resistant to cleavage, and both strands of the double-stranded structure store the same biological information. Biological information is replicated as the two strands are separated. A significant portion of DNA (more than 98% for humans) is non-coding. DNA is the material that makes up genes and is found in all living cells. It holds the information that is used by cells to build proteins and other molecules. These proteins and molecules then carry out processes that keep the cell alive. Through sexual reproduction, DNA is passed from parents to offspring, which in turn receive a mixture of DNA from both their parents. DNA codes for all of the genes found in a cell which can be more than 100,000 genes in some species.

Within cells, DNA is organized into structures called chromosomes, and is used by researchers as a molecular tool to explore physical laws and theories, such as the ergodic theorem and the theory of elasticity. The unique material properties of DNA have made it an attractive molecule for material scientists and engineers interested in micro- and nano-fabrication. Among notable advances in this field are DNA origanic and DNA-based hybrid materials (Mashaghi and Katan 2013).

For animals, plants, fungi and other eukaryotic organisms the major part of DNA is found within the cell nucleus within chromosomes. We also have DNA in our mitochondria, which are specialized structures within cells responsible for the process of respiration. The DNA within the mitochondria is specifically related to the functions of the mitochondria. In organisms which have prokaryotic cells, such as bacteria, DNA can be found coiled together anywhere within the cell in a nucleus-like structure called a nucleoid.

A major concern is of the adverse effects of DNA exposure to microwaves field is that it causes cancer promotion/induction, as the majority of cancers are initiated by damage to a cell's genome, by a possible pathway as shown in Fig. 3.2.

During cell division the chromosomes are duplicated in the process of DNA replication, providing each cell its own complete set of chromosomes. Eukaryotic organisms (animals, plants, fungi, and protists) store most of their DNA inside the cell nucleus and some of their DNA in organelles, such as mitochondria or chloroplasts (Russell 2001). In contrast, prokaryotes (bacteria and archaea) store their DNA only in the cytoplasm. Within the chromosomes, chromatin proteins such as histones compact and organize DNA. These compact structures guide the interactions between DNA and other proteins, helping control which parts of the DNA are transcribed.

The backbone of the DNA strand is made from alternating phosphate and sugar residues. The sugar in DNA is 2-deoxyribose, which is a pentose (five-carbon) sugar. The sugars are joined together by phosphate groups that form phosphodiester bonds between the third and fifth carbon atoms of adjacent sugar rings. These asymmetric bonds mean a strand of DNA has a direction. In a double helix the direction of the nucleotides in one strand is opposite to their direction in the other strand, hence the strands are antiparallel. The asymmetric ends of DNA strands are called the 5′ (five prime) and 3′ (three prime) ends, with the 5′ end having a terminal phosphate group and the 3′ end a terminal hydroxyl group. One major difference between DNA and RNA is the sugar, with the 2-deoxyribose in DNA being replaced by the alternative pentose sugar ribose in RNA (Berg et al. 2002).

3.3 Electromagnetic Fields and DNA: Biophysical Concepts

The basic question is: can microwaves disrupt the covalent bonds of DNA. The fundamentals of thermodynamics and physics suggest that this is not possible. The energy level of a microwave photon is ~10^{-5} eV, whereas the energy required to break a covalent bond is ~10 eV, or a million times greater. Based on this fact, it has been stated in the literature that "microwaves are incapable of breaking the covalent bonds of DNA" (Fujikawa et al. 1992; Jeng et al. 1987). Studies have shown that in the absence of water or moisture, biocidal effects of microwaves are severely diminished, or require considerably longer exposures (Jeng et al. 1987; Vela and Wu 1979). DNA has a dielectric dispersion, where microwaves are readily absorbed, at much lower frequencies than water (Takashima et al. 1984). However, it is demonstrated (Kakita et al. 1995) that the microwave effect is distinguishable from external heating by the fact that it is capable of extensively fragmenting viral DNA, something that heating to the same temperature does not accomplish. There

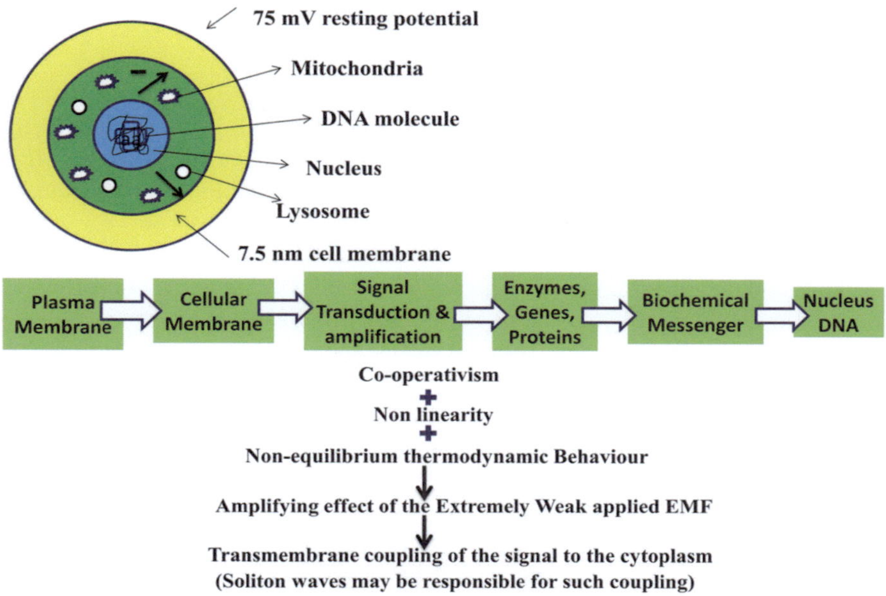

Fig. 3.4 An idealized excitable cell at rest in the human body .75 mv across a thickness of 705 nm corresponds to an electric field of ten million Vm^{-1}

is, in fact, plenty of evidence to indicate that there are alternate mechanisms for causing DNA covalent bond breakage without invoking the energy levels of radiations. A possible mechanism is shown in Fig. 3.4. It would seem there is reason to believe that the microwave effect does indeed exist, even if it cannot yet be adequately explained, and probably some more data are needed.

Apurinic/apyrimidinic sites are common lesions in DNA and are formed either spontaneously or as intermediates during the course of the normal repair of oxidized, deaminated, or alkylated bases (Lindahl 1993). Apurinic/apyrimidinic sites are one of the major types of damage generated by reactive oxygen species (ROS). Hydroxyl radical attack on the deoxyribose moiety can cause the release of free bases from DNA (Atamna et al. 2000). Apurinic/apyrimidinic sites can be mutagenic or they can cause cell death (Atamna et al. 2000; Kingma and Osheroff 1997). It has been estimated that endogenous ROS can cause approximately 50,000–200,000 AP sites per mammalian cell per day, and that brain cells contain the most AP sites (Atamna et al. 2000). In some organs, the level of AP sites seems to be age dependent, with higher levels found in older animals (Atamna et al. 2000).

Since nerve cells do not divide and are not likely to become cancerous, more likely consequences of DNA damage in nerve cells are changes in function and cell death, which could either lead to or accelerate the development of neurodegenerative diseases.

3.4 Double Strand Break (DSB)

DNA double-strand breaks (DSBs) arises due to the lesions formed when both strands of the DNA duplex are broken (Figs. 3.3 and 3.5a). A prominent example of such occurrence is due to the exposure by ionizing radiations (IR). The major endogenous source occurs when DNA replication forks encounter unrepaired DNA lesions, triggering the phenomena (Pfeiffer et al. 2000). Indeed, the frequency of replication-associated DSBs is so high that both leading- and lagging-strand replication in vivo is considered to be discontinuous (Lehmann and Fuchs 2006; Sonoda et al. 1998). DSBs are highly toxic and can cause genome rearrangements and cell death. Indeed, defective DSB repair is associated with various developmental, immunological, and neurological disorders, and is a major source of cancer promotion (Jackson and Bartek 2009; McKinnon 2009).

Unrepaired DNA double-strand breaks (DSBs) cause genetic instability that leads to malignant transformation or cell death. Cells respond to DSBs with the ordered recruitment of signaling and repair proteins to the sites of DNA lesions. To prevent this, organisms have evolved a range of DSB repair mechanisms.

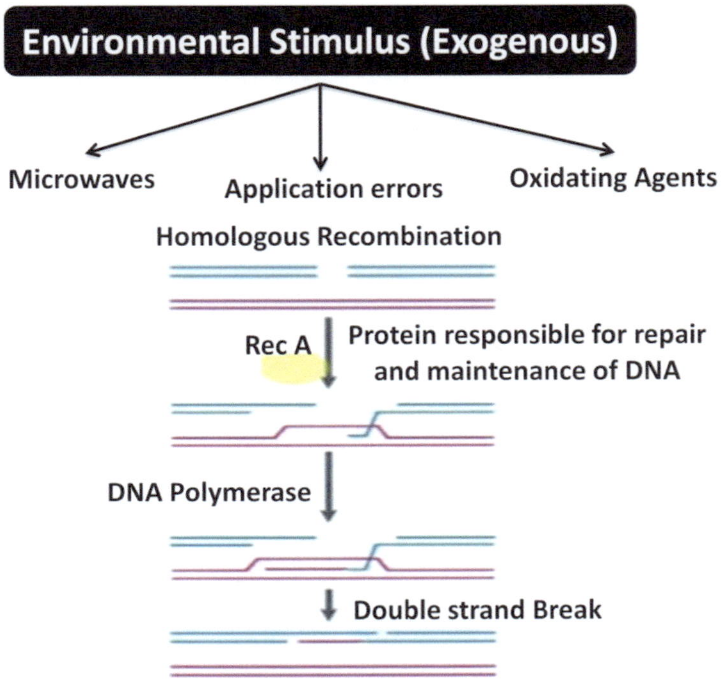

Fig. 3.5a Occurrence of double strand breaks under exogenous environmental stimulus

3.4.1 Mechanisms and Effect of DNA Double-Strand Break

Mailankot et al. (2009) tested the hypothesis that RF-EMR from mobile phones results in oxidative stress and decreases semen quality. Authors found a striking reduction in the percentage of motile sperm in rats exposed to RF-EMR, as well as a significantly elevated Lipid Peroxidation (LPO) and decreased Glutathione (GSH) content in the testis and epididymis.

The size of a double stranded DNA molecule is measured by the number of base pairs it contains while a single strand is measured by the number of nucleotides it has. Humans have 6×10^9 base pairs in all of their chromosomes. In eukaryotic cells, DNA is found mostly in the nucleus. In prokaryotic cells it can be found coiled together anywhere in the cell. If there are not enough antioxidants available or if the rate of DNA damage exceeds the rate of repair, genetic regulation processes or protein expression become impaired, leading to possible pathogenic ramifications. Thus, the likelihood of cancer formation increases (Trush and Kensler 1991; Wiseman and Halliwell 1996) because the processes of initiation and promotion of carcinogenesis are promoted by DNA damage mediated by Reactive Oxygen Species/Reactive Nitrogen Species (ROS/RNS) (Janssen et al. 1993; Takabe et al. 2001). Oncogenes are also activated in this process (Shibutani et al. 1991; Cheng et al. 1992). Damage to the DNA in mitochondria is particularly fatal (mtDNA, mitochondrial DNA) because this is ten times more susceptible to oxidative stress than DNA in the cell nucleus (nDNA). Further mtDNA is not protected by histone proteins and does not possess any effective repair mechanism (Hruszkewycz and Bergtold 1988; Druzhyna et al. 2008). Mitochondria can be damaged so severely by damage to their mtDNA that (i) the various steps of respiration can no longer proceed as usual, and more ROS are generated and (ii) the energy production will fall below a critical threshold, as a result of which the cell will die (apoptosis) (Kremer 2002; Kuklinski and van Lunteren 2005). In case the mechanisms of apoptosis are blocked, the cell will become transformed into a cancer cell (Kremer 2002) and, at the same time, its energy production process will be converted as well: from an oxygen dependent ATP production in the mitochondria to a non-oxygen enzymatic ATP production in the cell plasma (Gatenby and Gillies 2004). This physiological switch of energy production is a counterregulation by the cell because during anaerobic glycolysis far fewer ROS/RNS are generated and, thus, the oxidative stress situation is defused (Brand and Hermfiess 1997; Kremer 2002). In healthy cells, energy production is also switched periodically (during late-stage cell division) in order to protect exposed chromosomes from ROS/RNS. This reaction is regulated by the mitochondrial permeability transition pore whose activity, in turn, is controlled by NO and O^{2-} (Kremer 2002). mtDNA damage mediated by ROS/RNS plays a significant role in the formation of cancer (Carew and Huang 2002; Copeland et al. 2002). The decisive factor of whether a cell is transformed into a cancer cell depends on the redox status of the mitochondria or the mitochondrial membrane potential, respectively (Chen 1988; Kremer 2002). This fact explains the observation that cells may also be transformed into cancer cells when the DNA of the nucleus (nDNA) is not damaged (Weaver and Gilbert 2004; Maffini et al. 2004). This results in an increased ROS/RNS production, which can have further adverse health impact: The resulting damage to proteins, lipids, and the DNA lead to adverse health effects, which may cause cancer and degenerative diseases.

3.5 Biological Effects Relating to EMF Exposure to DNA

It is customary to look for tumor initiating/promoting tendencies due to external environmental perturbations, more so due to low level microwave exposure (in present case) which is pervading the environment with varying intensity. Since the energy of microwave quantum is much less than the energy of covalent/hydrogen bonds, this field exposure can at the most have a tumor promoting effect.

There are reports, (Paulraj and Behari 2002, 2006) indicting adverse health effects of cell phones which emit electromagnetic radiation. It is estimated that a maximum value of 50% of their energy being deposited when held close to the head. Also there is a likelihood that fields emanating from base stations (though very weak), would create a health hazard, though the issue is still being investigated.

In most studies, genotoxic effects have been investigated after short-term exposure (Moulder et al. 1999; Vijayalaxmi and Obe 2004). Diem et al. (2005) reported DNA strand breaks (by comet assay) in human diploid fibroblasts and cultured rat granulose cells after RF field exposure (1800 MHz, SAR 1.2 or 2Wkg^{-1}; different modulations; duration 4, 16 and 24 h; intermittent 5 min on/10 min off or CW). However it is not clear if continuous exposure of nonmodulated or modulated 1800 MHz was used. A statistically significant increase in micronucleus formation and in chromosomal aberrations were observed in fibroblasts as well. Speit et al. (2007) used CW with intermittent exposure (1800 MHz, SAR 2Wkg^{-1}), applied using the same cell system and reported negative results. Nikolova et al. (2005) reported, that after 6 h, but not after 48 h RF field exposure, a low and transient increase in DNA strand breaks in embryonic stem-cell-derived neural progenitor cells. Lai and Singh (1995, 1996) reported that rats exposed to pulsed (two-second pulses, 500 pulses per second) or CW 2.45 GHz fields with SARs of 0.6 or 1.2 Wkg^{-1} for 2 h increased the number of single and double-strand breaks in brain cell DNA when assayed 4 h after RF exposure. Lai and Singh (1997) also reported that treatment of rats immediately before or after exposure with either melatonin (1 mg kg^{-1}) or PBN (N -t-butylphenylnitrone, 100 mg kg^{-1}) blocks the formation of DNA breaks by RF fields. These experiments suggest that RF fields are unable to break molecular bonds. Epigenetic events may be involved. Cleary (1990) reported that RF radiation at 2.45 GHz and 27 MHz increase cell proliferation. However, Malyapa et al. (1998) reported that no DNA damage in cells of the rat cerebral cortex or the hippocampus after a 2 or 4 h exposure to 2450 MHz (CW).

In a relatively long term study, mice were exposed to 2.45 GHz fields at an SAR of 1.18 Wkg^{-1} for 2 h per day for 120, 150 and 200 days, whereby structural genomic re-arrangement were observed in brain and testes cells (Sarkar et al. 1994). Paulraj and Behari (2006) studied the same phenomena at other frequencies and low dose level, and confirmed DNA strand break; and concluded that the hippocampus is the targeted site of such interactions. 2.45 GHz radiation causes significant increase in ornithine decarboxylase activity and a decrease in protein kinase activity in rat brain after chronic microwave exposure (Paulraj and Behari 2002). Kilgallon and Simmons (2005) concluded that a decrease in sperm concentration has been found due to keeping cell phones close to waist.

Phillips et al. (1998) reported single-strand DNA breaks exposed to cellular telephone frequencies 813.5 MHz and 836.5 MHz at low SAR (average 2.4 and 24 $\mu W/g^{-1}$). These authors used the same comet assay techniques used by Lai and Singh (1997). They also postulated that DNA-repair rates may be affected by exposure to RFR (Phillips et al. 1998). Large electron flows are possible within the stakked base pairs of the double helix of DNA molecules.

Several findings viz. protein kinase C (PKC), apoptosis and sperm count (Paulraj and Behari 2002, 2006; Kesari et al. 2010, 2011) are suggestive that their levels are affected due to mobile phone radiation exposures. Such type of alterations may have implications leading to infertility. It is well known that PKC plays a key role in a variety of pathologic states including oncogenesis (Harvey and French 2000; Giwercman et al. 2003) and in mediating cellular responses to extracellular stimuli involved in proliferation, differentiation, apoptosis, and exocytotic release in a number of non-neuronal and sperm cells (Ohkusu et al. 1986). It was found to be localized mainly in the equatorial segment of the human sperm (Rotem et al. 1990a, b). Lee et al. (2004) reported that EMF may induce cell death (apoptosis) in several in vivo studies mostly on mice and rats. Fejes et al. (2005) also observed negative effects on sperm motility due to prolonged use of cell phones.

Paulraj and Behari (2004) and Kesari and Behari (2009) reported a decrease in the activity of PKC in rats exposed to 147 MHz amplitude modulated and 50 GHz microwave exposure respectively in developing rat brain. This has been earlier corroborated in our study (Kesari and Behari 2008). Findings, at such lower power show a significant decrease in sperm count and an increase in apoptosis are also in support of the findings of Agarwal et al. (2007).

Yan et al. (2007) reported with 6 h of daily cellular phone exposure for 18 weeks exhibited a significant higher incidence of sperm cell death than control group rats. De Iuliis et al. (2009) have also reported that the radio frequency-electromagnetic radiations emitted from mobile phones enhances mitochondrial reactive oxygen species generation by human spermatozoa, decreasing the motility and vitality of these cells and stimulating DNA fragmentation. These authors (Yan et al. 2007; De Iuliis et al. 2009) suggest that carrying cell phones close to reproductive organs could negatively affect male fertility and health, confirming that targeting central nervous system has also adverse effect on reproductive system.

3.5.1 Chromosome Aberrations and Micronuclei

Garaj-Vrhovac et al. (1999) reported that exposure to microwave radiation of only 10–20 μW/cm^2 in workers chronically exposed to 1250–1350 MHz was sufficient to cause an increase in micronuclei (an aberrant form of DNA). Vijayalaxmi et al. (1997, 1998) reported that the frequency of micronuclei was increased in the peripheral blood and bone marrow of cancer-prone mice chronically exposed to 2450 MHz microwave radiofrequency radiation. Maes et al. (1993) exposed human peripheral blood lymphocytes to microwaves at 2450 MHz. A marked increase in the frequency of chromosome aberrations and micronuclei (the formation of abnormal chromosome fragments) was reported at nonthermal levels. Chromosome aberrations increased with increasing time exposure (a dose-response). The results on this count (nonionizing radiations) are consistent with those obtained by the exposure due to ionizing radiations (Garaj-Vrhovac et al. 1991).

Maes et al. (1995) reported that whole blood exposed to the radiating antenna of a GSM base station showed increased chromosome aberrations when placed within a distance of 5 cm or less with 2 h exposures. Combined effects of 954 MHz radiofrequency radiation and the chemical mutagen mitomycin C were studied by these authors using human lymphocytes. Blood samples were exposed to AM radiation from a GSM base station at an estimated SAR of 1.5 W/Kg. Microwave exposure enhanced the harmful effect of the chemical mutagen and showed a clear increase in a form of chromosome aberration (sister chromatid exchange). Single strand DNA breaks were also reported.

3.6 Biology of Reproduction Process

Mammalian spermatogenesis occurs within U-shaped seminiferous tubules lined by sertoli cells, which support germ cells in varying stages of development. In mammalian testes, germ cells expand clonally through many rounds of mitosis before undergoing the differentiation steps that result in mature spermatozoa. Protein kinases encodes approximately 2% of eukaryotic genes and represent one of the major classes of cell regulatory molecules (Hunter and Plowman 1997). Manning et al. (2002) reported that the human gonome carries 518 protein kinase genes (~1.7% of all genes). A majority (~90%) of these protein kinases belong to the eukaryotic protein kinase superfamily, defined on the basis of a homologous kinase catalytic domain (Hanks and Hunter 1995) and its functions to transfer the c-phosphate of ATP onto the hydroxyl group of amino acid residue within the peptide substrate. The maturation promoting factor is a biologically recognized factor responsible for driving the cell cycle from the G to the M phase (Nurse 1990; Meikrantz and Schlegel 1992), as measured by cytoplasmic injection or cell fusion (Jung et al. 1993). In general, initiation of mitosis (M phase) requires a protein kinase complex (maturation promoting factor) consisting of a catalytic subunit

(Cdc2 protein kinase) (Dunphy et al. 1988; Gautier et al. 1988) and a regulatory subunit (cyclin B) (Labbe et al. 1989).

Catalytic activity of a specific protein kinase plays an important role in elucidating signal transduction pathways that affect cell behavior. These radiations enhance the overproduction of ROS, which may lead to genotoxicity and carcinogenicity. The decrease in histone kinase activity may serve as a measure of the ability of EMF to affect spermatogenesis and cell cycle in sperm. These results also substantiate that the activity of histone kinase is closely related to the G_2/M transition during the cell cycle (Pawse et al. 1971). The electromagnetic field that decreases the activity of histone kinase also decreases the G2/M phase of cell cycles. Levels of histone kinase phosphorylation are usually lowest in the G1 phase, continuously increase during the S phase and mitosis. Lowering of phosphorylation of histone kinase may affect permatogenesis and thereby causing infertility.

Kesari and Behari (2010) reported the effect of 2.45 GHz radiation on reproductive system of male rats. Chronic exposure to these radiations produced the formation of apoptotic cell in testis as well as alteration in antioxidant capacity of sperm cell. The data suggested that prolonged exposure of microwave radiation may induce genetic damage in reproductive system of male rats due to overproduction of ROS. Kesari et al. (2011) analysed effect of mobile phones (900 MHz having SAR 0.9 W/kg) on fertility pattern of male Wister rats. Antioxidant enzymes glutathione peroxidase ($p < 0.001$) and superoxide dismutase ($p < 0.007$) showed a decrease, while an increase in catalase ($p < 0.005$) was observed. Malondialdehyde ($p < 0.003$) showed an increase in the exposed group. Micronuclei also showed significant decrease (0.002) in the exposed group. A significant change in sperm cell cycle of G_0–G_1 ($p = 0.042$) and G_2/M ($p = 0.022$) were recorded. Generation of free radicals was recorded to be significantly increased ($p = 0.035$).

A study aimed to find out the effect of chronic exposure of radiation. Data show that prolonged (45 days or more) exposure to microwave radiation cause double-strand DNA break in brain cells (Fig. 3.5b). In the qualitative picture of DNA double-strand break, it is shown that more tail migration takes place in the exposed brain as compared to control. It is variously reported that the prolonged chronic exposure to EM radiations cause reproducible increase in double-strand DNA breaks in brain cells of rats in all the exposed-group animals. It shows that there is a significant increase in the head and tail length of DNA, and tail migration was also recorded as compared to the control group. A comparison for the control group, the average values of head length (93.26 ± 0.41 μm), tail length (66.26 ± 0.26 μm), and tail migration (19.63 ± 0.70 μm), was recorded, whereas for the exposed group, it was head length (109.17 ± 1.13 μm; $p < 0.01$), tail length (176.56 ± 2.35 μm; $p = 0.0023$), and tail migration (122.23 ± 2.19 μm; $p < 0.005$). At the same time, the total average value of head and tail intensity was also recorded. A significant decrease was observed in head intensity, whereas significant increase was observed in tail intensity of the exposed group. For the control group, the average value head intensity (87.05 ± 1.2%) and tail intensity (12.94 ± 1.20%) was scored.

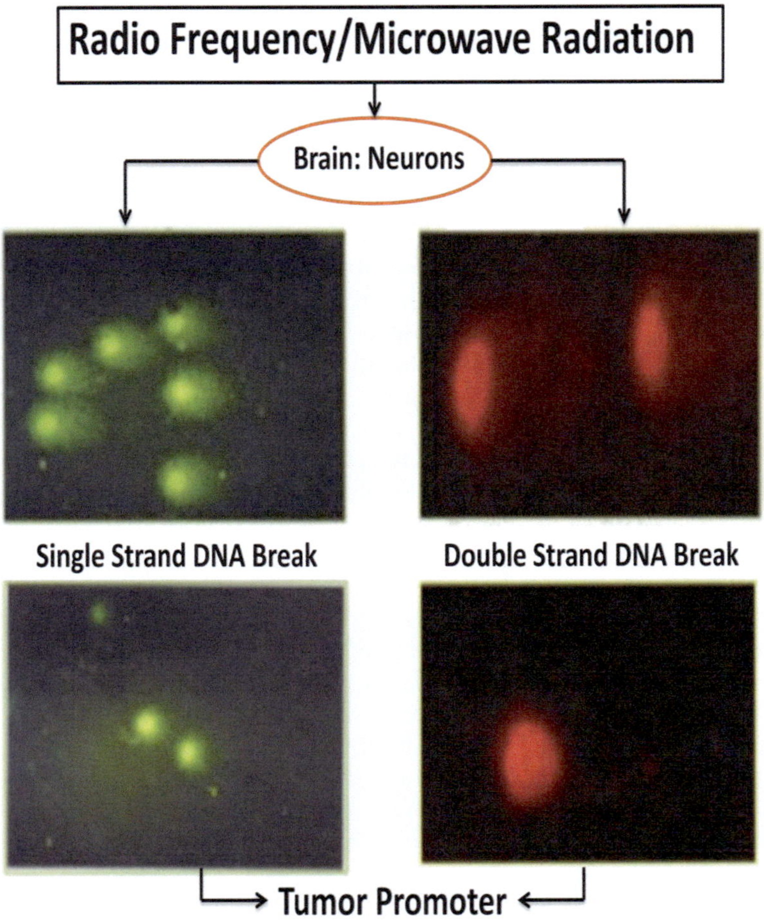

Fig. 3.5b Effect of radiofrequency/microwave radiation on neuronal cells

Alternatively, in tail exposed group, the average values in head intensity ($49.76 \pm 1.04\%$, $p < 0.03$) and tail intensity ($48.24 \pm 2.00\%$; $p < 0.02$) were recorded (Kesari and Behari 2009).

3.7 Reactive Oxygen Species: Generation and Role in DNA Fragmentation

Dasdag et al. (2008) reported the effect of 900 Mhz microwave radiation (0.0784 mW/cm^2) on Wistar albino rats with increased MDA and ROS levels. They stated that RF radiation from GSM cellular phone may play a role to induce oxidative damage by increasing lipid peroxidation and oxidative stress.

The balance between production and neutralization of ROS is maintained by concerned action of enzymatic and nonenzymatic defense systems. When unbalanced, it may lead to oxidation of polyunsaturated fatty acids in lipids, amino acids in proteins, and damage to the DNA. The ROS may cause a G_2/M arrest, as reported by several workers (Thorn et al. 2001; Zhang et al. 2003; Bijur et al. 1999). When ROS come in contact with DNA, it leads to oxidative damage to DNA and might result in mutagenesis (Ames et al. 1993; Cooke et al. 2003). Reports suggest that electromagnetic field increase the concentration of free radicals which may enhance the probability of damage to the biological system (Zmyslony and Jajte 1998; Yoshikawa et al. 2000; Eveson et al. 2000). The EMFs exhibit their effects on the cells through an elaborate intracellular signal transduction network, which ultimately results in progression through the cell cycle or alternatively to cell cycle arrest, cell differentiation, or apoptosis. Free radicals can cause the phosphorylation and activation of numerous signaling proteins, besides histone kinase and PKC (Finkel and Holbrook 2000). Earlier studies established a decrease in activity of protein kinase C due to EMF exposure at 2.45 GHz (Paulraj and Behari 2006; Kesari and Behari 2010). It is suggested that activity of histone kinase (maturation promoting factor) is closely related to the G_2/M transition during the cell cycle. The electromagnetic field that decreases the activity of histone kinase also decreases the G_2/M phase of the cell cycle. While the making and breaking of strand breaks is constantly happening, the presence of electromagnetic field acts as perturbant to disturb this dynamic equilibrium thereby causing SSBs and/or DSBs, governed by law of probability.

3.7.1 Mutations

All DNAs have hereditary coding meaning thereby that it is passed from parent to offspring or child. The only way new genetic material can be created is through genetic mutations within cells. Mutations can occur through a number of ways and either be positive, negative or make no difference at all.

DNA damage seems more alarming as it may lead to increased mutation frequency and could be linked to gene mutations. A gene is a section of DNA containing the information needed to make a particular protein or enzyme. There is also a section that can turn the gene on or off in response to outside signals. The growth of an organism from a fertilized egg involves a hugely complex pattern of switching genes on and off that regulates growth, cell division and differentiation into specific tissues. A damage to DNA (male/female or both) can impair reproduction pattern in an unpredicted manner (Fig. 3.6). Most of the time mutations don't lead to any significant change but in worse case scenarios mutations can lead to cancerous growths. In other circumstances, mutations can be beneficial and make an individual better suited to their environment which, through natural selection, leads to the evolution of new species. DNA damage can sometimes give

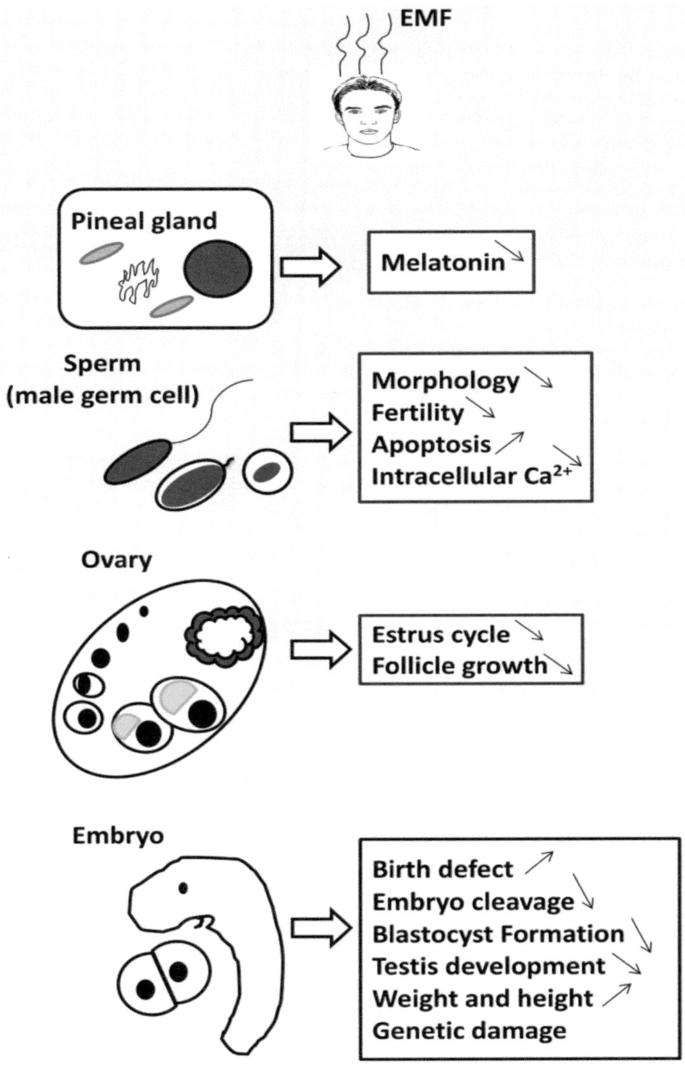

Fig. 3.6 EMF exposure may lead to genetic damage

unregulated growth to form tumors. However, the effect may not be immediate and may take years before it is fully replicated.

Halgamuge (2013) showed the significance disruption of melatonin due to exposure to weak EMFs, which may possibly lead to long term health effects in humans. Melatonin has also been identified as natural anti-cancer chemical (Reiter and Robinson 1995). If endogenous concentration of melatonin is reduced, the natural ability of the body to fight against cancer is decreased. Melatonin hormone works as most potent free radical scavenger opposing the most toxic free radicals

(Zang et al. 1998). It was reported that melatonin can protect sperm mitochondria from ROS through its effective antioxidant potential (Shang et al. 2004).

Chromosomes consist mainly of DNA, histones and nonhistone proteins and are organized during interphase in domains. In most cells chromosomes are visible only during the process of cell division (Cremer et al. 1996; Savage 1996; Bickmore and Craig 1997; Kreth et al. 1998; Wolffe 1998). It is believed that mostly G_0/G_1 chromosomes contain one DNA molecule which is replicated in the S phase of the mitotic cell cycle to yield two sister chromatids, each of which consists of one single DNA molecule. Sometimes, alterations in the normal chromosome structure (chromosomal aberrations (CA)), are observed. CA can be classified as intra- and inter-chromosomal aberrations. The first class comprises aberrations within a single chromosome, such as terminal and interstitial deletions and inversions; the second class comprises rearrangements between two or more chromosomes, such as translocations and dicentrics.

3.8 Bioindicaters of Cancer Promotion: Role of Melatonin

There are various indicators of tumor promotion, due to electromagnetic field exposure:

1. Increase in ODC and PKC activity.
2. Increase in Ca^{2+} efflux.
3. Altered anti oxidant enzyme activity(GX, GSH, Catalase)
4. DNA strand break (single and double)
5. Na^{+}/k^{+} Phosphatase activity
6. Decreased melatonin levels
7. Histone kinase activity

↓

Cancer promotion (Stress Development/RF/MW Syndrome)/effecting reproduction parameters.

Epidemiological studies reported an increased risk of brain tumors among analogue cellular phone users (Hardell et al. 2003). EMF-induced DNA damage could be blocked by treating rats with antioxidants, including melatonin, immediately before and after exposure. Melatonin is a hormone secreted from the brain's pineal gland (Fig. 3.7). As a potent antioxidant, it effectively eliminates free radicals inside cells.

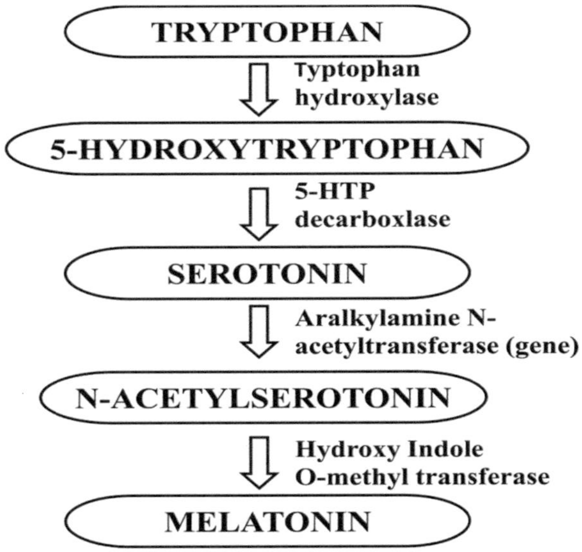

Fig. 3.7 Pathway of melatonin release from Pineal gland in brain

As per the investigation so far, melatonin is primarily synthesized in the pinealocytes of pineal gland in all vertebrates at night (Karasek 2006). Imbalances in the serotonin/melatonin cycle (Fig. 3.4b) may cause depression experienced by people living near high intensity of electromagnetic field. According to the melatonin hypothesis of Stevens and Davis (1996), the primary effect of exposure is reduction of the synthesis of melatonin which finally results in secondary endocrine effects which in turn promotes growth of cancer. If the synthesis of melatonin is low against higher amounts of oxygen radicals produced by electromagnetic radiation, these effects may lead to DNA damage which would increase cancer risk (Reiter et al. 1999).

Reiter (1993) stated that the activity of pineal gland in brain is quickly altered by electromagnetic fields. Melatonin is synthesized by pea shaped pineal gland located in the middle of the brain (Fig. 3.7). They also stated that exposure to sinusoidal electric and/or magnetic fields as well as pulsed static magnetic fields often reduce pineal melatonin production. Melatonin plays a central role in fine tuning circadian rhythms and seasonal changes through its daily nocturnal increase in the blood (Reiter 1991; Malpaux et al. 2001). Melatonin is a neuro-hormone that controls sleep cycle, sex hormones and reproduction. It serves as a compass (it detects changes in the geomagnetic field), a clock (it sense changes in visible light, a part of the EMF spectrum and induces sleep), and a calendar (it senses changes in photoperiod and induces hibernation as well as ovulation and thus controls reproductive cycles in seasonal breeding animals) (Havas 2004). The role of melatonin in reproduction in many species is related to seasonal reproductive cycles and in man it modulates its activity through reproductive neuroendocrine axis (Awad et al. 2006).

Damage to body occurs when intracellular antioxidant mechanism are overwhelmed by ROS (Zmyslony and Jajte 1998). GSH, CAT, and SOD protect cells against ROS. As indicative of damage to body systems, free radicals are identified as one of cause. Free radicals are chemical constituents that have an unpaired electron. If an electron is added to O_2 then the superoxide anion radical O_2^- is formed. O_2^- is reduced by superoxide dismutase to H_2O_2 which is toxic at high concentrations and can be reduced to •OH. The hydroxyl radical (•OH) damage cells. Melatonin is an efficient neutralizer of •OH. When melatonin interacts with the initial •OH, it by necessity, becomes a radical, in this case the indolyl (melatonyl) radical. This reactant has very low toxicity so there is a net gain when melatonin scavenges the •OH, a highly toxic reactant is replaced by a radical with low toxicity. After some molecular rearrangement the indodyl radical scavenges a second •OH to form cyclic 3-hydroxymelatonin (Reiter et al. 2000).

Melatonin donates an electron to hydroxyl radical which is relatively unreactive indolyl cation radical, which presumably scavenges a $•O_2^-$ thereby further reducing the likelihood of formation of the highly toxic OH^- (Hardeland et al. 1993). The findings, indicated that melatonin restricts lipid peroxidation by preventing the initiating events as well as interrupting the chain reaction.

A decrease in the *Superoxide dismutase* (SOD) activity was found in steel workers exposed to electromagnetic field which resulted in the accumulation of superoxide anion-radicals in RBCs (Kula et al. 2000). However, CAT activity was higher in worker and Glutathione peroxidase (GPx) inactivated H_2O_2 and organic hydroxides together with glutathione reductase and glucose-6-dehydrogenase with reduced glutathione. In their study, a statistically significant increased level of MDA was found which may indicate increased lipid peroxidation caused by the EMF. All these changes lead to some adaptative responses due to activation of systems controlling the body oxidative mechanism balance under the influence of electromagnetic fields. Kula et al. (2002) reported statistically significant decreases in red blood cells SOD and GSH-Px activities, a CAT activity increase and plasma MDA increased in workers exposed for 3–10 years and longer than 10 years to electromagnetic fields. Authors also concluded that all changes observed among workers exposed to electromagnetic field result in adaptive responses by activating systems controlling the balanced of body oxidative mechanisms.

Stopczyk et al. (2005) demonstrated that microwaves produced by mobile phones significantly depleted SOD-1 activity after 1, 5, and 7 min of exposure and increased after 3 min in comparison with the control test. There was a significant increase in the concentration of MDA after 1, 5, and 7 min and decrease after 3 min of exposure as compared with the control test. On the grounds of these results they concluded that oxidative stress after exposure to microwave may be the reason for many adverse changes in cells and may cause a number of systemic disturbances in the human body.

Under the influence of mounting evidences, it is now generally accepted that reactive oxygen species (ROS) production in sperm suspensions, lipid peroxidation and DNA oxidation availability of exogenous antioxidants, the parameters of exposure (e.g., generated power, duration of exposure and wave shape), and the oxidative damage is cumulative.

The findings on antioxidant, malondialdehyde, histone kinase, micronuclei and sperm cell cycle are clear indications of an infertility pattern, initiated due to an overproduction of reactive oxygen species. It can be concluded that radiofrequency electromagnetic wave from commercially available cell phones might affect the fertilizing potential of spermatozoa.

It has been proposed that microwave effects superconductive tunneling in carbon films at room temperature, suggest the possibility that non-thermal effects of microwaves on biological systems may occur by a superconductive mechanism (Cope 1976).

While the thermal effects of microwaves are not commonly observed, non-thermal effects, if undetected, may adversely effects human health. It is understood that RF electromagnetic fields (EMFs) from base stations to which humans are exposed (far-field exposures) are well below the allowed values. In another survey, it was reported that around 4% of people claim that they are sensitive to RF EMFs to some extent (Eltiti et al. 2007). The biological effects also depend upon the intermittency of the exposure, may be accumulative, and may include membrane interactions that affect the ion fluxes, the modulation of neuronal impulse activity and possibly induce arrhythmia in isolated heart. Behavioural effects and those on reproductive patterns are also possible at low levels of microwave expo-sure. There are important biomedical considerations associated with long-term exposure to any environmental factor capable of tissue interactions. These include effects specifically attributable to wave type, duration, intermittency, frequency of recurrent exposure, interactions involving simultaneous exposure to multiple factors, age at onset of exposure and may also reflect ethnicity. While examining the biological effects, it is realized that it is not only the external field, but the induced fields within the body, that are important. Despite the fact that many workers have claimed adverse effect of microwave frequency exposure on human being, there is no uniform guidelines of safe exposure have been evolved internationally. This may be because of difficulty in defining a suitable microwave background for all the exposures. Though this may be an empirical approach but it indicates an attempt to make the exposure level bit more conservative to reach a level of safe exposure to humans and other living species. However it will remain a challenge to adopt methodologies to monitor this exposure level.

While damage to DNA strands has been confirmed by several workers (Lai and Singh 1995, 1996; Paulraj and Behari 2004, 2006), it is argued that its repair is an ongoing process and the damaged chromosomes can be reconstituted. However, this proposition is not without risk. It is not certain that there will be a faithful replicate in the manner they were originally present. Pieces may be left out (deletions), joined in backwards (inversions), swapped between different parts of the chromosome (translocations) or even attached to the wrong chromosome. The quantum of effect can also be frequency dependent. In most cases, the new arrangement can work for a while, if most of the genes are still present, and any metabolic deficiencies can often be made good by the surrounding cells. However, things may be different when it comes to meiosis. During meiosis, the chromosomes line up in pairs (one from each original parent) along their entire length so that corresponding parts are adjacent and can be exchanged. Malformed pairs are torn apart in the later stages of meiosis so that eggs or sperms that have an incomplete or unbalanced set of genes may not function properly, and so reduce fertility and other physiological functioning. There is a possibility that this may lead to permanent genetic damage, which although may not be visible in the first generation, may be evident in subsequent births.

It is quite intriguing that the mode of EMF interaction with biological system is multifold. There is ample of evidence that low level electromagnetic fields causes DNA strand break in brain cells and sperm. On the other hand, the effect of RF radiation on DNA could conceivably be more significant on neurons than on other cell types, because these nerve cells have a low capability for DNA repair. An application of pulsed electromagnetic field on induced fracture and osteoporosis in rats causes an accelerated healing (Behari 2009) and deceleration (Fig. 3.5b) of osteoporosis in ovariectomised and neuroectomised rat (bones) (Jayanand and Behari 2005).

3.9 Method for Identifying DNA Damage

3.9.1 Comet Assay

The Comet assay, also known as single cell gel electrophoresis, is a sensitive technique for the detection of DNA damage at the level of the individual eukaryotic cell. DNA damage can lead to changes in cellular functions and may cause cell death. DNA damage, reflected as single-strand breaks, double-strand breaks, and crosslinks. The damage is represented by an increase in DNA fragments that have migrated out of the nucleus of the cell. This approach displays the DNA damage as a characteristic streak similar to the tail of a comet. The DNA damage is measured as DNA single strand breaks, double strand breaks, strand breaks induced by alkali-labile sites, DNA adducts, oxidative damage and cross-linking agents. The length and fragment content of the comet tail is directly proportional to the amount of DNA damage. This assay can be performed on tissue from both mice and rats and has been validated with the following tissues: liver, blood, bone marrow for mouse; blood, liver, kidney, spleen, glandular stomach, duodenum, jejunum, bone marrow for rat.

Comet assay provides a method to ascertain Genotoxicity mechanism and for screening DNA damage repair inhibitors.

3.9.2 Procedure for Comet Assay

1. Cells mixed with low melting agarose at 37 °C (LM Agarose)
2. Immobilize cells on ComerSlide™
3. Treat cells with lyrics Solution (removes membranes and histones from the DNA
4. Samples treated with alkali (unwinds and denature DNA)
5. Samples stained with intercalating dye and visualized by epiflurescence microscopy following alkaline electrophoresis, which reveals DNA breaks

3.9.3 Exposure Chamber for Animals

In a typical exposure setup used for multiple number of animals exposed at the same time be as follows. Six rats at a time are placed in a Plexiglass cage, which is quite ventilated with holes of 1-cm diameter. The dimension of the exposure cage was made in such a way that animals moved freely. The chamber is lined with radar-absorbing material (attenuation, 40 db) to minimize the possibility of any reflections. At far field distance from the horn antenna, the field is homogeneous in the vertical plane midline of the beam. Rats are exposed through the antenna, for a specified days and duration (Fig. 3.8a). The animals faced E filed orientation parallel to the exposure chamber. Since the free space wavelength at 2.45 GHz is 12.24 cm and within the medium is approximately 1.74 cm, it is assumed that the exposure is limited to the subcutaneous surface. In Fig. 3.8a variable number of small size animals can be placed for microwave exposure. For one animal to be exposed at a given time the setup shown in Fig. 3.8b is adopted. Another near field exposure setup used by different workers to examine possible effects of electromagnetic exposure on the central nervous system at the frequency bands (800–900 MHz), whereby ten adults rat are grouped around a symmetrical dipole, where the distance between feed point and snout is kept around 3 cm is also used in Kuster et al. (1997).

3.10 Conclusion

Since microwave exposure has been identified as a pollutant which has adverse effect on human health, there is a call to restrict its level in the environment. Several recommendations (*American Conference of Governmental Industrial Hygienists,* Federal Communications Commission and Institute of Electrical and Electronics Engineers) for criteria of field exposure have been proposed. The most liberal of this (ICNIRP) has been adopted by India at the microwave frequency corresponding to 900 MHz, while reducing it by a factor of 10. However there are inherent difficulties in defining the field intensity at a given point, more so in an urban environment. It is therefore apparent that the adverse effect of these radiations will

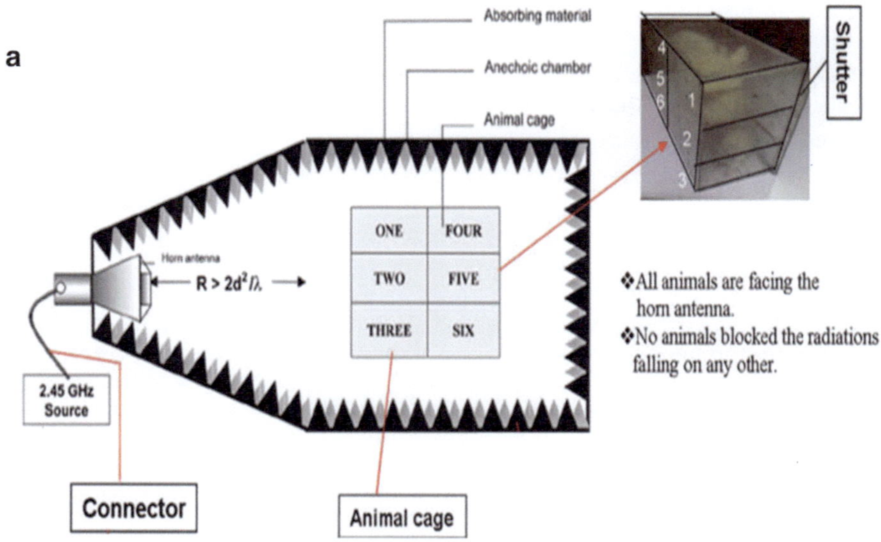

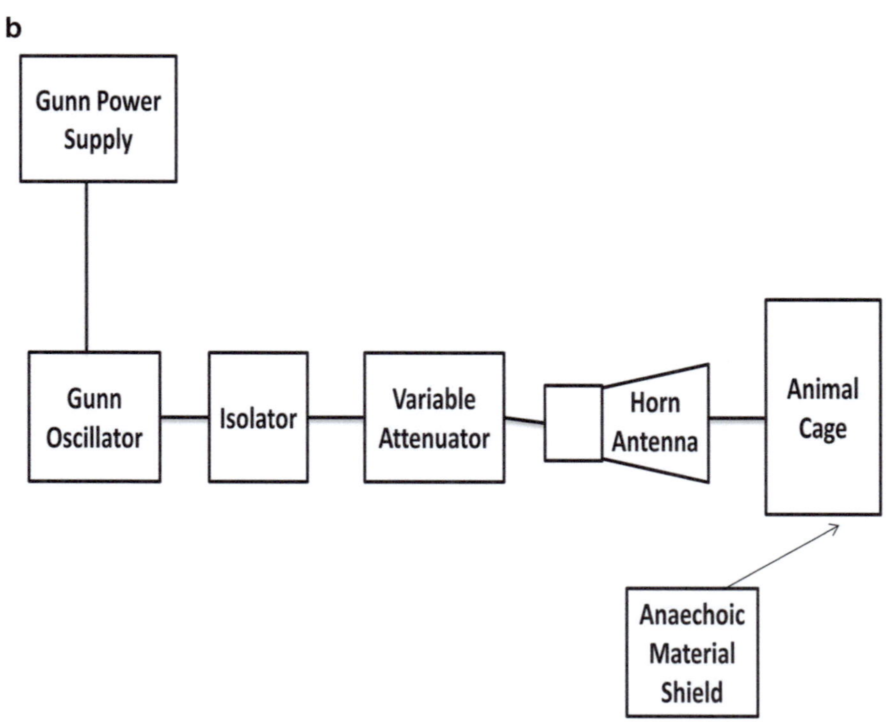

Fig. 3.8 (**a**) An experimental setup for the exposure of rats to microwave frequency at 2.45 GHz. (**b**) Schematic diagram for microwave radiation on rats

continue to haunt all living species for long time to come. However in ICNIRP report (ICNIRP Report 2004) on possible effects of long term exposure to RF radiation state that there is no convincing evidence of relation between RF and any adverse health effects. Even then most of the countries in Europe and America have adopted criteria for safe exposure.

References

Agarwal A, Deepinder F, Sharma RK, Ranga G, Li J (2007) Effect of cell phone usage on semen analysis in men attending infertility clinic: an observational study. Fertil Steril 89:124

Alberts, B., Johnson, A., Lewis, J., Raff, M., Roberts, K., & Walters, P. (2002). Molecular biology of the cell, 4th edn.. Garland Science, New York/London, ISBN 0–8153–3218-1. OCLC 145080076 48122761 57023651 69932405

Ames BN, Shigenaga MK, Hagen TM (1993) Oxidants, antioxidants, and the degenerative diseases of aging. Proc Natl Acad Sci U S A 90:7915–7922

Atamna H, Cheung I, Ames BH (2000) A method for detecting abasic sites in living cells: age-dependent changes in base excision repair. Proc Natl Acad Sci U S A 97:686–691

Awad H, Halawa F, Mostafa T, Atta H (2006) Melatonin hormone profile in infertile males. Int J Androl 29:409–413

Behari J (2009) Biophysical bone behavior principles and applications. John Willey & Sons (Asia) PTE LTD, Singapore

Berg J, Tymoczko J, Stryer L (2002) Biochemistry, W. H. Freeman and Company, New York, ISBN 0–7167–4955-6

Bickmore W, Craig J (1997) Chromosome bands: patterns in the genome. Springer, Heidelberg

Bijur GN, Briggs B, Hitchcock CL, Williams MV (1999) Ascorbic acid-dehydroascorbate induces cell cycle arrest at G2/M DNA damage checkpoint during oxidative stress. Environ Mol Mutagen 33:144–152

Brand KA, Hermfiess U (1997) Aerobic glycolysis by proliferating cells: a protective strategy against reactive oxygen species. FASEB J 11:388–395

Brusick D, Albertini R, McRee D, Peterson D, Williams G, Hanawalt P, Preston J (1998) Genotoxicity of radio frequency radiation DNA/Genetox Expert Panel. Environ Mol Mutagen 32:1–16

Carew JS, Huang P (2002) Mitochondrial defects in cancer. Mol Cancer 1(9). http://www.molecular-cancer.com/content/pdf/1476-4598-1-9.pdf

Chen LB (1988) Mitochondrial membrane potential in living cells. Annu Rev Cell Biol 4:155–181

Cheng KC, Cahill DS, Kasai H, Nishimura S, Loeb LA (1992) 8-Hydroxyguanine, an abundant form of oxidative DNA damage, causes G>T and A>C substitutions. J Biol Chem 267 (1):166–172

Cleary SF (1990) Biological effects of radiofrequency electromagnetic fields. In: Gandhi OP (ed) Biological effects and medical applications of electoromagnetic energy. Prentice Hall, Englewood Cliffs, pp 236–255

Cooke MS, Evans MD, Dizdaroglu M, Lunec J (2003) Oxidative DNA damage: mechanisms, mutations and disease. FASEB J 17:1195–1214

Cope FW (1976) Superconductivity--a possible mechanism for non-thermal biological effects of microwaves. J Microw Power 11(3):267–270

Copeland WC, Wachsman JT, Johnson FM, Penta JS (2002) Mitochondrial DNA alterations in cancer. Cancer Investig 20(4):557–569

Cremer C, Münkel CH, Granzow M, Jauch A, Dietzel S, Eils R, Guan X-Y, Meltzer PS, Trent JM, Langowski J, Cremer T (1996) Nuclear architecture and the induction of chromosomal aberrations. Mutat Res 366:97–116

Dasdag S, Bilgin HM, Akdag MZ, Celik H, Aksen F (2008) Effect of long term mobile phone exposure on oxidative-antioxidative processes and nitric oxide in rats. Biotechnol Biotechnol Equip 22:992–997
De Iuliis GN, Newey RJ, King BV, Aitken RJ (2009) Mobile phone radiation induces reactive oxygen species production and DNA damage in human spermatozoa *in vitro*. PLoS One 4:6446
Diem E, Schwarz C, Adlkofer F, Jahn O, Rüdiger HW (2005) Non-thermal DNA breakage by mobile phone radiation (1,800 MHz) in human fibroblasts and in transformed GFSH-R17 rat granulosa cells in vitro. Mutat Res 583:178–183
Druzhyna NM, Wilson GL, LeDoux SP (2008) Mitochondrial DNA repair in aging and disease. Mech Ageing Dev 129(7–8):383–390
Dunphy WG, Brizuela L, Beach D, Newport J (1988) The Xenopus cdc2 protein is a component of MPF, a cytoplasmic regular of mitosis. Cell 54:423–431
Eltiti S, Wallace D, Zougkou K, Russo R, Joseph S, Rasor P, Fox E (2007) Development and evaluation of the electromagnetic hypersensitivity questionnaire. Bioelectromagnetics 28:137–151
Eveson RW, Timmel CR, Brocklehurst B, Hore PJ, McLauchlan KA (2000) The effects of weak magnetic fields on radical recombination reactions in micelles. Int J Radiat Biol 76:1509–1522
Fejes I, Zavaczki Z, Szollosi J, Zavaczki Z, Pal A (2005) Is there a relationship between cell phone use and semen quality? Arch Androl 51:385
Finkel T, Holbrook NJ (2000) Oxidants, oxidative stress and the biology of ageing. Nature 408:239–247
Friedrich Miescher (1869). DNA with associated proteins, from cell nuclei. http://www.dnaftb.org/15/bio.html
Fujikawa H, Ushioda H, Kudo Y (1992) Kinetics of Escherichia coli destruction by microwave irradiation. Appl Environ Microbiol 58:920–924
Garaj-Vrhovac V, Horvat D, Koren Z (1991) The relationship between colony forming ability, chromosome aberrations and incidence of micronuclei in V70 Chinese hamster cells exposed to microwave radiation. Mutat Res 263:143–149
Garaj-Vrhrovac V et al (1999) Micronucleus assay and lymphocyte mitotic activity in risk assessment of occupational exposure to microwave radiation. Chemosphere 39(13):2301–2312
Gatenby RA, Gillies RJ (2004) Why do cancers have high aerobic glycolysis? Nat Rev Cancer 4:891–899
Gautier J, Norbury C, Lohka M, Nuese P, Mailer J (1988) Purified maturation-promoting factor contains the product of a Xenopus homolog of the fission yeast cell cycle control gene cdc2. Cell 54:433–439
Giwercman A, Richthoff J, Hjollund H, Bonde JP, Jepson K, Frohm B, Spano M (2003) Correlation between sperm motility and sperm chromatin structure assay parameters. Fertil Steril 80:1404
Gregory SG, Barlow KF, McLay KE, Kaul R, Swarbreck D, Dunham A et al (2006) The DNA sequence and biological annotation of human chromosome 1. Nature 441(7091):315–321
Halgamuge MN (2013) Pineal melatonin level disruption in humans due to electromagnetic fields and ICNIRP limits. Radiat Prot Dosim 154:405–416
Hanks SK, Hunter T (1995) The eukaryotic protein kinase superfamily: kinase (catalytic) domain structure and classification. FASEB J 9:576–596
Hardeland R, Reiter RJ, Poeggeler B, Tan DX (1993) The significance of the metabolism of the neurohormone melatonin: antioxidative protection and formation of bioactive substances. Neurosci Biobehav Rev 17:347–357
Hardell L, Mild KH, Carlberg M (2003) Further aspects on cellular and cordless telephones and brain tumors. Int J Oncol 22:399–407
Harvey C, French PW (2000) Effects on protein kinase C and gene expression in a human mast cell line, HMC-1, following microwave exposure. Cell Biol Int 23:739
Havas M (2004) Biological effects of low frequency electric and magnetic fields. In: Clements-Croome D (ed) Electromagnetism and health. Taylor & Francis Publication, London, p 25

Hruszkewycz AM, Bergtold DS (1988) Oxygen radicals, lipid peroxidation and DNA damage in mitochondria. Basic Life Sci 49:449–456
Hunter T, Plowman GD (1997) The protein kinase of budding yeast: six score and more. Trends Biochem Sci 22:18–22
ICNIRP Report (2004) Epidemiology of health effects of radiofrequency. Environ Health Perspect 112(17):1741–1754
Iliakis G (1991) The role of DNA double strand breaks in ionizing radiation induced killing of eukaryotic cells. Bio Essays 13:641–648
Irobalieva, R. N., Fogg, J. M. Catanese Jr, D. J., Sutthibutpong, T, Chen, M, Barker, A. K., Ludtke, S. J., Harris, S. A., Schmid, M. F. (2015). Structural diversity of supercoiled DNA. Nat Commun, 6:8440. doi:10.1038/ncomms9440. PMC 4608029. PMID 26455586
IUPAC-IUB Commission on Biochemical Nomenclature (CBN) (2006) Abbreviations and symbols for nucleic acids, polynucleotides and their constituents. Eur J Biochem. Retrieved 3 January 2006
Jackson SP, Bartek J (2009) The DNA-damage response in human biology and disease. Nature 461:1071–1078
Janssen Y, Van Houten B, Borm P, Mossman B (1993) Biology of disease. Cell and tissue responses to oxidative damage. Lab Investig 69:261–274
Jayanand, Behari J (2005) Proceedings of International Union of Radio Science (URSI) held in October 13–16, New Delhi
Jeng DKH, Kaczmarek KA, Woodworth AG, Balasky G (1987) Mechanism of microwave sterilization in the dry state. Appl Environ Microbiol 53:2133–2137
Jung T, Moor RM, Fulka J (1993) Kinetics of MPF and histone H1 kinase activity differ during the G_2- to M-phase transition in mouse oocytes. Int J Dev Biol 37:595–600
Kakita Y, Kashige N, Murata K, Kuroiwa A, Funatsu M, Watanabe K (1995) Inactivation of Lactobacillus bacteriophage PL-1 by microwave irradiation. Microbiol Immunol 39:571–576
Karasek M (2006) Melatonin in human physiology and pathology. In: Columbus F (ed) Frontiers in chronobiology research. Nova Science Publication, Haupage, pp 1–43
Kesari KK, Behari J (2008) Comparative study of 900 MHz and 2.45 GHz radiation effect on reproductive system of male rats. In: Sharma RS, Rajanna A, Rajalakshmi M (eds) Recent advances and challenges in reproductive health research. ICMR Publication, New Delhi, p 363
Kesari KK, Behari J (2009) Fifty-gigahertz microwave exposure effect of radiations on rat brain. Appl Biochem Biotechnol 158:126
Kesari KK, Behari J (2010) Effects of microwave at 2.45 GHz radiations on reproductive system of male rats. Toxicol Environ Chem 92:1135–1147
Kesari KK, Kumar S, Behari J (2010) Mobile phone usage and male infertility in Wistar rats. Indian J Exp Biol 48:987–992
Kesari KK, Kumar S, Behari J (2011) Effects of radiofrequency electromagnetic waves exposure from cellular phone on reproductive pattern in male Wistar rats. Appl Biochem Biotechnol 164:546–559
Kilgallon SJ, Simmons LW (2005) Image content influences men's semen quality. Biol Lett 1:253
Kingma PS, Osheroff N (1997) Spontaneous DNA damage stimulates topoisomerase II-mediated DNA cleavage. J Biol Chem 272:7488–7493
Kremer H (2002) Die stille Revolution der Krebs- und Aidsmedizin. Ehlers Verlag, Wolfratshausen
Kreth G, Münkel C, Langowski J, Cremer T, Cremer C (1998) Chromatin structure and chromosome aberrations: modelling of damage induced by isotropic and localized irradiation. Mutat Res 404:77–88
Kuklinski B, van Lunteren I (2005) Neue Chancen zur naturlichen Vorbeugung und Behandlung von umweltbedingten Krankheiten. J. Kamphausen Verlag, Bielefeld
Kula B, Sobczak A, Kuska R (2000) Effects of static and EMF magnetic fields on free-radical processes in rat liver and kidney. Electromagn Biol Med 19:99–105

Kula B, Sobczak A, Kuska R (2002) Effects of electromagnetic field on free-radical processes in steelworker. Part I: magnetic field influence on the antioxidant activity in red blood cells and plasma. J Occup Health 44:226–229
Kuster N, Balzano Q, Lin JC (1997) Mobile communications safety. Chapman & Hall, London, p 23
Labbe JC, Capony JP, Caput D, Cavadore JC, Derancourt J, Kaghad M et al (1989) MPF from starfish oocytes at first meiotic metaphase is a heterodimer containing one molecule of cdc2 and one molecule of cyclin B. EMBO J 8:3053–3058
Lai H, Singh NP (1995) Acute low intensity microwave exposure increases DNA single-strand breaks in rat brain cells. Bioelectromagnetics 16:207–210
Lai H, Singh NP (1996) Single and double strand breaks in rats brain cells after acute exposure to radio frequency electromagnetic radiation. Int J Radiat Biol 69:513–521
Lai H, Singh NP (1997) Melatonin and a spin-trap compound block radiofrequency electromagnetic radiation induced DNA strand breaks in rat brain cells. Bioelectromagnetics 18:446
Lee JS, Ahn SS, Jung KC, Kim YW, Lee SK (2004) Effects of 60 Hz electromagnetic field exposure on testicular germ cell apoptosis in mice. Asian J Androl 6:29
Lehmann AR, Fuchs RP (2006) Gaps and forks in DNA replication: rediscovering old models. DNA Repair (Amst) 5:1495–1498
Lindahl T (1993) Instability and decay of the primary structure of DNA. Nature 362:709–715
Maes A, Verschaeve L, Arroyo A, De Wagter D, Vercruyssen L (1993) In vitro cytogenetic effects of 2450 MHz waves on human peripheral blood lymphocytes. Bioelectromagnetics 14:495–501
Maes A, Collier M, Slaets D, Verschaeve L (1995) Cytogenetic effects of microwaves from mobile communication frequencies (954 MHz). Electro- Magnetobiol 14:91–98
Maffini MV, Soto AM, Calabro JM, Ucci AA, Sonnenschein C (2004) The stroma as a crucial target in rat mammary gland carcinogenesis. J Cell Sci 117:1495–1502
Mailankot M, Kunnath AP, Jayalekshmi H, Koduru B, Valsalan R (2009) Radiofrequency electromagnetic radiation (RF-EMR) from GSM (0.9/1.8 GHz) mobile phones induces oxidative stress and reduces sperm motility in rats. Clinics 64:561–565
Malpaux B, Migaud M, Tricoier H, Chemineau P (2001) Biology of mammalian photoperiodism and the critical role of the pineal gland and melatonin. J Biol Rhythm 16:336–347
Malyapa RS, Ahern EW, Chen BI, Straube WL, LaRegina M, Pickard WF, Roti Roti JL (1998) DNA damage in rat brain cells after in vivo exposure to 2450 MHz electromagnetic radiation and various methods of euthanasia. Radiat Res 149:637–645
Mandelkern M, Elias JG, Eden D, Crothers DM (1981) The dimensions of DNA in solution. J Mol Biol 152(1):153–161
Manning G, Whyte DB, Martinez R, Hunter T, Sudarsanam S (2002) The protein kinase complement of the human genome. Science 298:1912–1934
Mashaghi A, Katan A (2013) A physicist's view of DNA. De Physicus 24e(3):59–61
McKinnon PJ (2009) DNA repair deficiency and neurological disease. Nat Rev Neurosci 10:100–112
Meikrantz W, Schlegel RA (1992) M-phase-promoting factor activation. J Cell Sci 101:475–481
Moulder JE, Erdreich LS, Malyapa RS, Merritt J, Pickard WF, Vijayalaxmi (1999) Cell phones and cancer: what is the evidence for connection? Radiat Res 151:513–531
Nikolova T, Czyz J, Rolletschek A, Blyszczuk P, Fuchs J, Jovtchev G, Schuderer J, Kuster N, Wobus AM (2005) Electromagnetic fields affect transcript levels of apoptosis-related genes in embryonic stem cell-derived neural progenitor cells. FASEB J 19:1686–1688
Nurse P (1990) Universal control mechanism regulating onset of M-phase. Nature 344:503–508
Ohkusu K, Isobe K, Hidaka H, Nakashima I (1986) Elucidation of the protein ki nase C-dependent apoptosis pathway in distinct of T lymphocytes in MRL-lpr/lpr mice. Eur J Immunol 25:3180
Paulraj R, Behari J (2002) The effect of low level continuous 2.45 GHz wave on brain enzymes of developing rat brain. Electro-Magn Biol Med 21(3):231–241

Paulraj R, Behari J (2004) Radio frequency radiation effects on protein kinase C activity in rats brain. Mutat Res 545:127–130
Paulraj R, Behari J (2006) Single strand DNA breaks in rat brain cells exposed to microwave radiation. Mutat Res 596:76–80
Pawse AR, Margery GO, Stocken LA (1971) Histone kinase and cell division. Biochem J 122:713–719
Pfeiffer P, Goedecke W, Obe G (2000) Mechanisms of DNA doublestrand break repair and their potential to induce chromosomal aberrations. Mutagenesis 15:289–302
Phillips JL, Ivaschuk O, Ishida-Jones T, Jones RA, Campbell-Beachler M, Haggren W (1998) DNA damage in Molt-4 T-lymphoblastoid cells exposed to cellular telephone radiofrequency fields in vitro. Bioelectrochem Bioenerg 45:103–110
Rao KS (1993) Genomic damage and its repair in young and aging brain. Mol Neurobiol 7 (23):23–48
Reiter RJ (1991) Pineal melatonin: cell biology of its synthesis and of its physiological interactions. Endocr Rev 12:151–180
Reiter RJ (1993) Interactions of the pineal hormone melatonin with oxygen centered free radicals: a brief review. Braz J Med Biol Res 26:1141–1155
Reiter RJ, Robinson J (1995) Melatonin: your body's natural wonder drug. Bantam Publication, New York
Reiter RJ, Tan DX, Cabrera J, D'Arpa D, Sainz RM, Mayo JC, Ramos S (1999) The oxidant/antioxidant network: role of melatonin. Biol Signals Recept 8:56–63
Reiter RJ, Tan DX, Osuna C, Gitto E (2000) Actions of melatonin in the reduction of oxidative stress. A review. J Biomed Sci 7:444–458
Rotem R, Paz GF, Homonnai ZT, Kalina M, Naor Z (1990a) Protein kinase C is present in human sperm: possible role in flagellar motility. Proc Natl Acad Sci U S A 87:7305
Rotem R, Paz GF, Homonnai ZT, Kalina M, Naor Z (1990b) Further studies on the involvement of protein kinase C in human sperm flagellar motility. Endocrinology 127:2571
Russell P (2001) iGenetics. Benjamin Cummings, New York. ISBN 0-8053-4553-1
Saenger W (1984) Principles of nucleic acid structure. Springer, New York. ISBN 0-387-90762-9
Sarkar S, Ali S, Behari J (1994) Effect of low power microwave on the mouse genome: a direct DNA analysis. Mutat Res 320:141–147
Savage JRK (1996) Insight into sites. Mutat Res 366:81–95
Shang X, Huang Y, Ye Z, Yu X, Gu W (2004) Protection of melatonin against damage of sperm mitochondrial function induced by reactive oxygen species. Zhonghua Nan Ke Xue 10:604–607
Shibutani S, Takeshita M, Grollman AP (1991) Insertion of specific bases during DNA synthesis past the oxidation-damaged base 8-oxodG. Nature 349:431–434
Sonoda E, Sasaki MS, Buerstedde JM, Bezzubova O, Shinohara A, Ogawa H, Takata M, Yamaguchi-Iwai Y, Takeda S (1998) Rad51-deficient vertebrate cells accumulate chromosomal breaks prior to cell death. EMBO J 17:598–608
Speit G, Schütz P, Hoffmann H (2007) Genotoxic effects of exposure to radiofrequency electromagnetic fields (RF-EMF) are not independently reproducible. Mutat Res 626:42–47
Stevens RG, Davis S (1996) The melatonin hypothesis: electric power and breast cancer. Environ Health Perspect 104:135–140
Stopczyk D, Gnitecki W, Buczynski A, Kowalski W, Markuszewski L, Buczynska M (2005) Effect of electromagnetic field produced by mobile phones on the activity of superoxide dismutase (SOD-1) and the level of malonyldialdehyde (MDA)-in vitro study. Med Pr 53:311–314
Takabe W, Niki E, Uchida K, Satoh K, Noguchi N (2001) Oxidative stress promotes the development of transformation: involvement of a potent mutagenic lipid peroxidation product, acrolein. Carcinogenesis 22:935–941
Takashima S, Gabriel C, Sheppard RJ, Grant EH (1984) Dielectric behaviour of DNA solution at radio frequency and microwave frequencies. J Biophys 46:29–34

Thorn T, Gniadecki R, Petersen AB, Vicanova J, Wulf HC (2001) Differences in activation of G2/M checkpoint in keratinocytes after genotoxic stress induced by hydrogen peroxide and ultraviolet radiation. Free Radic Res 35:405–416
Trush M, Kensler T (1991) Role of free radicals in carcinogen activation. In: Sies H (Hrsg) Oxidative stress. Oxidants and antioxidants S. Academic Press, London, pp 277–317
Vela GR, Wu JF (1979) Mechanism of lethal action of 2450 MHz radiation on microorganisms. Appl Environ Microbiol 37:550–553
Vijayalaxmi, Obe G (2004) Controversial cytogenetic observations in mammalian somatic cells exposed to radiofrequency radiation. Radiat Res 162:481–496
Vijayalaxmi et al (1997) Frequency of micronuclei in the peripheral blood and bone marrow of cancer-prone mice chronically exposed to 2350 MHz radiofrequency radiation. Radiat Res 147 (4):495–500
Vijayalaxmi et al (1998) Correction of an error in calculation in the article in. Res Radiat 149 (3):199–202
Watson JD, Crick FH (1953) A structure for deoxyribose nucleic acid. Nature 171(4356):737–738
Weaver VM, Gilbert P (2004) Watch thy neighbor: cancer is a communal affair. J Cell Sci 117:1287–1290
Wiseman H, Halliwell B (1996) Damage to DNA by reactive oxygen and nitrogen species: role in inflammatory disease and progression to cancer. Biochem J 313:17–29
Wolffe A (1998) Chromatin structure and function. Academic, San Diego
Yan JG, Agresti M, Bruce T, Yan YH, Granlund A, Matloub HS (2007) Effects of cellular phone emissions on sperm motility in rats. Fertil Steril 88:957
Yoshikawa T, Tanigawa M, Tanigawa T, Imai A, Hongo H, Kondo M (2000) Enhancement of nitric oxide generation by low frequency electromagnetic field. Pathophysiology 7:131–135
Zang LY, Cosma G, Gardner H, Vallyathan V (1998) Scavenging of reactive oxygen species by melatonin. Biochim Biophys Acta 1425:469–477
Zhang Z, Leonard SS, Huang C, Vallyathan V, Castranova V, Shi X (2003) Role of reactive oxygen species and MAPKs in vanadate induced G2/M phase arrest. Free Radic Biol Med 34:1333–1342
Zmyslony M, Jajte JM (1998b) The role of free radicals in mechanisms of biological function exposed to weak, constant and net magnetic fields. Med Pr 49:177–186
Zmyslony M, Jajte JM (1998a) The role of free radicals in mechanisms of biological function exposed to weak, constant and net magnetic fields. Med Pr 49:177–186

Chapter 4
Modified Health Effects of Non-ionizing Electromagnetic Radiation Combined with Other Agents Reported in the Biomedical Literature

Ronald N. Kostoff and Clifford G.Y. Lau

Abstract Ionizing and non-ionizing electromagnetic field (EMF) radiation, either stand-alone or in combination with other agents, exert health effects on biological systems. The present chapter examines the scope of non-ionizing EMF radiation combined effects; i.e., identifies effects on biological systems from combined exposure to non-ionizing electromagnetic fields/radiation and at least one other agent. Only articles in which the presence of non-ionizing EMF radiation had some effect (beneficial or adverse) on the biological system were selected. A comprehensive and novel query was developed using an iterative hybrid approach, whereby articles related by common text and by citation linkages were retrieved. This retrieved literature was: (1) clustered algorithmically into 32 biomedical sub-themes (assigned by the authors); (2) grouped through factor analysis into 32 factors; and (3) subsequently grouped manually (by the authors) into an effects-based taxonomy. The common principles within each thematic cluster/group that accounted for the combined effects were identified.

Non-ionizing EMF radiation plays a supportive role in a wide range of beneficial and adverse effects. Major beneficial effects include (1) accelerated healing of wounds and injuries in concert with other agents and (2) treatment of cancer by combining chemotherapy with radiation. Major adverse effects, on the other hand, include (1) enhanced carcinogenesis, (2) enhanced cellular or genetic mutations, and (3) teratogenicity. It should be noted that community consensus (unanimity among papers published in peer-reviewed journals) does not exist on these potential effects, either beneficial or adverse, although there is substantial credible scientific evidence supporting the above effects (as described in this chapter).

R.N. Kostoff (✉)
Georgia Institute of Technology, Atlanta, GA, USA

13500 Tallyrand Way, Gainesville, VA 20155, USA
e-mail: ronald.kostoff@pubpolicy.gatech.edu; rkostoff@gmail.com

C.G.Y. Lau
Institute for Defense Analyses, Alexandria, VA, USA

C.D. Geddes (ed.), *Microwave Effects on DNA and Proteins*,
DOI 10.1007/978-3-319-50289-2_4

In daily living, the body is exposed to multiple external agents simultaneously, e.g., myriad non-ionizing EMF radiations, pesticides, food additives, heavy metal, legal and illegal drugs, ionizing radiation, and air pollution. The number of combinations of potential external agents is large. Each combination could potentially have synergistic or antagonistic, and beneficial or adverse, effects. However, non-ionizing EMF radiation exposure safety standards are based primarily on stand-alone radiation exposures. When combined with other agents, the adverse effects of non-ionizing EMF radiation on biological systems may be more severe. Much work remains to be done before definitive statements about non-ionizing EMF radiation exposure safety can be made.

Keywords EMF • Electromagnetic fields • Magnetic fields • Radio frequency radiation • Microwave radiation • Interactive effects • Combined effects • Synergistic effects • Additive effects • Antagonist effects • Potentiative effects • Co-promotion • Co-mutagenic • Co-carcinogenic • Combined exposure • Combined treatment • DMBA • TPA • Text mining • Document clustering

4.1 Introduction

4.1.1 Electromagnetic Spectrum Definitions (See Table 4.1)

The electromagnetic spectrum encompasses the entire span of electromagnetic radiation. The spectrum includes: ionizing radiation (gamma rays, x-rays, and the extreme ultraviolet, with wavelengths below ~10^{-7} m and frequencies above ~3×10^{15} Hz); non-ionizing visible radiation (wavelengths from ~4×10^{-7} m to ~7×10^{-7} m and frequencies between ~4.2×10^{14} Hz and ~7.7×10^{14} Hz); non-ionizing non-visible radiation (short wavelength radio waves and microwaves, with wavelengths between ~10^{-3} m and ~10^{5} m and frequencies between ~3×10^{11} to ~3×10^{3} Hz; long wavelengths, ranging between ~10^{5} m and ~10^{8} m and frequencies ranging between 3×10^{3} and 3 Hz).

In the present study of non-ionizing EMF radiation health effects, interest is focused on the frequency spectrum ranging from 3 Hz to 3000 GHz. The low frequencies (3 Hz–300 KHz) are used for electrical power line transmission (60 Hz in the U.S.) as well as maritime and submarine navigation and communications. Medium frequencies (300 KHz–900 MHz) are used for AM/FM/TV broadcasts in North America. Lower microwave frequencies (900 MHz–5 GHz) are used for telecommunications such as microwave devices/communications, radio astronomy, mobile/cell phones, and wireless LANs. Higher microwave frequencies (5 MHz–300GHz) are used for radar and proposed for microwave WiFi. Terahertz frequencies (300 GHz–3000 GHz) are used increasingly for imaging to supplement X-rays in some medical and security scanning applications.

Table 4.1 Electromagnetic spectrum

Region of EMF spectrum	Wavelength range (Meters)	Frequency range (Hz)
Ionizing	$<10^{-7}$	$>3 \times 10^{15}$
Non-ionizing visible	4×10^{-7}--->7×10^{-7}	4.2×10^{14}--->7.7×10^{14}
Non-ionizing non-visible shortwave	10^{-3}--->10^{5}	3×10^{11}--->3×10^{3}
Non-ionizing non-visible longwave	10^{5}--->10^{8}	3×10^{3}--->3×10^{0}

4.1.2 Modern Non-ionizing EMF Radiation Exposures

In ancient times, sunlight and its lunar reflections provided the bulk of the visible spectrum for human beings (with fire a distant second and lightning a more distant third). Now, many varieties of artificial light (incandescent, fluorescent, and light emitting diode) have replaced the sun as the main supplier of visible radiation during waking hours. Additionally, EMF radiation from other parts of the non-ionizing spectrum has become ubiquitous in daily life, such as from wireless computing and telecommunications. In the last two or three decades, the explosive growth in the cellular telephone industry has placed many residences in metropolitan areas within less than a mile of a cell tower. Health concerns have been raised about non-ionizing EMF radiation from (1) mobile communication devices, (2) occupational exposure, (3) residential exposure, (4) wireless networks in homes, businesses, and schools, and other non-ionizing EMF radiation sources such as ‘smart meters’ and ‘Internet of Things’.

4.1.3 Beneficial and Harmful Effects of Non-ionizing EMF Radiation Exposure

4.1.3.1 Beneficial Effects

The effects of non-ionizing electromagnetic radiation on health can be therapeutic; e.g.,

- “ELF-EMF modulates chemokine production and keratinocyte growth through inhibition of the NF-kappa B signalling pathway and thus may inhibit inflammatory processes. ELF-EMF could represent an additional therapeutic approach in the treatment of skin injury.” (Vianale et al. 2008).
- “ELF-EMF could augment the cell apoptosis effects of low doses of [radiotherapeutic] X-ray irradiation on [liver cancer cell line] BEL-7402 cells in a synergistic and cumulative way” (Jian et al. 2009).
- Treatment of cancer and other serious chronic diseases using chemotherapy combined with (typically) pulsed EMF was the largest category of combined effects from our database retrievals; e.g.,

- "A significant synergizing antitumor effect was seen when EL4 tumor-bearing mice were simultaneously exposed to EMF-BEMER and treated with suboptimal dose of synthetic HPMA copolymer-based doxorubicin, DOXHYD-HPMA." (Rihova et al. 2011)
- "These results demonstrated that MW radiation exposure and Gemcitabine treatment have a synergistic effect on apoptotic activity of Raji cells." (Canseven et al. 2015)

- Diathermy by low level RF radiation is used in physical therapy to deliver moderate heat directly to pathology lesions in deep tissues of the body. In surgery, extreme heat produced by diathermy (aka ablation) can be used to destroy neoplasms, warts, and infected tissues, and to cauterize blood vessels to prevent excessive bleeding. For a cutting-edge example, "This implantable magnetic nanofiber device can be exploited to apply hyperthermia with an alternating magnetic field and to achieve cancer cell-specific drug release to enable synergistic cancer therapy" (Sasikala et al. 2016)
- Bone regeneration using non-ionizing EMF radiation also was mentioned in a number of retrieved papers; e.g.,

  - "PEMFs may be considered a possible tool to improve autologous cell-based regeneration of bone defects in orthopedics" (Ongaro et al. 2014)
  - "Osteogenic differentiation of ASCs was accelerated by multiple-combination biophysical stimulation in vitro. However, both single stimulation and double-combination stimulation were sufficient to accelerate bone regeneration in vivo, while the osteogenic marker expression of those groups was not as high as that of triple-combination stimulation in vitro" (Kang et al. 2014)

4.1.3.2 Harmful Effects

But the effects of non-ionizing EMF radiation can also be potentially harmful:

- "Extremely Low Frequency-Magnetic Fields (ELF-MF) are possible carcinogens to humans and some data suggest that they can act as promoters or progressors." (Gobba et al. 2009).
- "The authors found that ELF-EMFs may be increase the risk of human breast cancer" (Zhao et al. 2014)
- Residents near a "60 kV electric distribution line" experienced the following: "Non statistically significant increases were observed for all and primary cancers; primary cancers were significantly increased among subjects with >30 years' residence and latency. A significant increase for all, primary, and secondary cancers, and a twofold increase for ischaemic diseases, was observed in subjects in the sub-area with the highest exposure" (Fazzo et al. 2009)

- Cellular Neoplastic Transformation Induced by 916 MHz Microwave Radiation (Yang et al. 2012)
- Heavy cell phone use; e.g.,
  - "These additional data support previous findings concerning a possible association between heavy mobile phone use and brain tumours" (Coureau et al. 2014)
  - "an increased risk was found for glioma and use of mobile or cordless phone. The risk increased with latency time and cumulative use in hours and was highest in subjects with first use before the age of 20" ["OR = 4.9"] (Hardell et al. 2011).
  - "Our result suggests that use of mobile phones can be related to the early spontaneous abortions." (Mahmoudabadi et al. 2015)
- Proximity to wireless transmission antennas/cell towers; e.g.,
  - Levitt and Lai performed a comprehensive study of biological effects from cell tower radiation, and found "headaches, skin rashes, sleep disturbances, depression, decreased libido, increased rates of suicide, concentration problems, dizziness, memory changes, increased risk of cancer, tremors, and other neurophysiological effects in populations near base stations" (Levitt and Lai 2010)
  - "We found an association between increased childhood leukaemia incidence and mortality and proximity to TV towers" (Hocking et al. 1996)
  - "The odds ratio for all types of leukemia was 2.15.....among children who resided within 2 km of the nearest AM radio transmitter as compared with those resided more than 20 km from it. For total RFR exposure from all transmitters, odds ratios for lymphocytic leukemia were 1.39.....and 1.59..... for children in the second and third quartiles" (Ha et al. 2007)
  - "There was an association between residential proximity to the television towers and decreased survival among cases of childhood leukemia in North Sydney, Australia." (Hocking and Gordon 2003)
  - "The risk of childhood leukemia was higher than expected for the distance up to 6 km from the radio station.....and there was a significant decline in risk with increasing distance both for male mortality.....and for childhood leukemia" (Michelozzi et al. 2002)
  - "the proportion of newly developing cancer cases was significantly higher among those patients who had lived during the past 10 years at a distance of up to 400 m from the cellular transmitter site.....compared to those patients living further away, and that the patients fell ill on average 8 years earlier..... after 5 years' operation of the transmitting installation, the relative risk of getting cancer had trebled for the residents of the area in the proximity of the installation compared to the inhabitants of Naila outside the area." (Eger et al. 2004)

- "A comparison of the relative risk revealed that there were 4.15 times more cases [of cancer] in area A [<350 m from the cell tower antenna] than in the entire population." (Wolf and Wolf 2004)
- A large Brazilian study showed relative risk for neoplasia decreasing as distance from the cell tower antenna increased, ranging from a high of 1.35 for distances between 0 and 100 m to a low of 1.03 for distances between 800 and 900 m. (Dode et al. 2011)

- Adverse effects on nervous system:
  - "Activation of VEGF/Flk-1-ERK pathway induced blood-brain barrier injury after microwave exposure" (Wang et al. 2015)
  - "extensive neurodegeneration on exposure to radio waves. Increased production of reactive oxygen species due to exhaustion of enzymatic and non-enzymatic antioxidants and increased lipid peroxidation indicate extensive neurodegeneration" (Saikhedkar et al. 2014)
  - "The study highlights the detrimental effects of mobile phone radiations on brain during young and adult ages." (Motawi et al. 2014)
  - "Fetal radiofrequency radiation exposure from 800 to 1900 Mhz-rated cellular telephones affects neurodevelopment and behavior in mice" (Aldad et al. 2012)
  - "Autism-relevant social abnormalities in mice exposed perinatally to extremely low frequency electromagnetic fields" (Alsaeed et al. 2014)
  - "Long-term, low-level microwave exposure may inhibit learning and memory by affecting protein and energy metabolic processes and signaling pathways relating to neurological functions or diseases" (Zhao et al. 2015)
  - "case study of brain and nervous system cancers.....The only exogenous risk factor consistently associated with higher incidence was the penetration rate of mobile/cellular telecommunications subscriptions" (de Vocht et al. 2013)
  - "low intensity microwave radiation induces oxidative stress, inflammatory response and DNA damage in brain by exerting a frequency dependent effect" (Megha et al. 2015)
  - Martin Pall has shown that electromagnetic fields activate voltage-gated calcium channels, allowing increased calcium fluxes into the cell, thereby disrupting synapse formation (Pall 2015a). He has also stated that toxic chemicals act in synergy with the electromagnetic fields to enhance the adverse effects, including contributing to autism, among others (Pall 2015b).

- Researchers at Kaiser-Permanente have instrumented humans with magnetic field meters, and tracked exposure over time. Some conclusions are:
  - "prenatal maximum magnetic field exposure above a certain level (possibly around 16 mG) may be associated with miscarriage risk" (Li et al. 2002).
  - "every 1-mG increase of maternal MF [magnetic field] level during pregnancy was associated with a 15% increased rate of asthma in offspring" (Li et al. 2011).

- Numerous studies have shown adverse effects on sperm quality, e.g.:
  - "Our study provides some evidence for the first time that MF exposure may have an adverse effect on sperm quality." (Li et al. 2010).
  - "Electromagnetic radiation at 900 MHz induces sperm apoptosis through bcl-2, bax and caspase-3 signaling pathways in rats" (Liu et al. 2015)
  - "21-day-old rat testicles exposed to 900-MHz EMF in the prenatal term may be adversely affected, and this effect persists after birth" (Hanci et al. 2013)
  - "Exposure to 60 Hz and 1mT EMF can disturb spermatogenesis and may produce subfertility or infertility." (Tenorio et al. 2012)
- Finally, other large-scale studies reviewing adverse effects include The Bioinitiative Report (The Bioinitiative Report 2012) and The Dart Report (Dart et al. 2013).

4.1.4 Importance of Effects of Non-ionizing EMF Radiation Combined with Other Agents

In daily living, we are exposed to myriad potentially toxic stimuli unknown to our ancestors of only a few generations ago. These include both ionizing and non-ionizing radiations of many frequencies, artificial foods, air pollutants, environmental pollutants, occupational pollutants, high technology drugs/surgeries/diagnostics, and sedentary lifestyle practices unknown to other species. Much research has shown that when potentially toxic stimuli are combined, their adverse effects can be increased substantially (see definitions below in Sect. 4.1.6). Unfortunately, most research into effects of potentially toxic stimuli examine them one at a time in isolation. This exposure to a singular stimulus tends to underestimate the toxic influence of combined stimuli.

4.1.5 Relation of Non-ionizing EMF Radiation Combined Effects Studies to Safety Studies

A major reason for conducting non-ionizing EMF radiation effects studies, whether EMF radiation in isolation or combined with other agents, is to help set non-ionizing EMF radiation exposure ceilings for safety and health protection. Federal Communications Commission (FCC) limits for maximum permissible exposure (MPE) is a function of power density, exposure time, frequency, and other variables. The MPE limits for occupational/controlled exposure are different from the population/uncontrolled exposure (Federal Communications Commission Office of Engineering and Technology 1997). As one example for an important frequency band, FCC MPE limits for EMF radiation at 900 Mhz cell tower frequencies are approximately 6.0×10^6 microwatts/m^2. As will be seen by some results in this chapter (Levitt and Lai 2010; The Bioinitiative Report 2012), these MPE limits are orders of magnitude above levels shown to cause serious disease. In addition, when other toxic stimuli are considered in combination with non-ionizing EMF radiation, the synergies tend to enhance the effects of each stimulus in isolation. In other words, combined exposure to toxic stimuli and non-ionizing EMF radiation translates into much lower levels of tolerance for each toxic stimulus alone. So, the exposure limits for non-ionizing EMF radiation when examined in combination with other potentially toxic stimuli would be far lower than those derived from non-ionizing EMF radiation exposures in isolation.

4.1.6 Goal of Present Chapter

The goal of this chapter is to examine the scope of the non-ionizing EMF radiation combined effects on biological systems; i.e., identify effects on biological systems from combined exposure to non-ionizing electromagnetic fields/radiation and at least one other agent. These interactive effects include additive effects, antagonistic effects, potentiative effects, and synergistic effects, and are defined and discussed in Appendix 1 (A.1.3.1). Other terminology is used in the documents, such as co-promotional, co-mutagenic, co-carcinogenic, etc., but these terms tend to be sub-sets of the more general terms defined above.

4.1.7 Previous Studies on Non-ionizing EMF Radiation Combined Effects

Background studies are discussed in Appendix 1 (A.1.1 and A.1.2). In summary, about 5–7% of total non-ionizing EMF radiation health/biological impact studies (probably less) are concerned with combination-type effects. Most of these previous studies that focused on combination effects of non-ionizing EMF radiation with other stimuli reflect the articles referenced in the body of this chapter, namely, selected non-ionizing EMF radiation ranges in the frequency spectrum combined with usually one or a few co-stimuli. However, none of these review articles in Appendix 1 (A.1.1 and A.1.2) covers the wide range of disciplines and impacts as the present chapter, as well as both adverse and beneficial combined effects. It is the scope of coverage and the comprehensiveness of retrieval using the hybrid search technique, as well as the integration across disciplines, which makes the present chapter unique.

4.1.8 Sources of Non-ionizing EMF Radiation Impacts Data

There are two main sources of data types supporting the above analysis: lab experiments and epidemiological studies. The few example citations shown in Sect. 4.1.3 illustrate the complexity of the problem, compounded further by the lack of comprehensive and long-term data (Kundi 2010) and the intrinsic limitations in the fundamental data.

4.1.8.1 Lab Experiments

One major data type is isolated laboratory experiments on seemingly uniform biological systems, where only one or a very few parameters are varied. Even under these controlled conditions, output response of the biological preparations can vary greatly. In other words, unlike many physical systems, biological specimens have varying sensitivity to the same or slight variations of the stimulus:

(1) Individuals or groups in a population, which would usually be regarded as uniform, may react to millimeter waves (MMW) in rather different or even opposite ways; (2) The factors that determine the MMW sensitivity of a specimen or a population are not yet known or controlled. Irradiation could increase antibiotic resistivity in one experiment and decrease it in the other; (3) The time duration of MMW effects may vary from specimen to specimen. Even robust MMW effects may be well reproducible for a limited time and then disappear; (4) MMW effects could often be revealed only in subjects that are experiencing some deviation from the 'normal' state; (5) Increased sensitivity and even hypersensitivity of individual specimens to MMW may be real (Pakhomov et al. 1998).

Even where laboratory results are not completely conclusive, and the uncertainties are relatively low, it is difficult to extrapolate these single or low parameter lab results to real world effects in human beings. People are exposed to multiple forms of radiation with potential synergistic/additive effects (e.g., "*adverse effects of gamma-rays on cellular functions are strengthened by EMF*" (Cao et al. 2009)), multiple drugs and pollutants, and other potentially damaging influences whose cumulative effects could be synergistic/additive. There have been no comprehensive multivariate sensitivity experiments and analyses to draw any conclusions. As Verschaeve and Maes (1998) state in their review article on genetic, carcinogenic, and teratogenic effects of radiofrequency fields: "*we believe that synergistic investigations deserve special attention. Indeed, people are exposed to many different influences, and theoretically it may well be that a RF-exposure alone is ineffective whereas this exposure might enhance the mutagenicity, carcinogenicity or teratogenicity of chemical or physical factors*". They provide the example of a synergistic effect from RF exposure ***preceding*** the mutagen mitomycin C in an investigation of 954-MHz waves emitted by the antenna of a GSM [Global System for Mobile Communications] base station (Verschaeve and Maes 1998).

These non-ionizing EMF radiation combined effects may go in multiple directions. As Whissel and Persinger (2007) state, in addressing the possibility of weak non-ionizing EMF as an enhancer of drug therapy: "*Very weak (microT range) physiologically-patterned magnetic fields synergistically interact with drugs to strongly potentiate effects that have classically involved opiate, cholinergic, dopaminergic, serotonergic, and nitric oxide pathways. The combinations of the appropriately patterned magnetic fields and specific drugs can evoke changes that are several times larger than those evoked by the drugs alone.*" Conversely, as reflected in the provocative article titled "Do electromagnetic fields enhance the effects of environmental carcinogens?" (Juutilainen 2008) and in a preceding article (Juutilainen et al. 2006), the non-ionizing EMF contribution could go in the opposite direction: "*ELF MFs have been reported to enhance the effects of known carcinogenic or mutagenic agents in a few animal studies and in several in vitro studies.....The majority of in vitro studies have reported positive findings..... Animal studies designed according to the classical initiation-promotion concept may not be sufficient for studying the co-carcinogenic effects of MFs, and further studies using novel study designs would be useful*".

There may even be a further complicating factor relative to non-ionizing EMF radiation combined effects. Most of these papers address non-ionizing EMF radiation combinations, but not their sequencing. For example, a paper published in Radiobiologia (Tikhonchuk et al. 1987) shows the critical importance of sequencing on the combined effect of microwave and gamma-ray radiation: "*Structural and functional changes in the central nervous system were shown to be the same with both microwave and ionizing radiation having different mechanism of action. When the two types of radiation were delivered in a combination the* ***sequence of delivery was of a significant importance****. Antagonism of the effects was noted when microwave radiation was delivered prior to gamma-radiation. The effect was synergistic when the exposure to microwaves followed gamma-irradiation.*" Thus, time sequencing of radiation exposure appears to have different effects. It seems that the residual effect of prior exposures can influence the response on subsequent exposures.

4.1.8.2 Epidemiological Studies

The other major data type is epidemiological studies, which approach the problem from the other end of the parameter spectrum. Here, highly integrated data are taken; they contain the influence of many different types of parameters, genetic make-ups, and sequencing, only a very few of which may be taken into account. Thus, what appears as a subtle effect over a large number of heterogeneous people may be a significant identifiable effect on a much smaller more homogeneous group if the effects of all the operational parameters and their sequencing were taken into account.

As stated above, the first author initially performed text mining-based retrievals and analyses of documents that mainly examined health effects of non-ionizing

EMF radiation in isolation. The results showed the myriad health impacts from a relatively high level vantage point. During the course of that text mining analysis, it became obvious that the combined effects of non-ionizing EMF radiation were extremely important, explained some of the ambiguous results from the non-ionizing EMF-only studies, and needed to be the focus of a much more detailed study. The previous study by the authors (Kostoff and Lau 2013) was the first step in developing a 'much more detailed study', and the present chapter updates and expands the previous study to produce that 'more detailed study'.

4.2 Background and Approach Summary

The approach in this chapter is to: (1) select the most credible global databases of research articles; (2) develop a query that will retrieve the relevant combined non-ionizing EMF radiation effects literature comprehensively; (3) identify the key biomedical thrusts in this retrieved literature; and (4) extract the mechanisms and principles that describe the influence of the non-ionizing EMF radiation component on the final combination effects.

Medline and the Science Citation Index (SCI) were selected as the source databases. An iterative relevance feedback technique was selected to generate the initial query for retrieving relevant articles, and it was augmented by examining the citation network (references, citing papers, records with at least one reference in common) of the papers retrieved by the initial query.

The key biomedical thrusts were identified through text mining approaches developed by the first author, including factor analysis and document clustering of the retrieved records. A final taxonomy based on the factor analysis and document clustering was generated manually. The retrieved records were then assigned manually to each taxonomy category. The non-ionizing EMF radiation-co-promoter-mechanism-disease 'signatures' were extracted by integrating the relevant factors, relevant clusters, and relevant sections of each record in the cluster.

Appendix 1 contains: (1) a brief overview of text mining; (2) a brief summary of past non-ionizing EMF radiation co-promoter studies; (3) the text mining methodology used to identify, retrieve, extract, and analyze the relevant non-ionizing EMF radiation co-promoter articles; (4) the details of the iterative relevance feedback component of the query used, and of the subsequent citation network traversing component of the query.

4.3 Analysis and Results

4.3.1 Taxonomy Structure

The factor analysis-based taxonomy was not significantly different from the document clustering-based taxonomy; only the latter will be summarized due to space limitations. The hierarchical document clustering-based taxonomy had 32 leaf (lowest level) clusters (categories), and each retrieved record was assigned to one cluster only.

The three highest levels in the taxonomy are shown in Fig. 4.1. They are self-explanatory, and will not be discussed further due to space limitations.

4.3.2 Manual Taxonomy

Based on the factor matrix results, the document clustering results, and other grouping studies, a manual taxonomy was generated as a framework for presenting the detailed results (see Fig. 4.2). It focuses on impacts, and incorporates the major impact categories. A brief summary of each category is presented first, ending with

LEVEL 1	LEVEL 2	LEVEL 3
Cluster 1A: Interactive effects from pulsed electric and magnetic fields (combined with other agents) on tumors; other types of interactive effects from electric fields	**Cluster 1A1**: Interactive effects from pulsed magnetic fields on tumors; especially pulsed EMF-induced hyperthermia	Cl 1A1a: Hyperthermia and chemotherapy or gene therapy for patient tum or therapy, especially with magnetic nanoparticles
		Cl 1A1b: Pulsed electromagnetic/magnetic fields and anti-cancer drugs/photodynamics for cell lines
	Cluster 1A2: Pulsed electric fields for improved transdermal delivery of chemotherapeutic agents and for suppressing bacterial growth	Cl 1A2a: Pulsed electric fields/ HIPEF, especially with Nisin,for inactivation, especially of foodborne microorganisms
		Cl 1A2b: Electric fields (especially pulsed), electroporation, electrochemo-therapy and agents for tumor treatment.
Cluster 1B: Interactive effects at the cellular and DNA level from low frequency alternating and static magnetic fields (combined with other agents), and microwaves	**Cluster 1B1**: Combined effects of low frequency electromagnetic fields and other agents on biological processes	Cl 1B1a: Magnetic fields and morphine-induced analgesia
		Cl 1B1b: Magnetic fields (especially ELF) and carcinogens impacts on rats
	Cluster 1B2: Combined effects of static magnetic fields and microwaves with other agents on cells, DNA, and apoptosis	**Cl 1B2a**: Damage from microwaves and other agents, especially on DNA
		Cl 1B2b: Impact of electromagnetic and static magnetic fields, especially with x-rays on oxidative stress

Fig. 4.1 Taxonomy of biomedical thrusts in non-ionizing EMF radiation health effects literature (Major category headings *bolded*)

Fig. 4.2 Manual taxonomy for presenting results (Major categories *bolded*)

CAT #	CATEGORY/SUB-CATEGORY TITLE Code: Shaded denotes mainly adverse effects;Unshaded denotes mainly beneficial effects
4.3.2.1	**Impact on cancer**
	Hyperthermia
	Chemotherapy
	Ionizing radiation and other treatments
	Chemical promoters
	Electromagnetic promoters
4.3.2.2	**Impact on neural system**
	Enhanced analgesia
	Enhanced performance and reduced seizures
	Reduced analgesia effectiveness
	Performance degradation
4.3.2.3	**Impact on circulatory system**
	Enhance heart, vascular
	Degrade heart, vascular
4.3.2.4	**Impact on immune system**
	Enhance immune system performance
	Degrade immune system performance
4.3.2.5	**Impact on endocrine system**
	Enhance endocrine system performance
	Degrade endocrine system performance
4.3.2.6	**Impact on skeletal system**
4.3.2.7	**Impact on genes**
	Positive genetic impacts
	DNA damage
	Mutations
	Teratogenicity
4.3.2.8	**Impact on cells**
	Permeation (beneficial effects)
	Apoptosis (cancer cells)
	Apoptosis (healthy cells)
	Cell proliferation
	Oxidative damage
	Permeation (adverse effects)
	Cell growth, differentiation, proliferation, morphology
4.3.2.9	**Impact on micro-organisms**
4.3.2.10	**Attenuation of EMF effect**
4.3.2.11	**Other**

one or more short illustrative examples. Where possible, integrative mechanisms for the category are discussed.

4.3.2.1 Impact on Cancer

This category includes non-ionizing EMF radiation combinations that help in the treatment of cancer, as well as combinations that promote the growth of cancer. Some of the later categories in this overall taxonomy, in particular those that deal with the impact of non-ionizing EMF radiation combinations on genes and cells, are related, since e.g. DNA damage or oxidative damage could eventually lead to cancer. These later categories were treated separately, since the cellular or genetic impacts were central to the research, not the cancer potential (although the latter was mentioned in many cases). The present category specifically includes non-ionizing EMF radiation for hyperthermia combined with other agents, and non-ionizing EMF radiation combined with chemotherapy, ionizing radiation, ultrasound, chemical promoters, and electromagnetic promoters from other segments of the spectrum. Generally, the non-ionizing EMF radiation that resulted in positive effects was short electric pulses (electroporation) or relatively short RF exposures for hyperthermia, and the non-ionizing EMF radiation that resulted in adverse effects was long power frequency exposures or relatively long RF exposures.

Hyperthermia

In the synergistic use of hyperthermia (generated by microwave heating of magnetic particles) combined with other agents to treat tumors/cancer, the high temperatures within the tumors enhance apoptosis (programmed cell death), and augment the apoptotic impacts of the other agents. ***While athermal (non-thermal) microwave effects cannot be ruled out***, the thermal effects were the main focus of the hyperthermia research reported here. Examples include combination of hyperthermia with: (a) doxorubicin to target controlled drug release for tumors (Purushotham and Ramanujan 2010); (b) TNF-alpha gene therapy for tumor growth reduction (Ito et al. 2001); (c) immunotherapy (hsp70 gene therapy) to reduce tumor growth and induce systemic antitumor immunity (Ito et al. 2003); (d) 3D–conformal radiotherapy for steep decrease of PSA values (Deger et al. 2002); (e) "When hyperthermia was performed immediately after application of electric pulses.....greater than additive antitumour effectiveness was observed..... ***Single treatment, application of electric pulses or hyperthermia had minor or no effect on tumour growth***." (Karner et al. 2004).

Chemotherapy

The synergistic use of non-ionizing EMF radiation with chemotherapy involves mainly electroporation (electric pulses that transiently permeabilize the cell membrane) combined with anticancer drugs that can enter the cells more efficiently in the short period when pores have been established.

> Permeability or permeabilization in the present context reflects the transient establishment of pores by non-ionizing EMF radiation, usually in the cell membrane. In its positive role, it is used to assist chemotherapy, as well as the delivery of other types of drugs. However, in its negative role, it may allow unwanted toxic materials to enter cells or allow toxic materials to penetrate e.g. the blood-brain-barrier through non-ionizing EMF radiation-loosened endothelial tight junctions. Lange and Sedmak (Lange and Sedmak 1991) present an interesting example of the lethality enhancement of Japanese Encephalitis Virus by microwave radiation assisting entry of the virus into the central nervous system. Since the present chapter was organized by systems rather than phenomena, concepts such as permeability appear in a number of different sections in myriad applications

Bleomycin tends to be the main anticancer drug used with electroporation for electrochemotherapy, although other drugs have appeared as well. Examples include: (a) electroporation and bleomycin for pancreatic adenocarcinomas (Jaroszeski et al. 1999); (b) electroporation and TLR-9 ligands, CpG oligodeoxynucleotides for a systemic antitumor response on a contra-lateral untreated tumor (Roux et al. 2008); (c) cisplatin for hamster oral fibrosarcoma (Fulimoto et al. 2005); (d) cisplatin and electrogene therapy with p53 on murine sarcomas (Grosel et al. 2006); (e) "with electroporation, the cytotoxicity of BLM [Bleomycin] in electroporated cells was increased by as much as 95.7-fold compared to that of non-electroporated MBT-2 cells..... the ***anticancer effect of BLM can be considerably potentiated by applying EP***." (Ogihara and Yamaguchi 2000).

Ionizing Radiation and Other Treatments

The combination of ionizing and non-ionizing radiation enhances the apoptotic effect of each modality on cancer cells, although multiple non-ionizing forms (such as EMF radiation and ultrasound) have been used successfully as well. These combinations are quite lethal on cells, so collateral damage on healthy cells needs to be avoided by precision targeting. Gamma and x-radiation appear to be the two main ionizing radiation modes used. While different non-ionizing EMF radiation forms were used overall, the majority were in the form of electric pulses, and these were the most successful in the combinations examined. Examples include: (a) electric pulses and Co-60 gamma radiation for subcutaneous glioma tumors, which destroyed the tumor vasculature and caused DNA-related damage from reactive oxygen (Engstrom et al. 2001); (b) electric pulses and ultrasound for significant retardation of mouse tumor growth (Haro et al. 2005); (c) "Studies on tumor growth kinetics have shown a significant growth delay (by 50% to that of control) 7 days after treatment of tumor with radiation [Co-60 gamma rays] and electroporation. The results suggest that ***radiocytotoxicity of tumor cells in vitro as well as in vivo were enhanced significantly by electric pulses***, which may offer a potentially improved treatment of cancer." (Shil et al. 2005).

Chemical Promoters

The three previous sub-sections addressed proactive use of non-ionizing EMF radiation combined with other agents for cancer treatment. The following sub-sections in this cancer section reflect modeling of reactions to environmental non-ionizing EMF radiation combined with other agents that might possibly promote cancer.

> There are two main components involved in the non-ionizing EMF radiation co-promotion of cancer or other diseases. One component actively induces biological changes that could result in disease, and this can be termed the proactive component (e.g., carcinogenic chemicals, other radiation forms). The other component reduces the biological resistance to the proactive component, and serves as a 'passive' promoter of disease. Thus, as one example in the present subsection shows, where EMF and DMBA-induced cancer was enhanced in one substrain of rat but not another, genetics could be viewed as a 'passive' promoter of non-ionizing EMF radiation-induced disease. As another example in a later section shows, where blood-brain-barrier permeation was enhanced in diabetic rats but not normal rats, the presence of existing disease could be viewed as a 'passive' promoter of non-ionizing EMF radiation-induced disease. While the co-promoters identified in the present chapter are the proactive component, the role of the 'passive' promoters should not be discounted.

The combinations of non-ionizing EMF radiation and tumor-promoting substances, such as DMBA and TPA, tend to dominate the papers in this sub-section. While the majority of non-ionizing EMF radiation used in the research are power frequency, there are a representative number in the RF range. The effects tend to be synergistic. Examples include: (a) EMF ELF and DMBA for increasing number and size of mammary tumors (Baum et al. 1995); (b) microwave radiation for a significant acceleration of the development of benzopyrene-induced skin cancer and in shortening of life span of the tumor-bearing hosts (Szudzinski et al. 1982); (c) EMF ELF to enhance the induction of mammary gland tumors in rats using nitrosomethyl urea and reduce the mean latent period of tumor development (Beniashvili et al. 1991); (d) solvents, lead, and pesticides/herbicides were only associated with glioma in workers also exposed to moderate or high levels of ELFMF (Navas-Acien et al. 2002); (e) ELF-MF exposure to strengthen all-trans-retinoic acid [neuronal differentiating agent] effects on neuroblastoma cells (Marcantonio et al. 2010); (f) "long-term exposure of DMBA-treated female Sprague-Dawley rats in an alternating MF of low flux density promotes the development and growth of mammary tumors, thus indicating that MF exposure exerts tumor-promoting and/or compromoting effects" (Mevissen et al. 1998); (g) "MF exposure significantly increased mammary tumor development and growth in SD1 [one substrain of Sprague-Dawley DMBA-exposed rats] but not SD2 [another substrain of Sprague-Dawley DMBA-exposed rats obtained from the same breeder] rats. These data indicate that ***the genetic background plays a pivotal role in effects of MF exposure***." (Fedrowitz et al. 2004); (h)" tumor-promoting effects of life-long RF-EMF exposure may occur at levels supposedly too low to cause thermal effects" following treatment of mice with the carcinogen ethylnitrosourea (Lerchl et al. 2015).

Electromagnetic Promoters

The combination of non-ionizing EMF radiation with other low frequency electromagnetic radiation dominated this section. The much larger promoter combination of non-ionizing EMF radiation with ionizing radiation tended to concentrate on cellular and genetic damage as a precursor to cancer, and is addressed in those later cellular and genetic sections. They are not included here to avoid redundancy, although one example is provided here linking ELF-EMF and gamma radiation to increased rates of cancer in animals. Examples include: (a) power frequency magnetic and electric fields for enhanced leukemia and lung cancer risk (Miller et al. 1996); (b) "Residing near TPP [thermoelectric power plant] or HPL [high-power lines] confers a higher risk of AML [acute myeloblastic leukemia], with synergistic effect between both factors" (Rodriguez-Garcia and Ramos 2012); (c) "The combination of weak steady-state and weak low-frequency alternating magnetic fields activates SOD in Ehrlich ascites carcinoma cells and catalase in liver cells by 3.7 and 1.3 times, respectively ($p < 0.05$), which can result from enhanced production of ROS induced by combined exposure to magnetic fields with the specified parameters." (Sergeeva et al. 2011); (d) "Although ***no association was found for childhood leukemia in relation to measured ELF or static magnetic fields alone***, an increasing trend of leukemia risk with measured ELF fields was found for subjects within these static field.....findings suggest that the risk of childhood leukemia may be related to the ***combined effects of the static and ELF magnetic fields***." (Bowman et al. 1995); (e) " Life-span exposure to sinusoidal-50 Hz magnetic field and acute low-dose γ radiation induce carcinogenic effects in Sprague-Dawley rats" (Soffritti et al. 2016).

4.3.2.2 Impact on Neural System

This category includes non-ionizing EMF radiation combinations that enhance neural system performance, as well as combinations that degrade neural system performance. The present category specifically includes non-ionizing EMF radiation-agent combinations for enhanced analgesia, enhanced performance and reduced seizures, enhanced pain, and performance degradation (memory, learning, motor activity, behavior, sensory perception). Compared to the previous cancer-related category, more use is made of complex physiologically patterned magnetic fields to influence the neural system in the present category.

Because of ion resonance effects related to the total magnetic field exposure, and the different types of ions present, results can vary greatly for different field combinations and resonance conditions. For example, DC and AC fields at the calcium cyclotron frequency lower the locomotor and exploratory activity of test subjects, whereas action of the fields at the magnesium cyclotron frequency enhances these forms of behavioral activity. Different field values can either reduce, have no effect on, or increase e.g. endogenous opioid mediated analgesia through releasing/inhibiting endogenous opioids or enhancing/decreasing the activity of opioid signalling pathways. Finally, in some cases, the behavioral effects of a drug can be modified by brief exposure to a low-level non-ionizing EMF field even when the radiation level alone has no apparent effects on the behavior.

Enhanced Analgesia

This subsection includes a broad range of non-ionizing EMF radiation frequencies combined with diverse analgesics, although the examples focus on radiofrequencies. These examples include: (a) microwaves to enhance the duration of phenobarbital-induced sleep (Benson et al. 1983); (b) "***In combination with either of the anaesthetics*** used [ketamine or chloral hydrate], mm waves increased the duration of anaesthesia by approximately 50%.....exposure of mice to mm waves in vivo releases endogenous opioids or enhances the activity of opioid signalling pathway." (Rojavin and Ziskin 1997)

Enhanced Performance and Reduced Seizures

This subsection covers a broad frequency range, and includes complex patterned fields as well. Some of the perfomance enhancements include sleep and memory, but there are also efforts aimed at reducing seizures and improving motor activity in the impaired. Examples include: (a) EMF ELF enhancement of morphine-induced conditional behaviors (Lei et al. 2005); (b) static magnetic field enhancement of anticonvulsant effects (on auditory-induced seizures in mice) in combination with phenytoin (McLean et al. 2003); (c) reduction of damage to lithium and pilocarpine-seized rats by early exposure to computer-generated LTP-patterned magnetic fields (Lagace et al. 2009); (d) "intermittent, AC pulsed applications of picotesla flux density EMFs improve Parkinsonian symptoms in part by ***enhancing the patient's response to levodopa***.....intermittent applications of AC pulsed EMFs of picotesla flux density reverse the course of chronic progressive PD." (Sandyk 1997).

Reduced Analgesic Effectiveness

This subsection focuses on pain exacerbation, or more specifically the reduction of pain blockage by analgesics. Most of the frequencies used are power level, although a very few radiofrequencies were included. This subsection, in juxtaposition with the previous section on 'enhanced analgesia' illustrates the complexity of co-promotional non-ionizing EMF radiation effects, especially in the effects on neural systems. Reading of many papers on the neural effects shows the existence of 'windows', where non-ionizing EMF radiation in one frequency range in combination with other co-promoter and environmental variables produces one type of result, while non-ionizing EMF radiation in another frequency range in combination with similar co-promoters and environmental variables can produce radically different results. This means any conclusions about the effects of non-ionizing EMF radiation and co-promoters on the neural system have to be conditioned on the specific non-ionizing EMF radiation, co-promoters, and environmental variables operable at the time of the experiment.

Morphine tended to be the main co-promoter in this subsection, although other substances were used as well. Examples include: (a) competitive antagonism of morphine by methylnaltrexone, where microwave energy might facilitate entry of methylnaltrexone into the central nervous system (Quock et al. 1986a); (b) "The magnetic field exposures ***inhibited the degree of morphine-induced analgesia*** in a field intensity-dependent manner.....these data demonstrated a functional relationship between the behavioral effects of morphine in mice and the strength of the 60-Hz magnetic field." (Ossenkopp and Kavaliers 1987).

Performance Degradation

This subsection focuses on neural system performance degradation, and contrasts with the previous section on neural system performance enhancement. The main frequency ranges were ELF, RF, and complex magnetic fields, and there was a wide range of co-promoters. Examples include: (a) potentiation by EMF ELF pre-exposure during morphine treatment of dopamine D2 receptor (D2R) density in the rat dorsal hippocampus following withdrawal (Wang et al. 2008); (b) linear summation of TBS patterned complex magnetic field treatment with the contextual fear learning impairment evoked by agmatine treatment alone (McKay and Persinger 2003); (c) "simultaneous exposure to lead and RF from mobile phone use was associated with increased ADHD symptom risk" (Byun et al. 2013); (d) co-application of EMF with iron overload increased lipid peroxidation as compared to EMF alone, while the increase in antioxidant defenses triggered by the sole iron overload was abolished, suggesting that EMF exposure may be harmful in young adults by impairing the antioxidant defenses directed at preventing iron-induced oxidative stress (Maaroufi et al. 2011); (e) "The exposure to the EMF also potentiated haloperidol catalepsy: ***it decreased the drug threshold dose and increased the catalepsy duration***. The EMF influence on the haloperidol effects was of a prolonged character" (Shtemberg et al. 2001); (f) "The dose of iron alone used in present study, was unable to induce any effect. However, the 128 mT SMF in the presence of iron ions in the body can induce disruption in the emotional behavior and can produce oxidative stress in brain tissue of rats." (Elferchichi et al. 2015).

4.3.2.3 Impact on Circulatory System (Heart, Blood, Vascular)

This category includes non-ionizing EMF radiation combinations that enhance cardiovascular system performance, as well as combinations that degrade cardiovascular system performance.

Enhance Heart, Vascular

Many of the enhancement components were variants on baroreflex sensitivity (BRS), performed by researchers who co-authored frequently. The baroreflex is responsible for maintaining a stable blood pressure (BP) despite changes in body positions and fails in many autonomic disorders. The baroreflex regulates BP by changing the heart rate (vagal component) and total peripheral resistance (adrenergic component). A substantial number of the papers in this section focused on the relation of non-ionizing EMF radiation, especially static magnetic fields, to changes in the baroreflex sensitivity. However, there were some papers that showed static magnetic fields enhanced the effectiveness of anti-cancer drugs, possibly by increasing membrane permeability. Examples include: (a) static magnetic field prevention of significant decrease of BRS induced by verapamil administration (Gmitrov and Ohkubo 1999); (b) static magnetic field induced increment in microcirculatory blood flow, and was counter-acted by geomagnetic disturbance (which also decreased BRS) (Gmitrov 2007); (c) "***SMF may enhance nicardipine-induced hypotension by more effectively antagonizing the Ca2+ influx through the Ca2+ channels compared with the nicardipine treatment alone***. In addition, the enhanced antihypertensive effects of the SMF on the nicardipine-treated rats might be, at least in part, related to the increased NOx, primarily due to the upregulation of inducible nitric oxide synthase." (Okano and Ohkubo 2006); (d) "application of a SMF could alter membrane permeability, increasing the flow of the anticancer drugs. This may be one of the reasons why ***SMF can strengthen***

the effect of anticancer drugs" (Liu et al. 2011); (e) "***SMF increased the effectiveness of paclitaxel-chemotherapy significantly***." (Gellrich et al. 2014).

Degrade Heart, Vascular

There were very few articles in this section, the main example being enhanced risk of acute myocardial infarction among ELF EMF exposed subjects with genetic susceptibility to the disease: "The authors evaluated the relation between occupational exposure to extremely low frequency (ELF) magnetic fields and mortality from cardiovascular diseases....***the risk of AMI*** [acute myocardial infarction] ***was strengthened among ELF magnetic field-exposed subjects with genetic susceptibility to the disease***...." (Hakansson et al. 2003).

4.3.2.4 Impact on Immune System

This category includes non-ionizing EMF radiation combinations that enhance immune system performance, as well as combinations that degrade immune system performance. A number of studies e.g. focused on modifying neutrophil activity by adjusting the carrier and modulation frequencies of the EMF for calcium ion resonance. This reflects the proactive use of non-ionizing EMF radiation for enhancing immune system performance. There could be other situations where non-ionizing EMF radiation is being used for other purposes, and it exerts adverse impacts on the immune system. Some of this is included in the later sections on cellular impacts.

Enhance Immune System Performance

This subsection covers a wide range of non-ionizing EMF radiation frequencies, ranging from power frequency to very high radiofrequencies, with a number of pulsed electric field frequencies as well. While a broad range of co-promoters is covered, some of the more exciting results involve the enhancement of DNA vaccines. Examples include: (a) use of electric pulses to administer a DNA and IL-12 adjuvant combination to obtain a tenfold increase in antigen-specific IFN-gamma(+) cells (Hirao et al. 2008); (b) pulsed EMFs potentiated the effect of A(2A) or A(3) agonists on cell proliferation in bovine chondrocytes and fibroblast-like synoviocytes (Varani et al. 2008); (c) electroporation and DNA as good adjuvants in promoting efficient Th1-directed responses during DNA vaccination (Gronevik et al. 2005); (d) "Intramuscular (i.m.) delivery of a plasmid encoding anthrax toxin protective antigen (PA) using electroporation (EP), a potent DNA delivery method, rapidly induced anti-PA IgG and toxin neutralizing antibodies within 2 weeks following a single immunization in multiple experimental species.....These results suggest that ***EP may be a valuable platform technology for***

the delivery of DNA vaccines against anthrax and other biothreat agents." (Luxembourg et al. 2008).

Degrade Immune System Performance

This small subsection includes some emphasis of non-ionizing EMF radiation effects on spleen and thymus, but does not seem to have the same theme commonality as most other sections. An example is: (a) "C3H mice have been used to investigate the effect of a combination of cyclophosphamide (CY) and electromagnetic fields (PEMF).....we found that the effect of PEMF is evident only if mice are exposed during the 24 h following CY injection. The data reported here indicates that ***PEMF exposure after CY injection increases the damage [bone marrow labeling index, spleen labeling index, spleen colonies] induced in mice by CY***." (Cadossi et al. 1991).

4.3.2.5 Impact on Endocrine System

This (small) category includes non-ionizing EMF radiation combinations that enhance endocrine system performance, as well as combinations that degrade endocrine system performance.

Enhance Endocrine System Performance

This small subsection emphasizes drug delivery and chick embryos. The drug delivery is accompanied by pulsed electric fields while the chick embryos are tested using power frequencies. Examples include: (a) synergistic enhancement of percutaneous absorption of insulin by a combined use of electric pulses and iontophoresis (Tokumoto et al. 2006); (b) protection of chick embryos against damage from UV light exposure by power frequency EM field exposures of appropriate duration (Dicarlo et al. 1999); (c) "A lipid formulation.....was tested as an in vivo enhancer for the transcutaneous delivery of insulin.....The formulation enhanced the transport of insulin through the epidermis by 40- to 100-fold.....Application of electroosmosis across the formulation-treated epidermis enhanced the transport of insulin by an additional tenfold.....***The synergistic application of anionic lipid formulation and electroosmosis offers a promising non-invasive technique to deliver insulin transcutaneously***." (Murthy et al. 2006).

Degrade Endocrine System Performance

This subsection is too small to have any unifying theme, although the following three examples exhibit strong clarity: (a) demonstration that prolonged use of cellular telephones may lead to reduced melatonin production, and elevated 60-Hz MF exposures may potentiate the effect (Burch et al. 2002); (b) Atrazine, an endocrine-disrupting compound, had a synergistic effect on degranulation of cutaneous mast cells at both low and high doses in combination with EMFs (Rajkovic et al. 2010); (c) "Environmental EMFs apparently lack the energy necessary to function as aneugens, but the possibility exists that EMFs could influence the incidence of aneuploidy synergistically because EMFs can activate the neuroendocrine system, and ovulation and oocyte meiotic maturation are under neurohormonal control..... The results support the hypothesis that ***EMF exposure can promote the occurrence of aneuploidy caused by an aneugen*** via a mechanism involving the neuroendocrine system." (Mailhes et al. 1997).

4.3.2.6 Impact on Skeletal System

This category includes non-ionizing EMF radiation combinations that enhance skeletal system performance. The emphasis is on enhancing bone growth (mainly for accelerated fracture repair) or preventing bone loss, with the main mechanism being stimulation of cell proliferation. The EMF range is concentrated in power frequencies and pulsed EMFs, with a wide range of co-promoters. Examples include: (a) synergy of daily rotary non-uniform magnetic field exposure with calcium supplementation to increase the indices of thigh bone density, energy absorption, maximum load, maximum flexibility, and elastic deformation in ovarectomized rats (Zhang et al. 2006); (b) combination of BMP-2 and PEMF stimulation for augmenting bone formation to a greater degree than treatment with either single stimulus (Selvarnurugan et al. 2007); (c) "BMC and PEMFs might have a separate effect on osteochondral regeneration, but it seems that they have a greater effect when used together (Veronesi et al. 2015); (d) "Osteogenic differentiation of ASCs was accelerated by multiple-combination biophysical stimulation in vitro. However, both single stimulation and double-combination stimulation were sufficient to accelerate bone regeneration in vivo, while the osteogenic marker expression of those groups was not as high as that of triple-combination stimulation in vitro." (Kang et al. 2014); (e) "pulsed electromagnetic field (PEMF) plus BMP-2 upregulates intervertebral disc-cell matrix synthesis more than either BMP-2 alone or PEMF alone" (Okada et al. 2013); (f) "These results demonstrate that ***PEMF enhances osteogenic effects of BMP-2 on MSCs*** [mesenchymal stem cells] ***cultured on calcium phosphate substrates***, suggesting that PEMF will improve MSC response to BMP-2 in vivo in a bone environment." (Schwartz et al. 2008).

4.3.2.7 Impact on Genes

This category includes non-ionizing EMF radiation combinations that have beneficial impacts on genes, as well as combinations that that have adverse impacts. While potential diseases are mentioned in some of these articles, genetic issues were the main focus of the studies, and these records were therefore classified under the genetic issues rather than the diseases. The main non-ionizing EMF radiation forms are ELF, RF, and some static magnetic fields. The main co-promoters are chemical, ionizing radiation, and thermal/heat shock.

Positive Genetic Impacts

The main non-ionizing EMF radiation frequencies in this small subsection are at the very low end of the spectrum, and the co-promoters are typically neither the chemicals nor the ionizing radiation. Examples include: (a) EMF ELF and mild heat shock to strongly enhance the expression of the lacZ reporter gene (Junkersdorf et al. 2000); (b) "It was found that ***transfection of Chinese hamster ovary cells by naked plasmid DNA was enhanced by combined exposure of the cells to ultrasound..... and a magnetic field***.....in the presence of one of two different microbubble/nanoparticle preparations." (Stride et al. 2009).

DNA Damage

This relatively large subsection includes two main frequency groups: radiofrequencies at cell phone and wireless Internet levels, and ELF power frequencies. Examples include: (a) enhancement of human lymphocyte DNA damage effects (induced by mitomycin and 4-nitroquinoline-1-oxide, UV-mimetic agent) by 1.8 GHz RFR (Wang et al. 2005); (b) enhancement of the number of apurinic/apyrimidinic sites (induced by methyl methane sulfonate or H_2O_2) by exposure to ELF magnetic fields, due to enhanced activity or longer lifetime of radical pairs (Koyama et al. 2008); (c) "....exposure of the cells to static or 50 Hz magnetic fields (MF) and simultaneous treatment with a known oxidant, ferrous chloride, may affect the oxidative deterioration of DNA molecules.....Lymphocyte exposure to MF at 7 mT did not increase the number of cells with DNA damage in the comet assay. Incubation of lymphocytes with 10 mug/ml FeCl2 did not produce a detectable damage of DNA either. However, ***when the FeCl2-incubated lymphocytes were simultaneously exposed to 7 mT MF the number of damaged cells was significantly increased*** and reached about 20% for static MF and 15% for power frequency MF. In the control samples about 97% of the cells did not have any DNA damage." (Zmyslony et al. 2000); (d) "Pre-treatment with MF enhanced menadione-induced DNA damage, DNA repair rate, and micronucleus formation in human SH-SY5Y neuroblastoma cells" (Luukkonen et al. 2011); (e) "Gd induces DNA damage and apoptotic cell death in human lymphocytes and that ELF-EMF enhances the cytotoxicity and genotoxicity of Gd." (Cho et al. 2014); (f) "872 MHz CW RF radiation at 5 W/kg might enhance chemically induced ROS production and thus cause secondary DNA damage" (Luukkonen et al. 2009); (g) "The frequency of micronucleated astrocytes in the 10mg/kg BW bleomycin plus EMF exposure group.....was 1.6 times higher than that in the 10mg/kg BW bleomycin plus sham-exposure group." (Miyakoshi et al. 2012); (h) "Exposure to a 60-Hz, 2 mT ELF-MF for 6 h produced increased gamma-H2AX expression, as well as gamma-H2AX foci production, a common DNA double-strand break (DSB) marker.....2 mT ELF-MF exposure potentiated the expression of gamma-H2AX and gamma-H2AX foci production when combined with IR [ionizing radiation], but not when combined with H_2O_2" (Yoon et al. 2014).

Mutations

This relatively large subsection is almost exclusively power frequency EMF, with half the co-promoters being ionizing radiation. Examples include: (a) "ELF EMF increases x-ray-induced mutations and alters the spectrum of mutations" (Koyama et al. 2005); (b) ELF-EMF is mutagenic as a single agent and can potentiate the mutagenicity of ionising radiation (Mairs et al. 2007); (c) "exposure to ELFMF may induce mutations and enhance X-ray-induced mutations, resulting from the inactivation of NF-kappa B through the inhibition of tyrosine phosphorylation" (Ding et al. 2000); (d) "peripheral blood cells exposed to both ionizing radiation and EMF ELF magnetic fields demonstrated an enhanced frequency of near tetraploid chromosome complements, a feature not observed following exposure to only ionizing radiation" (Hintenlang 1993); (e) "The co-exposure of cells to BLM [bleomycin] and ELF-EMF led to a significant increase in the frequencies of MN [micronuclei] and aneuploidy compared to the cells treated with BLM alone.....ELF-EMF enhances the cytotoxicity of BLM" (Cho et al. 2007); (f) "The interaction of extremely low frequency electromagnetic fields (ELF-EMF) on the frequency of micronuclei (MN) and sister chromatid exchange (SCE) induced by benzo(a)pyrene (BP) in human lymphocytes was examined.....The co-exposure of cells to BP and 0.8 mT ELF-EMF for 24 h, followed by BP exposure for 48 h led to significant increases in the frequencies of MN and SCE compared to BP treatment for 72 h..... The obtained results suggest ***that low density ELF-EMF could act as an enhancer of the initiation process of BP*** rather than as an initiator-of mutagenic effects in human lymphocytes." (Cho and Chung 2003).

Teratogenecity

This modest-sized subsection focuses on RF EMF frequencies, uses mainly 2-methoxyethanol and cytosine arabinoside as co-promoters, and targets fetus development and malformations in the research. Examples include: (a) concurrent RF radiation exposure changed the shape of the dose-effect (dose-related developmental toxicity [external and skeletal malformation]) curve of 2-methoxyethanol (Nelson et al. 1997); (b) combination of microwave-induced hyperthermia and gamma radiation was highly teratogenic, indicating a mutual potentiation of the embryotoxic action of these two teratogens (Roux et al. 1986); (c) microwave-radiation enhanced the teratogenic effect of cytosine-arabinoside in mice

(Marcickiewicz et al. 1986); (d) "the study results appear to suggest that the combined use of nicotine and cell phones might result in more pronounced detrimental effects on the health of smokers" (Boga et al. 2015); (e) "The incidence of resorption and dead fetuses was not affected by PMF [pulsed magnetic field] but was increased by ara-C [a cytosine arabinoside] injection.....the incidence of CP [cleft palate] and/or CL [cleft lip] in the PMF group is not significantly greater than that in the control group. A significantly higher incidence of CP and/or CL was found in the PMF + ara-C group (49%) than the ara-C alone group (26.1%). These data suggest that PMF might enhance the development of ara-C-induced CP and/or CL.....it is concluded that the very ***weak embryotoxic effects of PMF exposure may be revealed and enhanced in combination with a teratogenic agent.***" (Chiang et al. 1995).

4.3.2.8 Impact on Cells (Permeation, Apoptosis, Oxidative Damage, Cell Growth, Cell Differentiation, Cell Proliferation)

This category includes non-ionizing EMF radiation combinations that have beneficial impacts on cells, as well as combinations that have adverse impacts. While potential diseases are mentioned in some articles, cellular issues were the main focus of the studies, and these records were classified under the cellular issues rather than diseases.

Beneficial Cellular Impacts

This subsection includes mainly pulsed EMF and power frequencies, with a broad variety of co-promoters. Three mechanisms are dominant: electroporation to allow drug entry into cells; enhanced apoptosis for destroying cancer cells; and enhanced cell proliferation for accelerating wound and fracture repair.

Permeation (Beneficial Effects)

This subsection covers mainly pulsed electric fields for electroporation, with the eventual goal of improved drug delivery. Examples include: (a) synergistic electroporation/iontophoresis transdermal delivery of indomethacin in contrast to iontophoresis or electroporation alone (Wang et al. 2007); (b) feasibility demonstration of electroporation to deliver and maintain the overall efficacy of an anthrax-plague DNA vaccine cocktail whose individual components have qualitative immunological differences when combined (Albrecht et al. 2012); (c) For Leishmaniasis, "use of electroporation and silver nanoparticles simultaneously can induce greater accumulation of particles in infected cells, besides higher toxicity.....In parasites and cells receiving both treatments, higher toxicity was observed in comparison to each treatment given individually, showing a synergic effect on promastigotes" (Dolat et al. 2015); (d) "The objective of this study was to enhance and optimize the skin permeation of MTX [methotrexate] using two physical techniques: an erbium:

yttrium-aluminum-garnet (Er:YAG) laser and electroporation.....Application of the laser and electroporation significantly enhanced the permeation of MTX. The enhancing effect was more pronounced after applying the laser.....***A combination of laser pretreatment and subsequent electroporation for 10 minutes resulted in a higher drug permeation than either technique alone***" (Lee et al. 2008).

Apoptosis (Cancer Cells)

This subsection covers pulsed EMF power and static magnetic fields, in that order. Most of the focus is accelerated destruction of cancer cells, and co-promoters tend to be photofrequency sources and drugs. Examples include: (a) pulsed magnetic fields (PMF) in combination with UVC radiation have the ability to augment the cancer cell killing effects of UVC radiation, and the effects appear to be greater when PMF and UVC are applied at the same time (Ruiz-Gomez and Martinez-Morillo 2005); (b) "This study was designed to test whether extremely low frequency electromagnetic field (ELF-EMF) could enhance the apoptosis-induction effect of X-ray radiotherapy on liver cancer cell line BEL-7402 in vitro....These findings suggested ***that ELF-EMF could augment the cell apoptosis effects of low doses of X-ray irradiation on BEL-7402 cells in a synergistic and cumulative way.***" (Jian et al. 2009).

Apoptosis (Healthy Cells)

The previous subsection on cellular apoptosis focused on non-ionizing EMF radiation-enhanced destruction of cancer cells. This subsection focuses on non-ionizing EMF radiation-enhanced destruction of healthy cells, and includes power frequency EMF, and pulsed and static electric fields. Co-promoters include ionizing radiation, photoradiation, and drugs. Examples include: (a) -dexamethasone-induced apoptosis but not spontaneous apoptosis was substantially increased in thymocytes from 60 Hz field-exposed animals (Ismael et al. 1998); (b) "Combined exposure of MF and MNP-SiO2 resulted in remarkable cytotoxicity and increased apoptosis in PC12 cells." (Jia et al. 2014); (c) "combined effect of EF [electric field] plus ionising radiation.....In cells exposed to EF, death increased substantially compared to irradiation alone.....***Application of an EF following irradiation greatly increases cell death***. The observation that the DNA repair shoulder in the survival curve of *C. albicans* is suppressed when cells are exposed to irradiation+EF suggests that EF likely inactivate cellular recovering processes. The result for the number of nuclei with -H2AX foci in MRC5 cells indicates that an ***EF interferes mostly in the DNA repair mechanisms***." (Arruda-Neto et al. 2009).

Cell Proliferation

This section covers mainly power frequency EMF, with a wide variety of non-radiation promoters and an emphasis on wound healing. Examples include: (a) combination of a 50-Hz sinusoidal magnetic field and fibrin glue has significantly favorable effects on healing of experimental colon anastomosis (Girgin et al. 2009); (b) "investigate the effects of topical application of an *Aloe vera* gel combined or not with microcurrent application on the healing of skin wounds surgically induced in Wistar rats.....the group treated ***with microcurrent plus Aloe vera presented an earlier onset of the proliferative phase*** compared to the control group and animals treated with *Aloe vera* gel alone.....Simultaneous application of *Aloe vera* gel and microcurrent is an excellent choice for the treatment of open wounds thus indicating a synergistic action of these two applications" (Mendonca et al. 2009).

Adverse Cellular Impact

This section focuses on oxidative damage, permeation, and cell growth/proliferation.

Oxidative Damage

The present subsection includes power frequency EMF, radiofrequency, and static magnetic field, in that order, and co-promoters that are mainly metals and chemicals. Examples include: (a) combination of CCl4 injection and SMF exposure caused an increase in lipid peroxidation in the liver exceeding that caused by either treatment alone (Watanabe et al. 1997); (b) exposure of intact erythrocytes incubated with an oxygen-radical generating system (Fe(II)/ascorbate) to a magnetic field induced a significant further decay in hexokinase activity as well as a twofold increase in methemoglobin production compared with red blood cells that were exposed to the oxidant system alone (Fiorani et al. 1997); (c) "single EMF exposure in 1800 MHz frequency significantly reduced antioxidant capacity both in healthy animals and those with paw inflammation. A certain synergic mode of action between applied electromagnetic fields and administered tramadol in rats treated with CFA" [Complete Freunds Adjuvant] was observed (Bodera et al. 2013); (d) demonstrated that the combined effect of static magnetic field and Cadmium increased oxidative damage in rat brain as compared with Cadmium-exposed rats (Amara et al. 2011); (e) "The effects were more pronounced after treatment with both Cd and EMF than at the treatment with each exposure alone..... This work concluded that combined exposure to Cd and EMFs might increase the risk of plasma damage via enhancing free radical generation and protein oxidation." (Hassan and Abdelkawi 2014); (f) "Effects of melatonin, extremely-low-frequency magnetic field (ELF-MF), and their combination on AT478 murine squamous cell carcinoma line were studied..... These results strongly suggest that ***ELF-MF***

attenuates antioxidative actions of melatonin on cellular level." (Zwirska-Korczala et al. 2004).

Permeation (Adverse Effects)

The previous subsection on positive impacts of cellular permeation showed the benefits of non-ionizing EMF radiation-enhanced permeation, especially for drug delivery. However, as shown in the present subsection, permeation can also allow toxic substances to penetrate cellular membranes and other barriers. This subsection covers power frequency and radiofrequency, focuses mainly on brain penetration, and covers mainly drug co-promoters. Examples include: (a) microwave irradiation facilitated central effects of domperidone (a drug which acts mainly in the periphery), by possibly altering the permeability of the blood-brain-barrier (BBB) and increasing the entry of domperidone to central sites of action (Quock et al. 1987); (b) methylatropine pretreatment and microwave irradiation resulted in a central anticholinergic action by methylatropine; microwave radiation may enhance passage of quaternary ammonium compounds like methylatropine across the BBB and B-CSFB (Quock et al. 1986b); (c) "investigated the effect of long-term exposure to modulation magnetic field (MF), insulin, and their combination on blood brain barrier (BBB) permeability in a diabetic rat model.....DM [diabetes mellitus] and MF [50 hz magnetic field] increase BBB permeability; ***in combination, they cause more increase in BBB permeability***, and insulin decreases their effect on BBB." (Gulturk et al. 2010).

Cell Growth, Differentiation, Proliferation, Morphology

This subsection covers mainly power frequency EMF and radiofrequency. The main co-promoter is TPA, but other drugs are used, as well as some ionizing radiation. Examples include: (a) combination of EMF and gamma-ray exposure to SHG44 cells resulted in a synergistic effect by triggering stress response, which increased reactive oxygen species (Cao et al. 2009); (b) effect of 60-Hz EMF at 1G on cell differentiation is approximately equivalent to treatment of the cells with 250–500 pg/ml TPA [tumor-promoting phorbol ester]; effect of both EMF and TPA treatment on differentiation is additive at low TPA concentrations (Tao and Henderson 1999); (c) "In NB69 simultaneous administration of MF and ROL induced an additive effect on cell proliferation, associated to increased DNA content" (Trillo et al. 2012); (d) "investigated whether exposure to 60-Hz sinusoidal magnetic fields (0.3–1.2 G for 3–72 h) would cause proliferation of human astrocytoma cells. Sixty-Hertz magnetic fields (MF) caused a time- and dose-dependent increase in proliferation of astrocytoma cells.....and strongly potentiated the effect of two agonists (the muscarinic agonist carbachol and the phorbol ester PIMA).....***These data indicate that MF can increase the proliferation of human astrocytoma cells and strongly potentiate the effects of two agonists***. These findings may provide a biological basis for the observed epidemiological associations between MF exposure and brain tumors." (Wei et al. 2000).

4.3.2.9 Impact on Micro-organisms

This category covers the use of (mainly) pulsed electric fields to destroy bacteria for improving food storage or developing superior strains. Examples include: (a) combination of nisin and mild pulsed electric field (applied to vegetative cells of Bacillus cereus) resulted in a reduction of 1.8 log units more than the sum of the reductions obtained with the single treatments, indicating synergy (Pol et al. 2000); (b) combined treatments of electric field and chemical mutagen N-methyl-N′-nitro-N-nitroso-guanidine (NTG) for the strain improvement of Saccharomyces sp. in ethanol production (1) increased the lethal effect and auxotrophic mutation rate of NTG, (2) increased the chances of obtaining superior yeast strains for the ethanol production from tapioca, (3) produced a higher number of improved clones (Kim and Lee 1998); (c) "in the clinical setting, combining systemic antibacterial therapy with PEF will yield a synergistic effect leading to improved eradication of mesh infections" (Khan et al. 2016); (d) "electric fields generated using insulated electrodes can inhibit the growth of planktonic *Staphylococcus aureus* and Pseudomonas aeruginosa and that the effect is amplitude and frequency dependent, with a maximum at 10 MHz. ***The combined effect of the electric field and chloramphenicol was found to be additive***.." (Giladi et al. 2008).

4.3.2.10 Attenuation of Non-ionizing EMF Radiation Effect

This category focuses on ameliorating non-ionizing EMF radiation effects through combinations. Researchers have identified adverse impacts resulting from exposure to EMF, and have searched for agents that will ameliorate these adverse effects. The EMF frequencies covered are split between power frequencies and radiofrequencies, mainly at cell phone level. While a number of adverse effects are targeted, the main focus is reducing oxidative damage, followed by reducing DNA damage. The main ameliorating agents are propolis, melatonin, Vitamins C and E, and zinc. Noise is used in many examples to cancel out the EMF field effects. Examples include: (a) L-carnitine seems to have protective effects on the 2.45-GHz-induced blood toxicity (oxidative damage) by inhibiting free radical supporting antioxidant redox system (Gumral et al. 2009); (b) bee venom is demonstrated to have a radioprotective (915 MHZ) effect against basal and oxidative DNA damage (Gajski and Garaj-Vrhovac 2009); (c) long-term exposure to low-frequency EMF increases lipid peroxidation in the brain, which may be ameliorated by Zinc supplementation (Bediz et al. 2006); (d) CAPE [Caffeic acid phenethyl ester, a major component of honeybee propolis] exhibits a protective effect on mobile phone-induced and free radical mediated oxidative renal impairment in rats (Ozguner et al. 2005); (e) significant inhibition of the increased human epithelial amnion cell proliferation when a noise field was superimposed on an EMF ELFfield (Raskmark and Kwee 1996); (f) "investigate the effects of 12 kV/m electric (E) field sourced by power lines on oxidative and nitrosative stress, and antioxidant status.....examine the protective effects of N-Acetyl-L-cysteine (NAC) and epigallocatechin-gallate (EGCG) in the liver tissues of guinea pigs against the possible detriments of electromagnetic field exposure.....extremely low frequency (ELF) electric field has potential harmful effects on the living organisms by enhancing the free radical production. ***NAC and EGCG might have hepatoprotective effects in ELF-E field induced oxidative and nitrosative stress***." (Guler et al. 2008).

4.3.2.11 Other

This category covered papers difficult to assign to any of the previous sections. There was a broad range of topics covered, and a broad range of combinations, with a few papers focusing on the impact of a static magnetic field combined with non-ionizing EMF radiation on ion resonances. Examples include: (a) magnetotherapy with the magnetic field inductor applied over a dressing with native naphthalane oil + routine therapy (prodectin, parmidin) enhanced recovery and is recommended for the treatment of trophic ulcers of the shins, complicated by bacterial eczemas (Guseinzade et al. 1991)); (b) extremely low-frequency magnetic field and aluminum solution synergistically enhanced the growth of spruce seedlings (*Picea abies*) (Ruzic et al. 2000); (c) "exposure to static magnetic field causes accumulation of reactive oxygen species in *V. faba* and natural radioactivity of soil exaggerates oxidative stress" (Jouni et al. 2012); (d) Microsecond pulsed magnetic field improves efficacy of antifungal agents on pathogenic microorganisms. (Novickij et al. 2014); (e) combination of CCl4 injection and static magnetic field exposure induced elevation of the hepatic metallothionein content exceeding that induced by either treatment alone (Satoh et al. 1996); (f) "Groups of male CBA/J mice were injected with Salmonella typhimurium lipopolysaccharide (LPS) and irradiated with 2450 MHz (CW) microwaves. The 50% lethal dose (LD50) of LPS was determined for mice irradiated at 30, 20, 10 and 5 mW/cm2 immediately following injection.. ***A significant decrease in the LPS dose required to kill 50% of the mice was observed at power densities of 20 and 30 mW/cm2***." (Riddle et al. 1982).

4.3.3 Non-ionizing EMF Radiation-Biological Mechanisms Matrix

In order to gain further insight to the relation between biological mechanisms and varieties of EMF, a non-ionizing EMF radiation-biological mechanisms matrix was generated. All Abstract phrases (with a frequency of three or greater) generated by the VP software were examined, and those that were biomedically related (~800) were extracted. Additionally, all of these Abstract phrases that could be related specifically to an EMF band, either textually or numerically (ELF [25–100 Hz]; SMF [0 Hz]; cellphone [~800–2000 MHz]; WiFi [~2450 MHz]; Radar [50 MHz–100 GHz], PEF, PEMF) were extracted, and combined by band. Two matrices of these two groups of phrases were generated, but will only be summarized due to space limitations.

4.3.3.1 Extremely Low Frequency

This section reflects the AC power frequency band, and the main focus of the research was adverse effects resulting from exposure to residential or occupational wiring and equipment. The main phrase grouping reflects enhancement of DMBA-induced mammary tumors by ELF. Another grouping reflects the activation of protein kinase C by the tumor-promoting phorbol esters, such as TPA, with a related grouping focused on enhancing the suppression of gap junctional intercellular communication, again by inhibiters such as TPA. There is a smaller group related to suppression of melatonin and its potential impact on carcinogenesis. There is a broader group with two components, one related to cell survival and enhanced apoptosis, and the other addressing enhanced cell proliferation and differentiation. Finally, a group shows enhanced DNA damage and mutagenicity from ELF combinations with ionizing radiation (x-ray, gamma) or chemicals/drugs (MMC, H_2O_2, Menadione).

4.3.3.2 Pulsed Electric Field (PEF)

This section reflects pulsed electric fields, and the research focus was proactive use of this technology for improved health and food preservation. The most fundamental grouping reflects increased permeation of cell membranes by PEF for improved drug delivery. Primary emphasis is on enhanced chemotherapy for tumor reduction, with a secondary emphasis on use of PEF as an adjuvant for DNA vaccines. Another thrust area is enhancement of [e.g. nisin] preservation factor-inactivated micro-organisms [e.g. Listeria].

4.3.3.3 Static Magnetic Field

The effect was mainly on circulatory system improvement. However, static magnetic fields were used copiously in the total research to modify the effects of the geomagnetic field or alternating magnetic fields. One group focused on the combined impact of artificial static and geomagnetic field on cardiovascular regulation, especially the regulation of the baroreflex sensitivity. A related group focused on the combined effect of static magnetic fields and blood pressure modifiers on blood pressure. There were effects of combined ELF EMF and static magnetic fields [sometimes combined with other agents] on genotoxicity, especially chromosomal damage and micronuclei. A fundamental underlying driver is the effect of the superposition of the static magnetic field on either the geomagnetic field or an alternating magnetic field to provide a net magnetic field that can allow or suppress ion resonances for selected ions.

4.3.3.4 Cell Phones

This section relates to effects of mobile communication frequencies. The focus of the research was identifying adverse health impacts from exposure to relevant radiofrequencies, and how to ameliorate the adverse impacts. Generation of reactive oxygen species and the subsequent oxidative damage are central to effects from radiofrequency radiation combined with other agents, in this frequency range. Enhancement of DNA damage was also evident. Some research was focused on synergistic teratogenic effects from combined administration of RF radiation and chemical agents, especially 2-methoxyethanol. A number of articles focused on agents that, when combined with this range of radiofrequency radiation, helped reduce the oxidative damage resulting from use of the radiofrequency radiation alone. These agents included caffeic acid phenethyl ester [CAPE]/honeybee propolis, melatonin, zinc, and xanthine oxidase, among others.

4.3.3.5 Pulsed EMF

This medical therapeutic section relates to pulsed electromagnetic fields. There is some overlap with pulsed electric fields in improving drug delivery and photodynamic therapy (especially for cancer therapy) by increasing cell membrane permeability transiently. Another key positive impact is the combined effect of bone morphogenetic proteins and pulsed electromagnetic fields in augmenting bone formation. Additionally, there is a combined effect of AC pulsed magnetic fields (typically employed in transcranial stimulation) and dopaminergic medications on stimulating dopaminergic neurons and dopamine receptors to enhance dopamine release and alleviate symptoms in dopamine-deficient diseases.

4.3.3.6 WiFi

This section relates to wireless communication frequencies, particulary wireless Internet. Unlike the effects shown in the previous PEMF sub-section, where the non-ionizing EMF radiation combined impacts tended to be positive, the non-ionizing EMF radiation combined impacts in this section are almost universally negative. The combined effect of 1) magnetic fields at these frequencies and

2) analgesics on the central nervous system tended to be negative, for example, in learning, memory, and stereotypy. In additions, these combinations were shown to promote tumors, enhance drug-induced lethality, enhance DNA damage and teratogenic effects, decrease LPS LD50 dose, and alter phagocytic activity (which could be attenuated by vitamins C and E).

4.3.3.7 Radar

This section covers a wide range of high frequency non-ionizing EMF radiation, and was meant to include those frequencies beyond WiFi. The combined effect data in this frequency region is minimal, and relates to the combined effect of modulated millimeter waves and phorbol ester [PMA] on neutrophil respiratory bursts.

4.4 Discussion and Conclusions

4.4.1 Overview

Non-ionizing EMF radiation can serve as an initiator of health effects, as well as serve as a co-promoter or potentiator of biochemical agents, both beneficial and adverse. However, as this chapter has shown, the numbers and types of impacts are increased substantially when non-ionizing EMF radiation functions in combination or as a co-promoter. There were many examples where either (1) non-ionizing EMF radiation by itself had no observable effect, but it enhanced the demonstrated individual effect of another agent or (2) non-ionizing EMF radiation by itself had no observable effect and other agents by themselves had no measurable effects, but in combination the two had a pronounced effect. In other words, there is an enhancement of the singular effects.

Generally, the combined effects occured in 'windows' for specific combinations of multiple variables, or stated another way, in specific regions of parameter space. Thus, combined effects could be observed at one dose rate but not at another, at one frequency but not at another, at one intensity but not at another, at one modulation pattern but not at another, at a different sequence of exposure, and so on. This large combinatorial parameter space makes drawing conclusions about combined effects difficult in many cases, since the effect will display only under the proper combination of variables and parameters within limited ranges of each.

4.4.2 Frequency Grouping

The overall results could be frequency-grouped into three major categories. One category contained pulsed electric fields and pulsed magnetic/electromagnetic fields. A second category contained extremely-low frequency electromagnetic fields, cell phone radiofrequency fields, and WiFi radiofrequency fields. The third smaller category contained static magnetic fields.

The first category, which could be termed the Treatment category, tended to use non-ionizing EMF radiation proactively to enhance therapeutic treatment, whether for enhanced drug delivery, accelerated wound and fracture healing, or bacterial inactivation for prolonged food storage. The second category, which could be termed the Environmental Exposure category, tended to identify the overwhelmingly detrimental reactive effects from exposures to non-ionizing EMF radiation used for non-treatment purposes, such as electrical appliances, residential wiring, and wireless communications. These reactive health effects included enhanced oxidative damage, enhanced DNA damage, enhanced mutagenicity, enhanced teratogenicity, and many others. The third category, which could be termed the Superposition category, superpositioned the static magnetic field with other magnetic fields (geomagnetic or alternating) to provide a net magnetic field that could have a multiplicity of positive or negative health consequences by allowing or suppressing ion resonances for selected ions.

The non-ionizing EMF radiation in the Treatment category tended to be relatively short-term, especially those that were pulsed, whereas the non-ionizing EMF radiation in the Exposure category that had adverse health impacts tended to require relatively long exposures. The static magnetic fields in the Superposition category, when used as co-promoters with other non-ionizing EMF radiation, required an intensity on the order of magnitude comparable to the other non-ionizing EMF radiation. Even with other non-EMF agents, the static magnetic fields tended to have relatively substantive intensities.

4.4.3 Windows in Parameter Space

The documents selected for this chapter required that the non-ionizing EMF radiation component of the combination have some impact. In many articles not selected for this chapter, the non-ionizing EMF radiation-agent combination may have been similar to the non-ionizing EMF radiation-agent combination selected for this chapter, but the presence of the non-ionizing EMF radiation component did not affect the outcome (especially in the Environmental Exposure category for the adverse health impacts). What could account for this difference? For a given combination, one or more of the studies could have had poor research, or one or more of the studies could have had preconceived bias. Or, the experimental conditions for the two studies were sufficiently different that the 'window' in parameter space required for the effect to be observed was not present in one study but was present in another study.

There appears to be sufficient data among diverse research groups that adverse health effects from non-ionizing EMF radiation combinations exist in at least selected 'windows' of parameter space. Overall, the number and extent of these 'windows' need to be identified, to ascertain their overlaps with the operational non-ionizing EMF radiation parameter space. This overlap would provide some indication or estimate of potential real-world health effects.

4.4.4 Resolving Differences

The first step in this process would identify major areas of disagreement, where strong adverse effects have been shown or predicted by the proponents, and typically no adverse effects have been shown or predicted by the opponents. The above studies, focused on the conditions that produced these adverse effects, would be re-done with multiple performers participating, representing diverse viewpoints. The study criteria would match objectives, methodology, and operational environment as closely as possible. Any differences in results could be examined on a uniform basis.

The second step would involve expanding the parameter values to understand the boundaries of the 'window' in parameter space in which adverse health impacts can occur. The third (and most difficult) step is the inclusion of other potential co-promoters to reflect more closely the following real-world conditions. People are not exposed only to non-ionizing EMF radiation in isolation, or non-ionizing EMF radiation combined with one potential co-promoter. People are exposed to many potentially harmful agents, either harmful in their own right, harmful only when combined with non-ionizing EMF radiation, harmful only when combined with non-ionizing EMF radiation and one or more other agents, and so on. For example, there could be three agents which, by themselves, would exhibit no harmful effects, and in any combination of two might exhibit no harmful effects, but in combination of three would exhibit harmful effects. Non-ionizing EMF radiation could provide the 'tipping point' of multiple potential harmful agents.

Unfortunately, to identify these potential harmful combinations experimentally would require astronomical levels of effort. In a recent eBook (Kostoff 2015), the first author identified ~800 pervasive foundational causes of disease; i.e., 800 tangible contributing factors to myriad diseases, and these results were viewed as an extremely conservative estimate. The number of potential combinations of these pervasive contributing factors to disease would be determined by the binomial coefficient. For example, the number of combinations of three potentially toxic stimuli from the list of 800 is 800!/(3!*797!), or approximately 85 million, and the number of combinations of two is approximately 320,000. Thus (in the latter case), we would need to perform 320,000 experiments to identify potentially harmful effects of any two toxic stimuli (from the ~800 identified in (Kostoff 2015)) combined with non-ionizing EMF radiation. Given the problem of 'windows' in parameter space described above, each experiment would be fairly complex, involving examination over large ranges of parameters such as non-ionizing EMF radiation frequency, intensity, duration, etc. Realistically, we would need to prioritize the pool of potential toxic stimuli, and then examine their effects in very small combinations with non-ionizing EMF radiation.

4.4.5 Limitations of Text-Based Analyses

A final observation. An exhaustive search process retrieved under 500 papers total for non-ionizing EMF radiation combined therapeutic and environmental adverse effects, with about 350 papers devoted to the environmental adverse effects of non-ionizing EMF radiation co-promotion published in the previous 30–40 years. Given the potential impact of non-ionizing EMF radiation, and the real-world importance of non-ionizing EMF radiation as a co-promoter, there appears to be a major disconnect between the magnitude of the problem and the research reported in the published literature available to understand and ameliorate the problem. What accounts for such a large disconnect: (1) underfunding of the research on adverse effects and/or (2) underreporting of the adverse effects? Appendix 2 addresses the issue of underreporting the adverse health effects of substances/ phenomena like non-ionizing EMF radiation.

Appendices

A.1 Appendix 1 – Text Mining Query Used for Information Retrieval

Appendix 1 contains: (a) brief overview of text mining; (b) brief summary of past EMF co-promoter studies; (c) text mining methodology used to identify, retrieve, extract, and analyze the relevant EMF co-promoter articles from the premier biomedical literature; (d) details of the iterative relevance feedback component of the query used, and of the subsequent citation network traversing component of the query.

A.1.1 Text Mining Overview

Text mining is the extraction of useful information from large volumes of text (Hearst 1999; Feldman et al. 1998; Kostoff et al. 2001a). Its component capabilities of computational linguistics and information retrieval were the main analytical techniques used in the present chapter. A typical text mining study of the published literature involves the development of a query for comprehensive information retrieval, an analysis of the database using computational linguistics and bibliometrics, and an integration of the processed information.

Computational linguistics identifies the main technical/medical themes of the database(s) being examined as well as the relationships among these themes. Computational linguistics has been used to enhance information retrieval and increase awareness of the global technical literature (Kostoff et al. 1997; Greengrass 1997), as well as to track the impact of a specific research area across time and applications areas (Davidse and VanRaan 1997; Kostoff et al. 2001b). Upgraded versions of these techniques were used for enhanced text-based and citation-based information retrieval in the present chapter.

Computational linguistics has been used in three modes to identify potential innovation and discovery: (a) co-occurrence of disease/technical problem and potential treatment/problem solution in the same article for identifying innovation (Kostoff and Briggs 2008; Kostoff et al. 2008b, c, d; Kostoff 2012); (b) linking disease/technical problem and disparate potential treatment/problem solution literatures directly (Kostoff and Briggs 2008; Kostoff et al. 2008a, b, c, d, e; Kostoff 2008, 2011, 2012; Swanson 1986; Swanson et al. 2001; Kostoff and Patel 2015; Kostoff and Los 2013); linking disease/technical problem and disparate potential treatment/problem solution literatures indirectly (Kostoff and Briggs 2008; Kostoff et al. 2008b, c; d). While these three modes each have their own distinctive features, they share the common requirement for an 'intelligent' query that targets documents with specific characteristics while filtering out 'noise'. That was also the main requirement for the present chapter, and the experiences from the above references were directly applicable to the requirements of the present chapter. The focus of the present chapter was on co-occurrence of two or more promoters/stimuli in the same article (mode a), but modes b) and c) were used informally to help further understand the mechanisms linking the promoters/stimuli to the health impacts. Modes b) and c) could be used to start with non-ionizing EMF radiation promoters/stimuli or combinations ab initio, and identify direct and indirect linkages/mechanisms to potential health impacts, beneficial or adverse.

A.1.2 EMF Co-promoter Studies

Most non-ionizing EMF radiation health/biological impact studies focus on one non-ionizing EMF radiation form, with no identifiable co-promoters/stimuli. The initial comprehensive non-ionizing EMF radiation retrieval performed by the first author used a text-based query to search the Medline/Science Citation Index databases for a wide range of EMF radiation records, and retrieved over 6000 documents. The present chapter used a more intense hybrid text-based and citation-based query to search for non-ionizing EMF radiation ***combinations*** and their effects, and retrieved under 500 records. Thus, perhaps 5–7% of non-ionizing EMF radiation health/biological impact studies (probably less) are concerned with combination-type effects.

There are relatively few comprehensive review articles covering the combination effects. Juutilainen (2008; Juutilainen et al. 2006) examines cocarcinogens. Stam (2010) examines non-ionizing EMF radiation effects on the blood-brain barrier; this study is typical of reviews that may examine non-ionizing EMF radiation effects on an organ or system. Lee et al. examine the carginogenic potential of 60 Hz ELF-MF (1 mT) alone or in combination with ionizing radiation (IR), hydrogen peroxide (H_2O_2), or c-Myc overexpression (Lee et al. 2012).

Most of these previous studies that focused on combination effects of non-ionizing EMF radiation with other stimuli are like the articles referenced in the body of this chapter, namely, selected non-ionizing EMF radiation ranges in the frequency spectrum combined with usually one or a few co-stimuli. However, none of these review articles covers the wide range of disciplines as the present chapter, as well as both adverse and beneficial combined effects. It is the scope of coverage and the comprehensiveness of retrieval using the hybrid search technique, as well as the integration across disciplines, which makes the present chapter unique.

A.1.3 Methodology: Document Identification, Retrieval, Extraction, Analysis

A.1.3.1 Definitions of Combined/Interactive Effects

The goal of this chapter is to examine the scope of the non-ionizing EMF radiation combined effects on biological systems; i.e., identify effects on biological systems from combined exposure to non-ionizing electromagnetic fields/radiation and at least one other agent. These interactive effects include:

A.1.3.1.1 Additive effects (the combined effect of two or more agents acting in the same general direction approximates the sum of the effects of the agents administered separately, subject to the maximum possible effects in biological systems);

A.1.3.1.2 Antagonistic effects (the combined effects of two agents acting in different/opposite directions are smaller than the effect of any one of them in stand-alone mode);

A.1.3.1.3 Potentiative effects (the increased effect of an agent by concurrent action of another agent that does not have a stand-alone effect); and

A.1.3.1.4 Synergistic effects (the combined effect of two or more agents is significantly greater than the sum of the effects of each agent administered alone, subject to the maximum possible effects in biological systems).

Other terminology is used in the documents, such as co-promotional, co-mutagenic, co-carcinogenic, etc., but these terms tend to be sub-sets of the more general terms defined above.

The approach in this chapter is to: (1) select the most credible global databases of research articles; (2) develop a query that will retrieve the relevant combined effects literature comprehensively; (3) identify the key biomedical thrusts in this retrieved literature; and (4) extract the mechanisms and principles that describe the influence of the non-ionizing EMF radiation component on the final combined effects. These four approach components are now described in more detail.

A.1.3.2 Select Global Databases

The two premier biomedical research article databases are the Web of Science (WOS-Science Citation Index/Social Science Citation Index/Arts and Humanities Citation Index-SCI/SSCI/A&HCI) and Medline. Each has its unique strengths.

WOS has the capability for citation linkages (references, citing papers, papers that share at least one reference), while Medline has a unique taxonomy/keyword structure called MeSH. Both databases were used in this chapter for query development.

A.1.3.3 Develop Retrieval Query

The first step in query development is to define the scope of the study topic. The scope selected was effects on biological systems from combined exposure to non-ionizing EMF radiation and at least one other agent. A novel hybrid iterative relevance feedback technique (based on Kostoff et al. 1997) was used to develop the query. The non-ionizing EMF radiation component of the full query (based on Medline and the SCI) used by the first author for the comprehensive non-ionizing EMF radiation retrievals was intersected with the terms "synerg* and 'combined effect*'" to form an initial test query (see A.1.3.6.1. for complete initial test query). This initial query was inserted into the SCI search engine, further filtering was performed by restricting to biomedical Subject Areas (each SCI record has one or more Subject Area keywords assigned to it), and about eighty records were retrieved. These were termed the *core records*. The query was then expanded by examining the local citation network for each of the core records.

First, all the records that cited the *core records* were retrieved (~700), and were examined manually to select relevant non-duplicative records (~90). These were termed the *core citing records*. Second, all the references to the *core records* and *core citing records* that were in the SCI were examined manually to select relevant non-duplicative records (~150). These were termed the *core record references* and *core citing record references*. Third, some of the records that shared references with the *core records* and *core citing records* were examined, as follows.

> The SCI has a feature called Related Records. For a record of interest, the Related Records feature will display all records in the SCI database that share at least one reference in common with the record of interest. These Related Records can be ordered by the number of references in common. The numbers of Related Records can range from zero (all the references in the record of interest were cited only once) to hundreds of thousands for a large block of highly cited references. Typically, the numbers of Related Records for a record of interest range from hundreds to tens of thousands. In practice, only a few of the Related Records for each record of interest can be examined for relevance, due to the large volumes involved.

The protocol used was to examine the twenty-forty records with the most shared references for each of the *core records* and *core citing records*, and extract those non-duplicative records deemed relevant (~75). These were termed the *core records related records* and the *core citing records related records*. Many of the Related Records tended to display repeatedly on the twenty-forty records with the most shared references, and the marginal utility of this approach decreased with time. This can be visualized as a well-connected network, where the same material is being accessed repeatedly. Like any network problem, the path to new information is less through complete-link type approaches and more through single-link type approaches. Unfortunately, the numbers of records with two or one shared references are large compared to the numbers of records with many shared references, and there were too many of these low shared reference records to examine manually.

Fourth, all the relevant records retrieved above were combined, and imported into the Vantage Point (VP) software (Vantage Point 2015). Text patterns in the Abstracts and Titles were examined, and were added to the initial test/text query. The additional query terms are shown in A.1.3.6.2. These terms were inserted into the SCI search engine, with further filtering done by Subject Area. About 500 records were retrieved and examined manually, and those deemed relevant (~135) were extracted. The total of about 530 records were then examined in detail, *stricter criteria were applied for relevance*, and 436 records with Abstracts were judged to be relevant. Obviously, the citation and text linkages could have been continued in an iterative manner, and more relevant records would have been found. However, the marginal utility of both approaches was beginning to decrease, especially for the citation linkages, and the manual selection approach was becoming infeasible.

A.1.3.4 Identify Key Biomedical Thrusts

The 436 retrieved records were inserted into the VP text mining software package and into the CLUTO document clustering software package (CLUTO 2015). A factor analysis was performed in VP using 32 factors, and a hierarchical taxonomy was generated in CLUTO using 32 clusters. Text mining was performed on each factor and cluster, to identify the key biomedical phrases representative of the group and the titles of papers in the group. Based on reading the titles and phrases, and reading of many paper Abstracts in each group as well, the theme of each group was identified.

A.1.3.5 Extract the Mechanisms and Principles that Describe the Influence of the Non-ionizing EMF Radiation Component on the Final Combined Effects

This is the key analytic step. Based on the two groupings identified by the clustering and factor analysis described above, and a reading of all the Abstracts, a final taxonomy was generated. The records were assigned manually to each taxonomy category. The non-ionizing EMF radiation-co-promoter-mechanism-disease 'signatures' were extracted by integrating the relevant factors, relevant clusters, and relevant sections of each record in the cluster. These relevant sections are displayed in the narrative section following the clustering summaries and taxonomy (Figs. 4.1 and 4.2).

The records were updated by repeating the above process for years 2012–2016, and adding the additional relevant records to the existing taxonomy categories.

A.1.3.6 Iterative Relevance Feedback Query

A.1.3.6.1 Initial Test Query

Topic=((EMF OR "Electromagnetic Field*" OR "Radio-Frequency Radiation" OR "Radio-Frequency Irradiation" OR "RF-Radiation" OR "RF-Irradiation" OR "Microwave Radiation" OR "Microwave Irradiation" OR "Mobile Phone*" OR "Cell* Phone*" OR "Wireless Phone*" OR "Cordless Phone*" OR "Mobile Telephone*" OR "Cellular Telephone*" OR "Wireless Telephone*" OR "Cordless Telephone*" OR "Base Station*" OR "RF-Transmission Tower*" OR "Cell Tower*" OR (("Magnetic Field*" OR "Electric Field*") AND ("Power Line*" OR "Low Frequency" OR "Power Frequency" OR "Intermediate Frequency" OR "Transmission Line*" OR "Electric Power Transmission"))) AND (synerg* OR "combined effect*")) AND Document Type=(Article OR Review)

Refined by: Subject Areas=(Biology OR Behavioral Sciences OR Biophysics OR Cell Biology OR Environmental Sciences OR Toxicology OR Public, Environmental & Occupational Health OR Orthopedics OR Radiology, Nuclear Medicine & Medical Imaging OR Health Care Sciences & Services OR Developmental Biology OR Pharmacology & Pharmacy OR Neurosciences OR Materials Science, Biomaterials OR Oncology OR Parasitology OR Biotechnology & Applied Microbiology OR Physiology OR Biochemical Research Methods OR Biochemistry & Molecular Biology OR Reproductive Biology OR Rheumatology OR Genetics & Heredity OR Surgery)

A.1.3.6.2 Additional Terms in Refined Query

Topic=(("magnetic field*" OR EMF* OR microwaves OR "microwave radiation" OR "microwave exposure*" OR "microwave irradiation" OR "electromagnetic field*" OR "RF radiation" OR ELF-MF* OR "mobile phone*" OR PEMF* OR EMR OR "Electromagnetic radiation" OR ELF OR "Radiofrequency field*" OR "radiofrequency radiation" OR ELF-EMF OR "cell* phone*" OR "electric power" OR "electromagnetic noise" OR IFC OR "pulsed magnetic field*" OR "CELL TOWER*" OR "static MF" OR "electric field*" OR electricity OR "electromagnetic EM field*" OR "electromagnetic radiation" OR "EM field exposure*" OR geomagnetic OR "power line*" OR electroporation OR electrofusion OR electrochemotherapy OR electropermeabilization))

AND ("combin* effect*" OR potentiat* OR synerg* OR "combin* exposure*" OR co-exposure OR "combin* treatment*")) AND Document Type=(Article OR Review)

Refined by: Subject Areas=(Biophysics OR Biology OR Neurosciences OR Biochemistry & Molecular Biology OR Radiology, Nuclear Medicine & Medical

Imaging OR Toxicology OR Oncology OR Cell Biology OR Public, Environmental & Occupational Health OR Medicine, Research & Experimental OR Physiology OR Genetics & Heredity OR Microbiology OR Immunology OR Endocrinology & Metabolism OR Medicine, General & Internal OR Clinical Neurology OR Surgery OR Behavioral Sciences OR Developmental Biology OR Hematology OR Orthopedics OR Psychiatry OR Urology & Nephrology OR Dermatology OR Infectious Diseases OR Psychology, Experimental OR Cardiac & Cardiovascular Systems OR Gastroenterology & Hepatology OR Health Care Sciences & Services OR Neuroimaging OR Respiratory System OR Rheumatology OR Veterinary Sciences OR Virology OR Dentistry, Oral Surgery & Medicine OR Medical Informatics OR Nutrition & Dietetics OR Obstetrics & Gynecology OR Parasitology OR Pathology OR Pediatrics OR Peripheral Vascular Disease OR Psychology OR Psychology, Multidisciplinary OR Reproductive Biology)

A.2 Appendix 2 – Underreporting of Adverse Effects of EMF Radiation

Most of the papers in this chapter (~70%) focused on adverse health effects of non-ionizing EMF radiation combined with other agents. There is a growing literature providing evidence that adverse events in the biomedical literature are underreported; some of this literature is summarized in Chapter 9 of (Kostoff 2015). References 71–125 of (Kostoff 2015) are a modest sample of studies showing how adverse events in the biomedical literature have been, and are being, underreported. What are the reasons for, and consequences of, such underreporting of adverse events from non-ionizing EMF radiation?

A.2.1 Full Reporting of Adverse Events

A literature in which adverse events were fully reported would have the following characteristics:

- all critical research problems necessary for credible policy are addressed/ funded;
- all research performed is credible and high quality;
- all research findings are submitted for publication;
- all papers are reviewed by unbiased experts before publication;
- all high quality research submissions are published;
- all published articles are available to the general public;
- all accessible articles are easily retrieved.

A.2.2 Underreporting of Adverse Events

Adverse events are not reported fully in part because of the following reasons.

A.2.2.1 Critical Research Not Funded

Some critical research problems are not addressed/funded, for myriad reasons:

- the combinatorics parameter space is too complex to address all the areas;
- the funds available to the sponsor organization are insufficient to cover all critical research areas;
- the process for setting funding priorities within the sponsor organization is poor;
- selection based on potential for favorable results;
- external pressures effectively limit what topics can be funded, including
  - industry pressure to suppress topics that may have commercial sensitivity, and/or
  - government pressure to suppress topics that may have political sensitivity.

  The pressures may operate intra-organizationally or inter-organizationally.

A.2.2.2 Research Not Submitted for Publication

Some research findings are not submitted for publication, for myriad reasons:

- national security classification, or classification for other reasons;
- organizationally proprietary;
- no organizational or individual publishing tradition, with equally little incentive to publish;
- costs associated with submissions for publication (time and money), which some organizations may not be willing to spend.

Most disturbing is the potentially deliberate suppression of research findings. This may result from:

- negative findings, which many organizations/journals/researchers are reluctant to publish;
- adverse events, which many industrial and governmental organizations in the biomedical community are reluctant to publish;
- commercial sensitivity, which industry would rather not be published;
- political sensitivity, which government would rather not be published; and,
- unethical research, whose performers would rather not be published, and whose quality may be relatively low due to lack of research oversight and lack of reproducibility.

A.2.2.3 Poor Research Published; Good Research Not Published

Some research that enters the literature may be of low quality, due to:

- poor peer review (where the peer review process and/or the peer reviewers are of low quality) or no peer review;
- contribution to the journal Editor's pre-determined agenda.

Some high quality research may not get published, due to:

- poor peer review or biased peer review;
- lacking editorial expertise to judge the quality of the research;
- non-contribution to the journal Editor's pre-determined agenda;
- not viewed as potentially contributing to increasing journal's Impact Factor

A.2.2.4 Manufactured Research

Finally, some/much research that enters the published literature may be deliberately distorted or skewed; this research can be termed "manufactured research". The purpose of this manufactured research is to both

- counter publications showing adverse effects from specific products and
- sow confusion among the public and decision-makers, not allowing the consensus required for policy.

The books Merchants of Doubt (Oreskes and Conway 2011) and Doubt is Their Product (Michaels 2008) describe this 'research manufacturing' process quite well. A few illustrative biomedical examples of some of the more egregious misrepresentations of science mentioned above, especially suppressed and manufactured research, are presented in Chapter 9 of (Kostoff 2015).

A.2.2.5 Published Research Not Easily Accessible

Some good published research may not be easily accessible to the public and the decision-makers, because of:

- publication in relatively obscure media or foreign languages only;
- publication behind high paywalls;
- poor search engines/algorithms.

A.2.2.6 Incentives for Inadequate Literature

There are many disincentives for an adequate literature, including.

- **Industry:** financial advantages for concealing the adverse effects of their products and services.
- **Government:** supporting corporate and large donor interests (through selective topic sponsorship and suppressed/distorted research findings) to lay the **groundwork for future industry employment.**
- **Journal Editors:** maintaining industry-sponsored professional society and/or advertising support through selective publication favorable to sponsors.
- **Research performers:** receiving and maintaining grants by working on topics of interest to, and producing results desired by, corporate and government sponsors; producing publications aligned with the interests of journal sponsors or advertisers in order to increase publication likelihood; laying the groundwork for future industry employment and/or consultancies by not publishing findings antithetical to the interests of industry.

In this section, we have identified the disincentives (for an adequate literature) for four classes: Industry, Government, Journal Editors, Research Performers. There are individuals who span multiple classes. For example, a person who works in government may also be a research performer and a journal Editor. The incentive (for an inadequate literature) associated with e.g. their government function may 'spill over' to their journal Editor and research performer roles. So, even though the journal may not have industry or government financial support as a source of potential bias, the potential biases arising from the government or research performer affiliations of the Editor could (in theory) influence the journal Editor role.

A.2.2.7 Policy Implications of Inadequate Literature

There are myriad important drivers of government policy; the three critical drivers of policy considered now will be technical literature, interests of political donors, and interests of the electorate.

Three options that relate policy to technical literature are:

- Option 1: The topical area is non-sensitive commercially or politically (e.g., weather satellite research, age of universe research). There is little incentive for much 'manufactured research' in these topical areas. Donors and voters would agree with, or be indifferent to, policy dictated by adequate literature; donors and voters agree with policy dictated by inadequate literature; policy reflects literature.
- Option 2: The topical area is sensitive commercially and/or politically (e.g., EMF health impacts). There is incentive for much 'manufactured research' in these topical areas, and our own studies have confirmed this. In this case, donors and voters would ***disagree*** with policy dictated by adequate literature. The donors are driven by profit, and the voters are addicted to the specific technology in this case (e.g., wireless communications). Thus, donors and voters agree with policy dictated by inadequate literature.

- In the case of EMF health impacts, the policy on EMF exposures that would be required as the result of an objective reading of the credible technical literature (severe restrictions on the use of wireless communications, etc) would not be acceptable to the vast majority of donors and voters. Thus, the policy in practice reflects the interests of the donors and voters, not the dictates of an adequate technical literature.
- Option 3: The topical area is sensitive commercially and/or politically (e.g., exposures/treatments that cause disease). There is incentive for much 'manufactured research' in this case, and our own studies have confirmed this. In this case, donors would disagree with policy dictated by an adequate literature, whereas the voters would ***agree*** with policy dictated by adequate literature. The donors are driven by profit, whereas the voters are driven by the benefits of technology in this specific case. ***Unlike the previous option, the voters are not addicted to the technology, since its application may be unpleasant in many cases.*** The donors still agree with policy dictated by inadequate literature, whereas the voters agree with policy dictated by inadequate literature, ***only because they believe it is adequate***. This means that some literatures may be highly manufactured to maintain voter support. The policy reflects donors, not adequate technical literature.

In conclusion, the published technical literature is inadequate for myriad reasons, and the degree of inadequacy is unknown and may be unknowable. The fraction of inadequacy due to deliberate misinformation is unknown, but may be large for topical areas with commercial or political sensitivity.

A.2.2.8 Alleged Under- and Distorted Reporting of Non-ionizing EMF Radiation Adverse Effects

In the course of the present research, we have received numerous oral and written statements by non-ionizing EMF radiation researchers, and have seen many anecdotes reported, describing how adverse effects from non-ionizing EMF radiation have been suppressed as topics of research by the sponsoring agencies, and suppressed or distorted in the published research literature. This is not surprising; many government operations depend on the use of wireless communications, many industrial organizations are involved in producing wireless communications devices and systems, and many utilities involve high non-ionizing EMF radiation as part of their daily operations. If non-ionizing EMF radiation levels were required to be drastically reduced for safety purposes, the effect on our economies and many government operations would be devastating.

Two examples will be discussed in this section: alleged publication bias in a leading radiation journal (Slesin 2006) and the Bioinitiative Report (2012).

A.2.2.8.1 Alleged Journal Bias on Publishing Adverse Effects

As the reader will see from the following example, obtaining data to support and validate allegations of journal bias is extremely difficult. Dr. Louis Slesin has been publishing a newsletter addressing myriad issues related to microwave radiation, and it is aptly entitled 'Microwave News' (MN). We came across this newsletter during the course of our initial non-ionizing EMF radiation health impacts study (Kostoff and Lau 2013), and found the MN articles were quite accurate in the areas where they overlapped our study.

In 2006, MN published an article entitled "Radiation Research" and The Cult of Negative Results (Slesin 2006). It was a unique study with major contributions from Dr. Henry Lai, a leading researcher in the technical area of the article. The study's focus was essentially to ascertain how reflective (of the microwave-induced genotoxicity publications in the larger technical literature) were those articles published in the journal *Radiation Research* on this topic.

In short, MN found that:

- [In the larger technical literature on microwave-induced genotoxicity] "There is just about an even split between effect and no-effect papers";
- "A clear—and disconcerting—pattern emerges: 32 of the 35 studies that were paid for by the mobile phone industry and the U.S. Air Force show no effect. They make up more than 75% of all the negative studies. You don't need to be a statistician to infer that money, more often than not, secures the desired scientific result";
- "A similar loss of balance occurs when you look at only the papers published in Radiation Research.....Over the last 16 years, only one positive paper on microwave genotoxicity has appeared in Radiation Research. During the same time, the journal has published 21 negative genotox papers. (Australia's Pam Sykes, the lead author of the lone positive paper, was denied money for a follow-up and soon moved on to other research areas.)....80% of the negative papers (17 out of 21) published in Radiation Research were paid for by either industry or the U.S. Air Force." (Kostoff and Lau 2013).

At this point, the statements in MN are only ***allegations***. There could be journal bias, or the best papers submitted to the journal happen to be the ones showing the absence of an impact of microwaves on genotoxicity. How could this issue be resolved?

One could (in theory) re-evaluate the original peer reviews of all the manuscripts submitted to the journal on this topic for bias. Unfortunately, we would then have the issue of determining the biases of the second group of reviewers, a difficult task. Additionally, even for reviewers who are unbiased, there is not always complete agreement. Scientists can sometimes have very differing opinions on the value of the same concept. Proving deliberate bias for a journal is extremely difficult, and may border on the impossible in practice.

Further, while funding source was used as an important metric to assess potential bias in the above study, it is by no means definitive. The determining factor is 'intent'; were the researchers incentivized to bias their findings? If a researcher had no observable funding sources that could be interpreted as predisposing to potential bias, but wanted to obtain either (1) funding from industry or (2) future employment with industry, then one could argue the researcher had some predisposition to bias in favor of industry. How would we ever know this? Even if the researcher eventually received a grant from industry, or employment with industry, how would we know whether the researcher 'intended' to pursue these funding opportunities at the time the research was being performed and published?

Finally, how well this particular example reflects all, or any, other technical/biomedical journals relative to the presence or absence of potential bias, is unknown.

A.2.2.8.2 The Bioinitiative Report

The Bioinitiative Report (2012) was written by a large number of world-class experts in science and public health policy in 2007, and updated in 2012. It assessed safety of non-ionizing EMF radiations at myriad frequencies. Its main message is that the present permissible non-ionizing EMF radiation exposure limits are orders of magnitude too high to protect against potential adverse health effects from non-ionizing EMF radiation. The Report recommends permissible chronic non-ionizing EMF radiation exposure limits of about one milligauss in the power frequency range (~60 Hz), and about ten microwatts/m^2 in the outdoor RF range. Contrast that with the ICNIRP (International Commission on Non-Ionizing Radiation Protection; an independent scientific organization) reference levels for general public exposure of about 800 milligauss (time varying magnetic field unperturbed rms value) in the 60 Hz power frequency range (ICNIRP n.d.), and the FCC (a regulatory agency) MPE power density limits of about 6.0×10^6 microwatts/m^2 at ~900 Mhz cell phone RF frequencies (Federal Communications Commission Office of Engineering and Technology 1997). These are three to six orders of magnitude higher than recommended in the Bioinitiative Report.

Why is there such a difference, and how does this difference relate to the underreporting of adverse health effects from non-ionizing EMF radiation? The answer is partially stated and partially implied within Section 3 of the Bioinitiative Report (The Existing Public Exposure Standards), as follows:

> Professional bodies from technical societies like IEEE and ICNIRP continue to support "thermal-only" guidelines routinely defend doing so a) by omitting or ignoring study results reporting bioeffects and adverse impacts to health and wellbeing from a very large body of peer-reviewed, published science because it is not yet "proof" according to their definitions; b) by defining the proof of "adverse effects" at an impossibly high a bar (scientific proof or causal evidence) so as to freeze action; c) by requiring a conclusive demonstration of both "adverse effect" and risk before admitting low-intensity effects should be taken into account; e) by ignoring low-intensity studies that report bioeffects and health impacts due to modulation; f) by conducting scientific reviews with panels heavily burdened with industry experts and under-represented by public health experts and independent scientists with relevant low-intensity research experience; g) by limiting public participation in standard-setting deliberations; and other techniques that maintain the status quo.

> Much of the criticism of the existing standard-setting bodies comes because their contributions are perceived as industry-friendly (more aligned with technology investment and dissemination of new technologies) rather than public health oriented.

There are many published studies that bolster the above conclusions, and address the influence of industry funding on scientific results. For example (Huss et al. 2007): "We examined the methodologic quality and results of experimental studies investigating the effects of the type of radiofrequency radiation emitted by handheld cellular telephones. We hypothesized that studies would be less likely to show an effect of the exposure if funded by the telecommunications industry, which has a vested interest in portraying the use of mobile phones as safe. We found that the studies funded exclusively by industry were indeed substantially less likely to report statistically significant effects on a range of end points that may be relevant to health. Our findings add to the existing evidence that single-source sponsorship is associated with outcomes that favor the sponsors' products....... Most previous studies of this issue were based on studies of the efficacy and cost-effectiveness of drug treatments. A recent systematic review and meta-analysis showed that studies sponsored by the pharmaceutical industry were approximately four times more likely to have outcomes favoring the sponsor's drug than studies with other sources of funding......The influence of the tobacco industry on the research it funded has also been investigated......To our knowledge, this is the first study to examine this issue in the context of exposure to radiofrequency electromagnetic fields."

Chapter 9 of reference (Kostoff 2015) shows this problem is endemic to many industries and government agencies. Given the difficulty of obtaining this type of information, much of which is provided by 'whistleblowers', we should not under-estimate the amount of distorted non-ionizing EMF radiation adverse effects findings published in the literature, nor under-estimate the amount of non-ionizing EMF radiation adverse effects findings suppressed from publication by government, industry, and the journals themselves.

To conclude, this sub-section has provided reasons for adverse health effects of myriad substances (including non-ionizing EMF radiation) being underreported in the premiere biomedical literature, or entering this literature in distorted form. Since there is no way to gauge the extent of this under/distorted-reporting, the quality and credibility of the 'premiere' biomedical/non-ionizing EMF radiation adverse effects literature is unknown. Therefore, any types of meta-analyses or scientometric analyses of this literature will have unknown quality and credibility. The most sophisticated scientometric analysis cannot compensate for a highly-flawed database.

Equally damaging is the effect of this flawed database on the larger scientific enterprise. Science can be viewed as a never-ending construction project, where the building blocks and support structures are the documents from past scientific studies. If some, or many, of these building blocks are flawed, the upper parts of the structure will be weakly supported, and may collapse. Through the citation process, the misleading findings at the lower parts of the structure are promulgated to the upper portions, and the dilution of quality increases. While the propensity for misconduct is greatest in areas of commercial and political sensitivity, the broad reach of basic science will have a 'spill-over' adverse impact on myriad directly and indirectly related areas of science.

Disclaimer The views in this chapter are solely those of the authors, and do not necessarily represent the views of Georgia Institute of Technology or any of its components, or of the Institute for Defense Analyses.

References

Albrecht MT, Livingston BD, Pesce JT, Bell MG, Hannaman D, Keane-Myers AM (2012) Electroporation of a multivalent DNA vaccine cocktail elicits a protective immune response against anthrax and plague. Vaccine 30(32):4872–4883

Aldad TS, Gan G, Gao X-B, Taylor HS (2012) Fetal radiofrequency radiation exposure from 800-1900 MHz-rated cellular telephones affects neurodevelopment and behavior in mice. Sci Rep 2:312

Alsaeed I, Al-Somali F, Sakhnini L, Aljarallah OS, Hamdan RMM, Bubishate SA, Sarfaraz ZK, Kamal A (2014) Autism-relevant social abnormalities in mice exposed perinatally to extremely low frequency electromagnetic fields. Int J Dev Neurosci 37:58–64

Amara S, Douki T, Garrel C, Favier A, Ben Rhouma K, Sakly M, Abdelmelek H (2011) Effects of static magnetic field and cadmium on oxidative stress and DNA damage in rat cortex brain and hippocampus. Toxicol Ind Health 27(2):99–106

Arruda-Neto JDT, Friedberg EC, Bittencourt-Oliveira MC, Cavalcante-Silva E, Schenberg ACG, Rodrigues TE, Garcia F, Louvison M, Paula CR, Mesa J, Moron MM, Maria DA, Genofre GC (2009) Static electric fields interfere in the viability of cells exposed to ionising radiation. Int J Radiat Biol 85(4):314–321

Baum A, Mevissen M, Kamino K, Mohr U, Loscher W (1995) A histopathological study on alterations in DMBA-induced mammary carcinogenesis in rats with 50 Hz, 100 mut magnetic field exposure. Carcinogenesis 16(1):119–125

Bediz CS, Baltaci AK, Mogulkoc R, Oztekin E (2006) Zinc supplementation ameliorates electromagnetic field-induced lipid peroxidation in the rat brain. Tohoku J Exp Med 208(2):133–140

Beniashvili DS, Bilanishvili VG, Menabde MZ (1991) Low-frequency electromagnetic radiation enhances the induction of rat mammary tumors by nitrosomethyl urea. Cancer Lett 61(1): 75–79

Benson EB, Lange DG, Fujimoto JM, Ishii TK (1983) Effects of acute microwave irradiation on phenobarbital sleep and disposition to brain in mice. J Toxicol Environ Health 11(2):261–274

Bodera P, Stankiewicz W, Zawada K, Antkowiak B, Paluch M, Kieliszek J, Kalicki B, Bartosinski A, Wawer I (2013) Changes in antioxidant capacity of blood due to mutual action of electromagnetic field (1800 Mhz) and opioid drug (tramadol) in animal model of persistent inflammatory state. Pharmacol Rep 65(2):421–428

Boga A, Emre M, Sertdemir Y, Akillioglu K, Binokay S, Demirhan O (2015) The effect of 900 and 1800 Mhz GSM-like radiofrequency irradiation and nicotine sulfate administration on the embryonic development of xenopus laevis. Ecotoxicol Environ Saf 113:378–390

Bowman JD, Thomas DC, London SJ, Peters JM (1995) Hypothesis: the risk of childhood leukemia is related to combinations of power-frequency and static magnetic fields. Bioelectromagnetics 16(1):48–59

Burch JB, Reif JS, Noonan CW, Ichinose T, Bachand AM, Koleber TL, Yost MG (2002) Melatonin metabolite excretion among cellular telephone users. Int J Radiat Biol 78(11): 1029–1036

Byun Y-H, Ha M, Kwon H-J, Hong Y-C, Leem J-H, Sakong J, Kim SY, Lee CG, Kang D, Choi H-D, Kim N (2013) Mobile phone use, blood lead levels, and attention deficit hyperactivity symptoms in children: a longitudinal study. PLoS One 8(3):e59742

Cadossi R, Zucchini P, Emilia G, Franceschi C, Cossarizza A, Santantonio M, Mandolini G, Torelli G (1991) Effect of low-frequency low-energy pulsing electromagnetic-fields on mice injected with cyclophosphamide. Exp Hematol 19(3):196–201

Canseven AG, Esmekaya MA, Kayhan H, Tuysuz MZ, Seyhan N (2015) effects of microwave exposure and gemcitabine treatment on apoptotic activity in Burkitt's Lymphoma (Raji) cells. Electromagnetic Biol Med 34(4):322–326

Cao Y, Zhang W, Lu M-X, Xu Q, Meng Q-Q, Nie J-H, Tong J (2009) 900-MHz microwave radiation enhances gamma-ray adverse effects on SHG44 cells. J Toxicol Environ Health A 72(11–12):727–732

Chiang H, Wu RY, Shao BJ, Fu YD, Yao GD, Lu DJ (1995) Pulsed magnetic field from video display terminals enhances teratogenic effects of cytosine arabinoside in mice. Bioelectromagnetics 16(1):70–74

Cho YH, Chung HW (2003) The effect of extremely low frequency electromagnetic fields (ELF-EMF) on the frequency of micronuclei and sister chromatid exchange in human lymphocytes induced by benzo (a) pyrene. Toxicol Lett 143(1):37–44

Cho YH, Jeon HK, Chung HW (2007) Effects of extremely low-frequency electromagnetic fields on delayed chromosomal instability induced by bleomycin in normal human fibroblast cells. J Toxicol Environ Health A 70(15–16):1252–1258

Cho S, Lee Y, Lee S, Choi YJ, Chung HW (2014) Enhanced cytotoxic and genotoxic effects of gadolinium following ELF-EMF irradiation in human lymphocytes. Drug Chem Toxicol 37(4):440–447

CLUTO (2015) http://glarosdtcumnedu/gkhome/views/cluto, University of Minnesota

Coureau G, Bouvier G, Lebailly P, Fabbro-Peray P, Gruber A, Leffondre K, Guillamo J-S, Loiseau H, Mathoulin-Pelissier S, Salamon R, Baldi I (2014) mobile phone use and brain tumours in the CERENAT case-control study. Occup Environ Med 71(7):514–522

Dart P, Cordes K Elliott A, Knackstedt J, Morgan J, Wible P (2013) Biological and health effects of microwave radio frequency transmissions: a review of the research literature. http://wwwnational-toxic-encephalopathy-foundationorg/wp-content/uploads/2012/01/Biological_and_Health_Effects_of_Microwave_Radio_Frequency_Transmissionspdf

Davidse RJ, VanRaan AFJ (1997) Out of particles: impact of CERN, DESY and SLAC research to fields other than physics. Scientometrics 40(2):171–193

de Vocht F, Hannam K, Buchan I (2013) Environmental risk factors for cancers of the brain and nervous system: the use of ecological data to generate hypotheses. Occup Environ Med 70(5):349–356

Deger S, Boehmer D, Turk I, Roigas J, Budach V, Loening SA (2002) Interstitial hyperthermia using self-regulating thermoseeds combined with conformal radiation therapy. Eur Urol 42(2):147–153

Dicarlo AL, Hargis MT, Penafiel LM, Litovitz TA (1999) Short-term magnetic field exposures (60 Hz) induce protection against ultraviolet radiation damage. Int J Radiat Biol 75(12): 1541–1549

Ding GR, Yaguchi H, Yoshida M, Miyakoshi J (2000) Increase in X-ray-induced mutations by exposure to magnetic field (60 Hz, 5 Mt) In NF-Kappab-inhibited cells. Biochem Biophys Res Commun 276(1):238–243

Dode AC, Leao MMD, Tejo F d AF, Gomes ACR, Dode DC, Dode MC, Moreira CW, Condessa VA, Albinatti C, Caiaffa WT (2011) Mortality by neoplasia and cellular telephone base stations in the Belo horizonte municipality, Minas Gerais State, Brazil. Sci Total Environ 409(19):3649–3665

Dolat E, Rajabi O, Salarabadi SS, Yadegari-Dehkordi S, Sazgarnia A (2015) Silver nanoparticles and electroporation: their combinational effect on leishmania major. Bioelectromagnetics 36(8):586–596

Eger H, Hagen KU, Lucas B, Vogel P, Voit H (2004) The influence of being physically near to a cell phone transmission mast on the incidence of cancer. Umwelt Medizin Gesellschaft 17:4

Elferchichi M, Maaroufi K, Sakly M, Abdelmelek H (2015) Effects of combined ferrous sulfate administration and exposure to static magnetic field on brain oxidative stress and emotional behavior. Arch Ital Biol 153(1):37–45

Engstrom PE, Persson BR, Brun A, Salford LG (2001) A new antitumour treatment combining radiation and electric pulses. Anticancer Res 21(3B):1809–1815

Fazzo L, Tancioni V, Polichetti A, Iavarone I, Vanacore N, Papini P, Farchi S, Bruno C, Pasetto R, Borgia P, Comba P (2009) Morbidity experience in populations residentially exposed to 50 hz magnetic fields: methodology and preliminary findings of a cohort study. Int J Occup Environ Health 15(2):133–142

Federal Communications Commission Office of Engineering & Technology. Evaluating compliance with FCC guidelines for human exposure to radiofrequency electromagnetic fields OET Bulletin 65, Edition 97–01, August 1997 https://transitionfccgov/Bureaus/Engineering_Technology/Documents/bulletins/oet65/oet65pdf

Fedrowitz M, Kamino K, Loscher W (2004) Significant differences in the effects of magnetic field exposure on 7,12-dimethylbenz (a) anthracene-induced mammary carcinogenesis in two substrains of sprague-dawley rats. Cancer Res 64(1):243–251

Feldman R, Dagan I, Hirsh H (1998) Mining text using keyword distributions. J Intell Inf Syst 10(3):281–300

Fiorani M, Biagiarelli B, Vetrano F, Guidi G, Dacha M, Stocchi V (1997) In vitro effects of 50 Hz magnetic fields on oxidatively damaged rabbit red blood cells. Bioelectromagnetics 18(2): 125–131

Fulimoto T, Maeda H, Kubo K, Sugita Y, Nakashima T, Sato E, Tanaka Y, Madachi M, Aiba M, Kameyama Y (2005) Enhanced anti-tumour effect of cisplatin with low-voltage electrochemotherapy in hamster oral fibrosarcoma. J Int Med Res 33(5):507–512

Gajski G, Garaj-Vrhovac V (2009) Radioprotective effects of honeybee venom (*Apis mellifera*) against 915-MHz microwave radiation-induced DNA damage in wistar rat lymphocytes: in vitro study. Int J Toxicol 28(2):88–98

Gellrich D, Becker S, Strieth S (2014) Static magnetic fields increase tumor microvessel leakiness and improve antitumoral efficacy in combination with paclitaxel. Cancer Lett 343(1):107–114

Giladi M, Porat Y, Blatt A, Wasserman Y, Kirson ED, Dekel E, Palti Y (2008) Microbial growth inhibition by alternating electric fields. Antimicrob Agents Chemother 52(10):3517–3522

Girgin S, Ozturk H, Gedik E, Akpolat V, Kale E (2009) Effect of a 50-Hz sinusoidal electromagnetic field on the integrity of experimental colonic anastomoses covered with fibrin glue. Adv Clin Exp Med 18(1):13–18

Gmitrov J (2007) Geomagnetic field modulates artificial static magnetic field effect on arterial baroreflex and on microcirculation. Int J Biometeorol 51(4):335–344

Gmitrov J, Ohkubo C (1999) Static magnetic field and calcium channel blocking agent combined effect on baroreflex sensitivity in rabbits. Electro Magnetobiol Med 18(1):43–55

Gobba F, Bargellini A, Scaringi M, Bravo G, Borella P (2009) extremely low frequency-magnetic fields (ELF-EMF) occupational exposure and natural killer activity in peripheral blood lymphocytes. Sci Total Environ 407(3):1218–1223

Greengrass E (1997) Information retrieval: an overview, national security agency, TR-R52–02-96

Gronevik E, Mathiesen I, Lomo T (2005) Early events of electroporation-mediated intramuscular DNA vaccination potentiate thi-directed immune responses. J Gene Med 7(9):1246–1254

Grosel A, Sersa G, Kranjc S, Cemazar M (2006) Electrogene therapy with P53 of murine sarcomas alone or combined with electrochemotherapy using cisplatin. DNA Cell Biol 25(12):674–683

Guler G, Turkozer Z, Tomruk A, Seyhan N (2008) The protective effects of N-acetyl-L-cysteine and epigallocatechin-3-gallate on electric field-induced hepatic oxidative stress. Int J Radiat Biol 84(8):669–680

Gulturk S, Demirkazik A, Kosar I, Cetin A, Dokmetas HS, Demir T (2010) Effect of exposure to 50 Hz magnetic field with or without insulin on blood-brain barrier permeability in streptozotocin-induced diabetic rats. Bioelectromagnetics 31(4):262–269

Gumral N, Naziroglu M, Koyu A, Ongel K, Celik O, Saygin M, Kahriman M, Caliskan S, Kayan M, Gencel O, Flores-Arce MF (2009) Effects of selenium and L-carnitine on oxidative stress in blood of rat induced by 245-GHz radiation from wireless devices. Biol Trace Elem Res 132(1–3):153–163

Guseinzade KM, Agakishiyev DD, Mejidov NM (1991) Magnetotherapy of trophic ulcers of the shin. Vestnik Dermatologii I Venerologii 11:61–64

Ha M, Im H, Lee M, Kim HJ, Kim B-C, Gimm Y-M, Pack J-K (2007) Radio-frequency radiation exposure from AM radio transmitters and childhood leukemia and brain cancer. Am J Epidemiol 166(3):270–279

Hakansson N, Gustavsson P, Sastre A, Floderus B (2003) Occupational exposure to extremely low frequency magnetic fields and mortality from cardiovascular disease. Am J Epidemiol 158 (6):534–542

Hanci H, Odaci E, Kaya H, Aliyazicioglu Y, Turan I, Demir S, Colakoglu S (2013) The effect of prenatal exposure to 900-Mhz electromagnetic field on the 21-old-day rat testicle. Reprod Toxicol (Elmsford, NY) 42:203–209

Hardell L, Carlberg M, Hansson Mild K (2011) Pooled analysis of case-control studies on malignant brain tumours and the use of mobile and cordless phones including living and deceased subjects. Int J Oncol 38(5):1465–1474

Haro AMR, Smyth A, Hughes P, Reid CN (2005) McHale AP electro-sensitisation of mammalian cells and tissues to ultrasound: a novel tumour treatment modality. Cancer Lett 222(1):49–55

Hassan NS, Abdelkawi SA (2014) Assessing of plasma protein denaturation induced by exposure to cadmium, electromagnetic fields and their combined actions on rat. Electromagnetic Biol Med 33(2):147–153

Hearst MA (1999) Untangling text data mining proceedings of ACL 99, the 37th annual meeting of the association for computational linguistics, University of Maryland, 1999, June 20–26

Hintenlang DE (1993) Synergistic effects of ionizing radiation and 60 Hz magnetic fields. Bioelectromagnetics 14(6):545–551

Hirao LA, Wu L, Khan AS, Hokey DA, Yan J, Dai AL, Betts MR, Draghia-Akli R, Weiner DB (2008) Combined effects of IL-12 and electroporation enhances the potency of DNA vaccination in macaques. Vaccine 26(25):3112–3120

Hocking B, Gordon I (2003) Decreased survival for childhood leukemia in proximity to television towers. Arch Environ Health 58(9):560–564

Hocking B, Gordon IR, Grain HL, Hatfield GE (1996) cancer incidence and mortality and proximity to TV towers. Med J Aust 165(11–12):601–605

Huss A, Egger M, Hug K, Huwiler-Muntener K, Roosli M (2007) Source of funding and results of studies of health effects of mobile phone use: systematic review of experimental studies. Environ Health Perspect 115(1):1–4

ICNIRP Guidelines (n.d.) http://wwwemclabit/attachment/ICNIRPGuidelinesPDF; http://wwwemfsinfo/limits/limits-organisations/comparison/

Ismael SJ, Callera F, Garcia AB, Baffa O, Falcao RP (1998) Increased dexamethasone-induced apoptosis of thymocytes from mice exposed to long-term extremely low frequency magnetic fields. Bioelectromagnetics 19(2):131–135

Ito A, Shinkai M, Honda H, Kobayashi T (2001) Heat-inducible TNF-alpha gene therapy combined with hyperthermia using magnetic nanoparticles as a novel tumor-targeted therapy. Cancer Gene Ther 8(9):649–654

Ito A, Matsuoka F, Honda H, Kobayashi T (2003) Heat shock protein 70 gene therapy combined with hyperthermia using magnetic nanoparticles. Cancer Gene Ther 10(12):918–925

Jaroszeski MJ, Illingworth P, Pottinger C, Hyacinthe M, Heller R (1999) Electrically mediated drug delivery for treating subcutaneous and orthotopic pancreatic adenocarcinoma in a hamster model. Anticancer Res 19(2A):989–994

Jia HL, Wang C, Li Y, Lu Y, Wang PP, Pan WD, Song T (2014) Combined effects of 50 Hz magnetic field and magnetic nanoparticles on the proliferation and apoptosis of PC12 cells. Biomed Environ Sci 27(2):97–105

Jian W, Wei Z, Zhiqiang C, Zheng F (2009) X-ray-induced apoptosis of BEL-7402 cell line enhanced by extremely low frequency electromagnetic field in vitro. Bioelectromagnetics 30(2):163–165

Jouni FJ, Abdolmaleki P, Ghanati F (2012) Oxidative stress in broad bean (*Vicia faba* L) induced by static magnetic field under natural radioactivity. Mut Res Gen Toxicol Environ Mutagen 741(1–2):116–121

Junkersdorf B, Bauer H, Gutzeit HO (2000) Electromagnetic fields enhance the stress response at elevated temperatures in the nematode *Caenorhabditis elegans*. Bioelectromagnetics 21(2):100–106

Juutilainen J (2008) Do electromagnetic fields enhance the effects of environmental carcinogens? Radiat Prot Dosim 132(2):228–231

Juutilainen J, Kumlin T, Naarala J (2006) Do extremely low frequency magnetic fields enhance the effects of environmental carcinogens? A meta-analysis of experimental studies. Int J Radiat Biol 82(1):1–12

Kang KS, Hong JM, Jeong YH, Seol YJ, Yong WJ, Rhie JW, Cho DW (2014) Combined effect of three types of biophysical stimuli for bone regeneration. Tissue Eng A 20(11–12):1767–1777

Karner KB, Lesnicar H, Cemazar M, Sersa G (2004) Antitumour effectiveness of hyperthermia is potentiated by local application of electric pulses to LPB tumours in mice. Anticancer Res 24 (4):2343–2348

Khan SI, Blumrosen G, Vecchio D, Golberg A, McCormack MC, Yarmush ML, Hamblin MR, Austen WG (2016) Eradication of multidrug-resistant pseudomonas biofilm with pulsed electric fields. Biotechnol Bioeng 113(3):643–650

Kim K, Lee JY (1998) Strain improvement of yeast for ethanol production using a combined treatment of electric field and chemical mutagen N-methyl-N '-nitro-N-nitrosoguanidine. J Microbiol Biotechnol 8(2):119–123

Kostoff RN (2008) Literature-related discovery (LRD): introduction and background. Technol Forecast Soc Chang 75(2):165–185

Kostoff RN (2011) Literature-related discovery: potential treatments and preventatives for SARS. Technol Forecast Soc Chang 78(7):1164–1173

Kostoff RN (2012) Literature-related discovery and innovation – update. Technol Forecast Soc Chang 79(4):789–800

Kostoff RN (2015) Pervasive causes of disease. Georgia Institute of Technology, Atlanta, GA. PDF http://hdlhandlenet/1853/53714

Kostoff RN, Briggs MB (2008) Literature-related discovery (LRD): potential treatments for parkinson's disease. Technol Forecast Soc Chang 75(2):226–238

Kostoff RN, Lau CGY (2013) Combined biological and health effects of electromagnetic fields and other agents in the published literature. Technol Forecast Soc Chang 80(7):1331–1349

Kostoff RN, Los LI (2013) Literature-related discovery techniques applied to ocular disease: a vitreous restoration example. Curr Opin Ophthalmol 24(6):606–610

Kostoff RN, Patel U (2015) Literature-related discovery and innovation: chronic kidney disease. Technol Forecast Soc Chang 91:341–351

Kostoff RN, Eberhart HJ, Toothman DR (1997) Database tomography for information retrieval. J Inf Sci 23(4):301–311

Kostoff RN, Toothman DR, Eberhart HJ, Humenik JA (2001a) Text mining using database tomography and bibliometrics: a review. Technol Forecast Soc Chang 68(3):223–253

Kostoff RN, del Rio JA, Humenik JA, Garcia EO, Ramirez AM (2001b) Citation mining: integrating text mining and bibliometrics for research user profiling. J Am Soc Inf Sci Technol 52(13):1148–1156

Kostoff RN, Briggs MB, Solka JL, Rushenberg RL (2008a) Literature-related discovery (LRD): methodology. Technol Forecast Soc Chang 75(2):186–202

Kostoff RN, Briggs MB, Lyons TJ (2008b) Literature-related discovery (LRD): potential treatments for multiple sclerosis. Technol Forecast Soc Chang 75(2):239–255

Kostoff RN, Solka JL, Rushenberg RL, Wyatt JA (2008c) Literature-related discovery (LRD): water purification. Technol Forecast Soc Chang 75(2):256–275

Kostoff RN, Block JA, Stump JA, Johnson D (2008d) Literature-related discovery (LRD): potential treatments for Raynaud's Phenomenon. Technol Forecast Soc Chang 75(2):203–214

Kostoff RN, Block JA, Solka JRL, Briggs MB, Rushenberg RL, Stump JA, Johnson D, Lyons TJ, Wyatt JR (2008e) Literature-related discovery (LRD): lessons learned, and future research directions. Technol Forecast Soc Chang 75(2):276–299

Koyama S, Nakahara T, Sakurai T, Komatsubara Y, Isozumi Y, Miyakoshi J (2005) Combined exposure of ELF magnetic fields and x-rays increased mutant yields compared with X-rays alone in Ptn89 plasmids. J Radiat Res 46(2):257–264

Koyama S, Sakurai T, Nakahara T, Miyakoshi J (2008) Extremely low frequency (ELF) magnetic fields enhance chemically induced formation of apurinic/apyrimidinic (AP) sites in A172 cells. Int J Radiat Biol 84(1):53–59

Kundi M (2010) The controversy about a possible relationship between mobile phone use and cancer. Cien Saude Colet 15(5):2415–2430

Lagace N, St-Pierre LS, Persinger MA (2009) Attenuation of epilepsy-induced brain damage in the temporal cortices of rats by exposure to LTP-patterned magnetic fields. Neurosci Lett 450(2):147–151

Lange DG, Sedmak J (1991) Japanese encephalitis virus (JEV): potentiation of lethality in mice by microwave radiation. Bioelectromagnetics 12(6):335–348

Lee WR, Shen SC, Fang CL, Zhuo RZ, Fang JY (2008) Topical delivery of methotrexate via skin pretreated with physical enhancement techniques: low-fluence erbium: YAG laser and electroporation. Lasers Surg Med 40(7):468–476

Lee HJ, Jin YB, Lee JS, Choi JI, Lee JW, Myung SH, Lee YS (2012) Combined effects of 60 Hz electromagnetic field exposure with various stress factors on cellular transformation in NIH3T3 cells. Bioelectromagnetics 33(3):207–214

Lei Y, Liu T, Wilson FAW, Zhou D, Ma Y, Hu X (2005) Effects of extremely low-frequency electromagnetic fields on morphine-induced conditioned place preferences in rats. Neurosci Lett 390(2):72–75

Lerchl A, Klose M, Grote K, Wilhelm AFX, Spathmann O, Fiedler T, Streckert J, Hansen V, Clemens M (2015) Tumor promotion by exposure to radiofrequency electromagnetic fields below exposure limits for humans. Biochem Biophys Res Commun 459(4):585–590

Levitt BB, Lai H (2010) Biological effects from exposure to electromagnetic radiation emitted by cell tower base stations and other antenna arrays. Environ Rev 18:369–395

Li D-K, Odouli R, Wi S, Janevic T, Golditch I, Bracken TD, Senior R, Rankin R, Iriye R (2002) A population-based prospective cohort study of personal exposure to magnetic fields during pregnancy and the risk of miscarriage. Epidemiology (Cambridge, MA) 13(1):9–20

Li D-K, Yan B, Li Z, Gao E, Miao M, Gong D, Weng X, Ferber JR, Yuan W (2010) exposure to magnetic fields and the risk of poor sperm quality. Reprod Toxicol (Elmsford, NY) 29(1):86–92

Li D-K, Chen H, Odouli R (2011) Maternal exposure to magnetic fields during pregnancy in relation to the risk of asthma in offspring. Arch Pediatr Adolesc Med 165(10):945–950

Liu Y, Qi H, Sun RG, Chen WF (2011) An investigation into the combined effect of static magnetic fields and different anticancer drugs on K562 cell membranes. Tumori 97(3): 386–392

Liu Q, Si T, Xu X, Liang F, Wang L, Pan S (2015) Electromagnetic radiation at 900 Mhz induces sperm apoptosis through Bcl-2, bax and caspase-3 signaling pathways in rats. Reprod Health 12:65

Luukkonen J, Hakulinen P, Maki-Paakkanen J, Juutilainen J, Naarala J (2009) Enhancement of chemically induced reactive oxygen species production and DNA damage in human SH-SY5Y neuroblastoma cells by 872 Mhz radiofrequency radiation. Mutat Res 662(1–2):54–58

Luukkonen J, Liimatainen A, Hoyto A, Juutilainen J, Naarala J (2011) Pre-exposure to 50 Hz magnetic fields modifies menadione-induced genotoxic effects in human SH-SY5Y neuroblastoma cells. Plos One 6(3):e18021

Luxembourg A, Hannaman D, Nolan E, Ellefsen B, Nakamura G, Chau L, Tellez O, Little S, Bernard R (2008) Potentiation of an anthrax DNA vaccine with electroporation. Vaccine 26(40):5216–5222

Maaroufi K, Save E, Poucet B, Sakly M, Abdelmelek H, Had-Aissouni L (2011) Oxidative stress and prevention of the adaptive response to chronic iron overload in the brain of young adult rats exposed to a 150 kilohertz electromagnetic field. Neuroscience 186:39–47

Mahmoudabadi FS, Ziaei S, Firoozabadi M, Kazemnejad A (2015) Use of mobile phone during pregnancy and the risk of spontaneous abortion. J Environ Health Sci Eng 13:34

Mailhes JB, Young D, Marino AA, London SN (1997) Electromagnetic fields enhance chemically-induced hyperploidy in mammalian oocytes. Mutagenesis 12(5):347–351

Mairs RJ, Hughes K, Fitzsimmons S, Prise KM, Livingstone A, Wilson L, Baig N, Clark AM, Timpson A, Patel G, Folkard M, Angerson WJ, Boyd M (2007) Microsatellite analysis for determination of the mutagenicity of extremely low-frequency electromagnetic fields and ionising radiation in vitro. Mutat Res 626(1–2):34–41

Marcantonio P, Del Re B, Franceschini A, Capri M, Lukas S, Bersani F, Giorgi G (2010) Synergic effect of retinoic acid and extremely low frequency magnetic field exposure on human neuroblastoma cell line BE(2)C. Bioelectromagnetics 31(6):425–433

Marcickiewicz J, Chazan B, Niemiec T, Sokolska G, Troszynski M, Luczak M, Szmigielski S (1986) Microwave-radiation enhances the teratogenic effect of cytosine-arabinoside in mice. Biol Neonate 50(2):75–82

McKay BE, Persinger MA (2003) Combined effects of complex magnetic fields and agmatine for contextual fear learning deficits in rats. Life Sci 72(22):2489–2498

McLean MJ, Engstrom S, Holcomb RR, Sanchez D (2003) A static magnetic field modulates severity of audiogenic seizures and anticonvulsant effects of phenytoin in DBA/2 mice. Epilepsy Res 55(1–2):105–116

Megha K, Deshmukh PS, Banerjee BD, Tripathi AK, Ahmed R, Abegaonkar MP (2015) Low intensity microwave radiation induced oxidative stress, inflammatory response and DNA damage in rat brain. Neurotoxicology 51:158–165

Mendonca FAS, Passarini JR, Esquisatto MAM, Mendonca JS, Franchini CC, dos Santos GMT (2009) Effects of the application of *Aloe vera* (L) and microcurrent on the healing of wounds surgically induced in wistar rats. Acta Cir Bras 24(2):150–155

Mevissen M, Haussler M, Lerchl A, Loscher W (1998) Acceleration of mammary tumorigenesis by exposure of 7,12-dimethylbenz a anthracene-treated female rats in a 50-Hz, 100-microt magnetic field: replication study. J Toxicol Environ Health A 53(5):401–418

Michaels D (2008) Doubt is their product: how industry's assault on science threatens your health, 1st edn. Oxford University Press, Oxford

Michelozzi P, Capon A, Kirchmayer U, Forastiere F, Biggeri A, Barca A, Perucci CA (2002) Adult and childhood leukemia near a high-power radio station in Rome. Ital Am J Epidemiol 155(12):1096–1103

Miller AB, To T, Agnew DA, Wall C, Green LM (1996) Leukemia following occupational exposure to 60-Hz electric and magnetic fields among ontario electric utility workers. Am J Epidemiol 144(2):150–160

Miyakoshi Y, Kajihara C, Shimizu H, Yanagisawa H (2012) Tempol suppresses micronuclei formation in astrocytes of newborn rats exposed to 50-Hz, 10-Mt electromagnetic fields under bleomycin administration. Mutat Res 747(1):138–141

Motawi TK, Darwish HA, Moustafa YM, Labib MM (2014) Biochemical modifications and neuronal damage in brain of young and adult rats after long-term exposure to mobile phone radiations. Cell Biochem Biophys 70(2):845–855

Murthy SN, Zhao YL, Hui SW, Sen A (2006) Synergistic effect of anionic lipid enhancer and electroosmosis for transcutaneous delivery of insulin. Int J Pharm 326(1–2):1–6

Navas-Acien A, Pollan M, Gustavsson P, Floderus B, Plato N, Dosemeci M (2002) Interactive effect of chemical substances and occupational electromagnetic field exposure on the risk of gliomas and meningiomas in Swedish men. Cancer Epidemiol Biomarkers Prev 11(12): 1678–1683

Nelson BK, Conover DL, Shaw PB, Snyder DL, Edwards RM (1997) Interactions of radiofrequency radiation on 2-methoxyethanol teratogenicity in rats. J Appl Toxicol: JAT 17(1):31–39

Novickij V, Grainys A, Svediene J, Markovskaja S, Paskevicius A, Novickij J (2014) Microsecond pulsed magnetic field improves efficacy of antifungal agents on pathogenic microorganisms. Bioelectromagnetics 35(5):347–353

Ogihara M, Yamaguchi O (2000) Potentiation of effects of anticancer agents by local electric pulses in murine bladder cancer. Urol Res 28(6):391–397

Okada M, Kim JH, Yoon ST, Hutton WC (2013) Pulsed electromagnetic field (PEMF) plus BMP-2 upregulates intervertebral disc-cell matrix synthesis more than either BMP-2 alone or PEMF alone. J Spinal Disord Tech 26(6):E221–E226

Okano H, Ohkubo C (2006) Elevated plasma nitric oxide metabolites in hypertension: synergistic vasodepressor effects of a static magnetic field and nicardipine in spontaneously hypertensive rats. Clin Hemorheol Microcirc 34(1–2):303–308

Ongaro A, Pellati A, Bagheri L, Fortini C, Setti S, De Mattei M (2014) Pulsed electromagnetic fields stimulate osteogenic differentiation in human bone marrow and adipose tissue derived mesenchymal stem cells. Bioelectromagnetics 35(6):426–436

Oreskes N, Conway EM (2011) Merchants of doubt: how a handful of scientists obscured the truth on issues from tobacco smoke to global warming. Bloomsbury Press, New York

Ossenkopp KP, Kavaliers M (1987) MORPHINE-induced analgesia and exposure to low-intensity 60-Hz magnetic-fields – inhibition of nocturnal analgesia in mice is a function of magnetic-field intensity. Brain Res 418(2):356–360

Ozguner F, Oktem F, Ayata A, Koyu A, Yilmaz HR (2005) A novel antioxidant agent caffeic acid phenethyl ester prevents long-term mobile phone exposure-induced renal impairment in rat – prognostic value of malondialdehyde, N-acetyl-beta-D-glucosaminidase and nitric oxide determination. Mol Cell Biochem 277(1–2):73–80

Pakhomov AG, Akyel Y, Pakhomova ON, Stuck BE, Murphy MR (1998) Current state and implications of research on biological effects of millimeter waves: a review of the literature. Bioelectromagnetics 19(7):393–413

Pall ML (2015a) Microwave frequency electromagnetic fields (EMFs) produce widespread neuropsychiatric effects including depression. J Chem Neuroanat 75:43–51 pii: S0891-0618(15) 00059-9 doi: 101016/jjchemneu201508001

Pall ML (2015b) Presentation to AutismOne Conference, 23 May 2015. https://wwwyoutubecom/watch?v=yydZZanRJ50

Pol IE, Mastwijk HC, Bartels PV, Smid EJ (2000) Pulsed-electric field treatment enhances the bactericidal action of nisin against bacillus cereus. Appl Environ Microbiol 66(1):428–430

Purushotham S, Ramanujan RV (2010) Modeling the performance of magnetic nanoparticles in multimodal cancer therapy. J Appl Phys 107(11):114701

Quock RM, Kouchich FJ, Ishii TK, Lange DG (1986a) Microwave facilitation of methylnaltrexone antagonism of morphine-induced analgesia in mice. J Bioelectricity 5(1): 35–46

Quock RM, Fujimoto JM, Ishii TK, Lange DG (1986b) Microwave facilitation of methylatropine antagonism of central cholinomimetic drug effects. Radiat Res 105(3):328–340

Quock RM, Kouchich FJ, Ishii TK, Lange DG (1987) Microwave facilitation of domperidone antagonism of apomorphine-induced stereotypic climbing in mice. Bioelectromagnetics 8(1):45–55

Rajkovic V, Matavulj M, Johansson O (2010) Combined exposure of peripubertal male rats to the endocrine-disrupting compound atrazine and power-frequency electromagnetic fields causes degranulation of cutaneous mast cells: a new toxic environmental hazard? Arch Environ Contam Toxicol 59(2):334–341

Raskmark P, Kwee S (1996) The minimizing effect of electromagnetic noise on the changes in cell proliferation caused by ELF magnetic fields. Bioelectrochem Bioenerg 40(2):193–196

Riddle MM, Smialowicz RJ, Rogers RR (1982) Microwave-radiation (2450-Mhz) potentiates the lethal effect of endotoxin in mice. Health Phys 42(3):335–340

Rihova B, Etrych T, Sirova M, Tomala J, Ulbrich K, Kovar M (2011) Synergistic effect of EMF-BEMER-type pulsed weak electromagnetic field and HPMA-bound doxorubicin on mouse EL4 T-cell lymphoma. J Drug Target 19(10):890–899

Rodriguez-Garcia JA, Ramos F (2012) High incidence of acute leukemia in the proximity of some industrial facilities in El Bierzo, Northwestern Spain. Int J Occup Med Environ Health 25(1):22–30

Rojavin MA, Ziskin MC (1997) Electromagnetic millimeter waves increase the duration of anaesthesia caused by ketamine and chloral hydrate in mice. Int J Radiat Biol 72(4):475–480

Roux C, Elefant E, Gaboriaud G, Jaullery C, Gardette J, Dupuis R, Lambert D (1986) Association of microwaves and ionizing radiation: potentiation of teratogenic effects in the rat. Radiat Res 108(3):317–326

Roux S, Bernat C, Al-Sakere B, Ghiringhelli F, Opolon P, Carpentier AF, Zitvogel L, Mir LM, Robert C (2008) Tumor destruction using electrochemotherapy followed by CpG oligodeoxynucleotide injection induces distant tumor responses. Cancer Immunol Immunother CII 57(9):1291–1300

Ruiz-Gomez MJ, Martinez-Morillo M (2005) Enhancement of the cell-killing effect of ultraviolet-C radiation by short-term exposure to a pulsed magnetic field. Int J Radiat Biol 81(7):483–490

Ruzic R, Vodnik D, Jerman I (2000) Influence of aluminum in biologic effects of ELF magnetic field stimulation. Electro- Magnetobiol 19(1):57–68

Saikhedkar N, Bhatnagar M, Jain A, Sukhwal P, Sharma C, Jaiswal N (2014) Effects of mobile phone radiation (900 MHz radiofrequency) on structure and functions of rat brain. Neurol Res 36(12):1072–1079

Sandyk R (1997) Treatment with AC pulsed electromagnetic fields improves the response to levodopa in parkinson's disease. Int J Neurosci 91(3–4):189–197

Sasikala ARK, Unnithan AR, Yun Y-H, Park CH, Kim CS (2016) An implantable smart magnetic nanofiber device for endoscopic hyperthermia treatment and tumor-triggered controlled drug release. Acta Biomater 31:122–133

Satoh M, Tsuji Y, Watanabe Y, Okonogi H, Suzuki Y, Nakagawa M, Shimizu H (1996) Metallothionein content increased in the liver of mice exposed to magnetic fields. Arch Toxicol 70(5):315–318

Schwartz Z, Simon BJ, Duran MA, Barabino G, Chaudhri R, Boyan BD (2008) Pulsed electromagnetic fields enhance BMP-2 dependent osteoblastic differentiation of human mesenchymal stem cells. J Orthop Res 26(9):1250–1255

Selvarnurugan N, Kwok S, Vasilov A, Jefcoat SC, Partridge NC (2007) Effects of BMP-2 and pulsed electromagnetic field (PEMF) on rat primary osteoblastic cell proliferation and gene expression. J Orthop Res 25(9):1213–1220

Sergeeva EY, Titova NM, Sherbinina AS, Sergeev NV, Shirokova AV (2011) Effect of magnetic fields on antioxidant system enzymes in mice with ehrlich ascites Carcinoma. Bull Exp Biol Med 150(3):365–367

Shil P, Sanghvi SH, Vidyasagar PB, Mishra KP (2005) Enhancement of radiation cytotoxicity in murine cancer cells by electroporation: in vitro and in vivo studies. J Environ Pathol Toxicol Oncol 24(4):291–298

Shtemberg AS, Bazyan AS, Shikhov SN, Chernyakov GM, Uzbekov MG (2001) Modulatory influence of ultralow-intensity electromagnetic fields on the effects of some psychotropic drugs. Zhurnal Vysshei Nervnoi Deyatelnosti Imeni I P Pavlova 51(3):373–377

Slesin L (2006) "Radiation research" and the cult of negative results, Microwave News. http://microwavenewscom/RRhtml, 31 July

Soffritti M, Tibaldi E, Padovani M, Hoel DG, Giuliani L, Bua L, Lauriola M, Falcioni L, Manservigi M, Manservisi F, Panzacchi S, Belpoggi F (2016) Life-span exposure to sinusoidal-50 Hz magnetic field and acute low-dose gamma radiation induce carcinogenic effects in sprague-dawley rats. Int J Radiat Biol 92(4):202–214

Stam R (2010) Electromagnetic fields and the blood-brain barrier. Brain Res Rev 65(1):80–97

Stride E, Porter C, Prieto AG, Pankhurst Q (2009) Enhancement of microbubble mediated gene delivery by simultaneous exposure to ultrasonic and magnetic fields. Ultrasound Med Biol 35(5):861–868

Swanson DR (1986) Fish oil, raynauds syndrome, and undiscovered public knowledge. Perspect Biol Med 30(1):7–18

Swanson DR, Smalheiser NR, Bookstein A (2001) Information discovery from complementary literatures: categorizing viruses as potential weapons. J Am Soc Inf Sci Technol 52(10): 797–812

Szudzinski A, Pietraszek A, Janiak M, Wrembel J, Kalczak M, Szmigielski S (1982) Acceleration of the development of benzopyrene-induced skin cancer in mice by microwave radiation. Arch Dermatol Res 274(3–4):303–312

Tao Q, Henderson A (1999) EMF induces differentiation in HL-60 cells. J Cell Biochem 73(2): 212–217

Tenorio BM, Jimenez GC, de Morais RN, Peixoto CA, de Albuquerque Nogueira R, da Silva VA Jr (2012) Evaluation of testicular degeneration induced by low-frequency electromagnetic fields. J Appl Toxicol 32(3):210–218

The Bioinitiative Report (2012) http://wwwbioinitiativeorg/

Tikhonchuk VS, Ushakov IB, Fedorov VP (1987) Structural and metabolic analysis of the reaction of the central nervous system to the combined action of microwave and ionizing radiations. Radiobiologiia 27(3):361–365

Tokumoto S, Higo N, Sugibayashi K (2006) Effect of electroporation and pH on the iontophoretic transdermal delivery of human insulin. Int J Pharm 326(1–2):13–19

Trillo MA, Martinez MA, Cid MA, Leal J, Ubeda A (2012) Influence of a 50 Hz magnetic field and of all-trans-retinol on the proliferation of human cancer cell lines. Int J Oncol 40(5):1405–1413

Vantage Point (2015) wwwtheVantagePointcom, Search Technology, Inc, Norcross, GA

Varani K, De Mattei M, Vincenzi F, Gessi S, Merighi S, Pellati A, Ongaro A, Caruso A, Cadossi R, Borea PA (2008) Characterization of adenosine receptors in bovine chondrocytes and fibroblast-like synoviocytes exposed to low frequency low energy pulsed electromagnetic fields. Osteoarthr Cartil 16(3):292–304

Veronesi F, Cadossi M, Giavaresi G, Martini L, Setti S, Buda R, Giannini S, Fini M (2015) Pulsed electromagnetic fields combined with a collagenous scaffold and bone marrow concentrate enhance osteochondral regeneration: an in vivo study. BMC Musculoskelet Disord 16:233

Verschaeve L, Maes A (1998) Genetic, carcinogenic and teratogenic effects of radiofrequency fields. Mutat Res 410(2):141–165

Vianale G, Reale M, Amerio P, Stefanachi M, Di Luzio S, Muraro R (2008) Extremely low frequency electromagnetic field enhances human keratinocyte cell growth and decreases proinflammatory chemokine production. Br J Derm 158(6):1189–1196

Wang BH, He JL, Jin LF, Lu DQ, Zheng W, Lou JL, Deng HP (2005) Studying the synergistic damage effects induced by 18 Ghz radiofrequency field radiation (RFR) with four chemical mutagens on human lymphocyte DNA using comet assay in vitro. Mut Res Fundam Mol Mech Mutagen 578(1–2):149–157

Wang RJ, Hung YB, Wu PC, Fang JY, Tsai YH (2007) The effects of iontophoresis and electroporation on transdermal delivery of indomethacin evaluated in vitro and in vivo. J Food Drug Anal 15(2):126–132

Wang X, Liu Y, Lei Y, Zhou D, Fu Y, Che Y, Xu R, Yu H, Hu X, Ma Y (2008) Extremely low-frequency electromagnetic field exposure during chronic morphine treatment strengthens downregulation of dopamine D2 receptors in rat dorsal hippocampus after morphine withdrawal. Neurosci Lett 433(3):178–182

Wang L-F, Li X, Gao Y-B, Wang S-M, Zhao L, Dong J, Yao B-W, Xu X-P, Chang G-M, Zhou H-M, Hu X-J, Peng R-Y (2015) Activation of VEGF/Flk-1-ERK pathway induced blood-brain barrier injury after microwave exposure. Mol Neurobiol 52(1):478–491

Watanabe Y, Nakagawa M, Miyakoshi Y (1997) Enhancement of lipid peroxidation in the liver of mice exposed to magnetic fields. Ind Health 35(2):285–290

Wei M, Guizzetti M, Yost M, Costa LG (2000) Exposure to 60-Hz magnetic fields and proliferation of human astrocytoma cells in vitro. Toxicol Appl Pharmacol 162(3):166–176

Whissell PD, Persinger MA (2007) Emerging synergisms between drugs and physiologically-patterned weak magnetic fields: implications for neuropharmacology and the human population in the twenty-first century. Curr Neuropharmacol 5(4):278–288

Wolf R, Wolf D (2004) Increased incidence of cancer near a cell-phone transmitter station. Int J Cancer 1(2):123–128

Yang L, Hao D, Wang M, Zeng Y, Wu S, Zeng Y (2012) Cellular neoplastic transformation induced by 916 MHz microwave radiation. Cell Mol Neurobiol 32(6):1039–1046

Yoon HE, Lee JS, Myung SH, Lee Y-S (2014) Increased gamma-H2AX by exposure to a 60-Hz magnetic fields combined with ionizing radiation, but not hydrogen peroxide, in non-tumorigenic human cell lines. Int J Radiat Biol 90(4):291–298

Zhang XY, Xue Y, Zhang Y (2006) Effects of 04 T rotating magnetic field exposure on density, strength, calcium and metabolism of rat thigh bones. Bioelectromagnetics 27(1):1–9

Zhao G, Lin X, Zhou M, Zhao J (2014) Relationship between exposure to extremely low-frequency electromagnetic fields and breast cancer risk: a meta-analysis. Eur J Gynaecol Oncol 35(3):264–269

Zhao YL, Li YX, Ma HB, Li D, Li HL, Jiang R, Kan GH, Yang ZZ, Huang ZX (2015) The screening of genes sensitive to long-term, low-level microwave exposure and bioinformatic analysis of potential correlations to learning and memory. Biomed Environ Sci 28(8):558–570

Zmyslony M, Palus J, Jajte J, Dziubaltowska E, Rajkowska E (2000) DNA damage in rat lymphocytes treated in vitro with iron cations and exposed to 7 Mt magnetic fields (Static Or 50 Hz). Mut Res Fundam Mol MechMutagen 453(1):89–96

Zwirska-Korczala K, Adamczyk-Sowa M, Polaniak R, Sowa P, Birkner E, Drzazga Z, Brzozowski T, Konturek SJ (2004) Influence of extremely-low-frequency magnetic field on antioxidative melatonin properties in AT478 murine squamous cell carcinoma culture. Biol Trace Elem Res 102(1–3):227–243

Chapter 5
Misleading Scientific Papers on Health Effects from Wireless Communication Devices

Luc Verschaeve

Abstract Extensive research has been conducted in recent years on how radiofrequency fields (RF) generated by mobile phones and other wireless communication devices might affect health. However, so far, no scientific consensus is reached. On the one hand there are reports from independent international expert groups who claim that there are no solid indications that such fields may impair human health, on the other hand there are numerous published scientific papers showing alarming effects on virtually all investigated endpoints. Members of 'action groups' and 'engaged citizens' are essentially aware of the frightening data and hence they do not understand how authorities and health agencies do not seem that much alarmed. We have performed a literature search over an approximately two year period on *in vitro* and *in vivo* experimental studies and found that 85% of the papers indeed reported adverse bio effects. However, many of the papers are deficient in one or many ways and are not seldom published in non-specialized journals with poor quality control. Looking at specialized peer reviewed scientific journals the number of papers reporting bio effects drops to about 50%. From these many report only minor (not necessarily adverse) effects, while others report effects at high exposure levels usually not found in our daily life. This explains the more reassuring attitude of expert groups and authorities. It should be realized yet that lack of consistency in scientific data does not means lack of effect and that a simple application of the precautionary approach is still needed.

Keywords Radiofrequency fields • Contradictory results • Cancer • Immunological effects • Neurological disorders • Reproduction • Electromagnetic hypersensitivity • Predatory journals

---

L. Verschaeve (✉)
Unit of Toxicology, Scientific Institute of Public Health, Brussels, Belgium

Department of Biomedical Sciences, University of Antwerp, Antwerp, Belgium
e-mail: Luc.Verschaeve@wiv-isp.be

C.D. Geddes (ed.), *Microwave Effects on DNA and Proteins*,
DOI 10.1007/978-3-319-50289-2_5

5.1 Introduction

Extensive research has been conducted in recent years on how radiofrequency fields (RF) generated by mobile phones and other wireless communication devices might affect health. Laboratory studies on cells, tissues, animals and human volunteers were conducted with respect to virtually all possible endpoints and many human epidemiologic studies were also performed, mainly with respect to cancer (Juutilainen et al. 2009; IARC 2013). However, so far, no scientific consensus is reached as some investigators claim that low-level (non-thermal) exposures to RF-fields may for example be carcinogenic, whereas others claim that there are no sufficiently strong indications whatsoever that such fields may impair human health (Repacholi et al. 2012). This is rather intriguing as all have access to the same scientific data, but apparently risk perception is different. This is true for many scientific disciplines as exemplified by the debate about glyphosate (Portier et al. 2016). This is an active ingredient in many popular herbicides that was classified in 2015 by the International Agency for research on Cancer in category 2A, which is 'probably carcinogenic to humans'. However the European Food Safety Authority concluded somewhat later in the same year that '*the overall weight of evidence indicates that glyphosate is unlikely to pose a human cancer risk*'). Different risk evaluations of an agent are thus not uncommon but may be even more comprehensive when non ionizing radiations (RF and other electromagnetic fields) are concerned, because no one is expert in all aspects of the subject which is per definition multidisciplinary. Medical doctors, biologists, biophysicists and engineers specialised in, e.g., dosimetry, should, amongst others, collaborate. This is not always the case and false claims of adverse effects may be due to unintentional errors or misinterpretations. Even when rectified later on only the previously reported adverse effect will most often be remembered by many scientists and the public. An example can be given by a series of papers showing induced DNA single- and double strand breaks at 1800 and 1950 MHz electromagnetic fields at relatively low SAR-values (Diem et al. 2005; Schwarz et al. 2008). However, these studies were severely criticized (Vijayalaxmi et al. 2006; Lerchl 2009) and no genotoxic effects were found in a study aimed at (partly) replicating the above mentioned investigations (Speit et al. 2007). For this reason the alarming papers can only be seen as possibly suggestive but not conclusive. 'Engaged citizens' and 'action groups' nevertheless often only refer to them and largely ignore the lack of replication and other evidence supporting the other, reassuring, conclusion. A paper of De Pomerai et al. (2000) on non-thermal RF-effects may be cited as another example. This highly quoted paper was published in *Nature* but retracted later on after the results of a follow-up study became available. This resulted in reinterpretation of the initial data and designation of subtle thermal effects caused by slight heating as an explanation for the observed biological effect (Chou 2013). It is however the original paper and original conclusions that are still cited as proof of non-thermal adverse effects, the reinterpretation being largely ignored. As a matter of fact, scientists may mislead themselves as they are no expert in all aspects of the research and be unaware of some essential aspects that may influence the results of a study, and readers may misinterpret research data for the same reason. To give an example one may refer to a number of quotes sometimes heard during lectures (even given by renowned scientists) or seen in newspapers where adverse effects of RF-radiation were claimed based on published data on extremely low frequency fields (ELF; another, yet considerably different type of non-ionizing radiations), or even worse, based on data from research on ionizing radiations, such as X-rays.

5.2 Reasons for Wrong or Controversial Conclusions

Actually there are several reasons why interpretation of scientific data or claims for adversity may be incorrect:

1. All studies are not equally important. An *in vitro* investigation will never be sufficient to prove adverse effects in humans (as is not seldom heard yet, especially from representatives of 'action groups' or 'engaged citizens'. An observed 'positive' effect in one single *in vitro* genotoxicity test on human lymphocytes is sometimes enough to let them conclude that mobile phone base stations do impair our health.
2. A single study can form the basis of an hypothesis, but does not provide the basis for hazard identification. Confirmation of the results of any study are needed through replication and/or supportive investigations. The resulting body of evidence forms the basis for science-based judgments by defining exposure levels for adverse health effects and no observable adverse effects. Often only those study results are referred to that coincide with one's conviction whilst other data are disregarded or simply ignored.
3. A weight of evidence approach is necessary but it should be realised that this is not a 'mathematical average' and hence more negative studies do not necessarily mean that the final conclusion will be *negative*. Such a weight of evidence approach is not always followed or not always properly done.
4. Contradictory results may simply be due to different experiments with other frequencies and exposure regimes, signal modulation, exposure times and intensities, different statistical power, over- or misinterpretation of the data, even 'bad science', etc. Results from different studies are often generalized and as a consequence, claims are made for exposure situations that are not usually found or that are unrelated to our daily life exposures. We previously conducted a literature search of human cytogenetic biomonitoring studies on RF-exposed subjects and concluded that despite the overwhelming alarming data no conclusion was possible because virtually all reported investigations were deficient at some level (Verschaeve 2009).
5. As already mentioned, results from electromagnetic fields other than radiofrequency fields are sometimes used to conclude about the potential health

effects of (e.g.,) mobile phone radiation due to poor knowledge of electromagnetic fields.

6. Use of different units in the estimation of the exposure level (e.g., W/m^2 or mW/cm^2) is sometimes misleading: 10 W/m^2 is often considered a high exposure and 1 mW/cm^2 a low exposure; actually they are identical. Results may be overestimated as a result of a wrong perception.
7. Some reported effects are presented as proof of adverse health effects but they may be no more than mild responses to an external stimulus without any consequence. Many results are over-interpreted.
8. Results at high (thermal) exposure levels cannot be extrapolated to low (non-thermal) exposure levels as the underlying molecular mechanisms are different. However this is often done, again by poor knowledge of electromagnetic fields.
9. A nocebo (negative expectation) effect, for example with respect to 'electromagnetic hypersensitivity', can be enforced by negative reports in the media or by other (societal) influences.

Above aspects may explain why different perceptions may exist even between scientists and why the public may have very different opinions about an agent's health consequences. People are not always aware of these aspects, consequently they do not take them into account and may reach wrong conclusions.

It is clear that so far no scientific consensus is reached. On the one hand there are reports from independent international expert groups who claim that there are no solid indications that such fields may impair human health (at least at 'normal' exposure levels), on the other hand there are numerous published scientific papers showing alarming effects on virtually all investigated endpoints. Members of 'action groups' and 'engaged citizens' are essentially aware of the frightening data (for example as highlighted by the media) and hence they do not understand how authorities and health agencies do not seem that much alarmed. For them this attitude is likely to be driven by a 'conflict of interest'.

5.3 Conclusions from Scientific Expert Group Reports

Let us first have a look at what scientific experts conclude.

Several international expert groups have evaluated and continue to evaluate the many scientific papers that were produced on potential adverse health effects of RF-radiation. A detailed overview of 'expert group reports' and their conclusions is given in Verschaeve (2012) and Vijayalaxmi and Scarfi (2014). More recent reports are for example those from respectively the Swedish Scientific Council on electromagnetic fields (SSM 2014) and the Scientific Committee on Emerging and Newly Identified Health Risks (SCENIHR 2015). It is not possible to give an exhausted overview of all discussed topics. Below a short summary is given of the main conclusions reached by most expert groups on the most cited or discussed scientific disciplines. We also included some of the more recent scientific papers that may enlarge the scientific debate. They were therefore also briefly discussed in this section.

5.3.1 Cancer

In recent years many studies have investigated whether RF fields, particularly those of mobile phones and their base station antennas, could cause or favour cancer development. Evaluating the existing data IARC classified in May 2011 RF fields into the group 2B, i.e. possibly carcinogenic for humans, essentially based on weak evidence with regard to glioma (Baan et al. 2011; IARC 2013). A study by Coureau et al. (2014) enforced this conclusion, but most other studies that were published since the announcement of the IARC classification indicated that, although the uncertainty with regard to the relationship of mobile phones and brain tumours remains, the available literature not currently points at an increased risk (SCENIHR 2015). Most recent studies did not show an increased risk of brain tumours and did not indicate an increased risk for other cancers of the head and neck region. Thus, overall, recent epidemiological studies did not confirm the association between RF exposures and glioma or other cancers. The only available epidemiologic study which addressed mobile phone use and brain tumours in children and adolescents also showed no association (Aydin et al. 2011). A census-based cohort study also suggested that there is no connection between calculated RF exposure from cell towers and childhood leukaemia. The results for central nervous system (CNS) tumours are less consistent, but could not be established in the most comprehensive analysis (Hauri et al. 2014). Furthermore, the overall brain cancer indices among the general population did not suggest an increasing trend after the introduction of mobile phones (Röösli et al. 2007; de Vocht et al. 2011; Deltour et al. 2012; Little et al. 2012), and in a prospective study even a decreased risk for glioma was found in mobile phone users (Benson et al. 2013). This raised the question whether an RF-induced adaptive response may play a role in reducing carcinogenesis, at least in some individuals (Vijayalaxmi and Prihoda 2014).

Further evidence for the absence of a carcinogenic effect was provided by a considerable number of well-performed experimental studies which investigated whether RF fields are able to induce cancer in laboratory animals (Elder 2010; Juutilainen et al. 2011; see also Annex D in Health Council of the Netherlands 2010). Overall, the results of these studies indicated no carcinogenic effects at exposure levels relevant to human exposure from mobile phones.

Studies assessing the potential of RF fields to cause genetic damage overall also did not shown such effects or provided only very weak evidence (Verschaeve et al. 2010; Vijayalaxmi and Prihoda 2012). The IARC classification was therefore not supported by a genotoxicity-based mechanism. Other potential endpoints were also investigated, such as cell death, expression of genes, cell proliferation etc. Most of the studies did not find any reproducible effect (IARC 2013).

Expert group reports thus remain somewhat cautious but nevertheless indicate that there is to date no compelling evidence to believe that RF fields from mobile phones are carcinogenic.

5.3.2 Immunological Effects

Reports on non-thermal effects from RF-fields were published since the 1970s, some studies reporting for example stress responses, stimulation of cytokines and other immune parameters in exposed cultured cells, whereas others did not find any such effects. In a literature review of possible effects of radiofrequency fields on humans related to cardiovascular, reproductive, immune and other systems, Jauchem (2008) reported some changes in immunoglobulin levels and peripheral blood lymphocytes of radar and radio/television-transmission workers. However, no unifying hypothesis of immune-system effects could be proposed and it was concluded that '*the majority of changes relating to each of the diseases or conditions were small and not significant*'. More recently Szmigielski (2013) reviewed the impact of weak RF-fields on various immune functions. He concluded that the available data clearly indicated that various shifts in the number and/or activity of immunocompetent cells are possible, but that the results were too much inconsistent to allow any firm conclusion. Actually, the only significant effects were associated with a rise in temperature of the body (thermal effects) due to exposures in which the whole-body SAR (= specific absorption rate) is greater than 4 W/kg.

We nevertheless cannot deny that many studies did show immunological effects, even at low exposure levels. This should not be surprising yet as the immune system is able to react in a measurable way to discrete environmental stimuli. Effects can often be interpreted as subtle, totally normal and harmless responses of cells to a mild external stress resulting in activation of the immune system, which may eventually even led to a protection against such stresses, for example infections. This may be so for RF- but also ELF-EMFs (e.g., Galeev 2000; Cuppen et al. 2007). Research on plants showed for example a stress response resulting in activation of certain immunological parameters but this can also be seen in case of other minor stresses such as a slight decreased or increased temperature, low mechanical stress or subtle osmotic disturbances. It is important to particularly emphasize that effects were often found to be more important in cells of plants, animals and humans when they were afflicted with some disease or deficiency (see also section on 'non-specific health symptoms'). An example is induced meristem production in flax (*Linus usitatissimum L.*) after exposure to RF-fields. This reflects a mild stress, especially in calcium deprived seedlings, which is also found when cells/organisms are exposed to other abiotic treatments such as a mild cold shock or simply touching (Tafforeau et al. 2004; Verschaeve 2013). Reported measurable molecular changes in tomato plants shortly after low-level EMF (RF) stimulation most probably also reflect a mild stress (Vian et al. 2007). Another example of a mild response to RF-radiation is the reported increased frequency of monocytes in RF-exposed rats

(Adang and Vander Vorst 2006) which, as pointed out at the International Conference on Electromagnetic Fields, Health & Environment (Madeira, Portugal, 2006), may fall within the normal variability expected in unexposed rats. The complete study showing effects of long term exposure to radiofrequency electromagnetic fields on the blood picture and on survival (Adang et al. 2009) attracted a lot of media attention and was presented as proof that RF-exposure lowers the survival rates of exposed animals and actually doubles the mortality rate. This study encountered a lot of criticism as a result of a number of flaws in the execution of the study and in the statistical analysis of the data (e.g., Health Council of the Netherlands, 2010) and because many other studies reached different conclusions (e.g., Elder 2003, 2010). However, the criticism did not reach the public.

Besides possible mild stress reactions which activate the immune system, low-level RF-fields also may (as a consequence) activate DNA repair mechanisms and hence result in lower DNA damage levels compared to non-exposed controls. We for example found less than background DNA damage in larvae and brain cells of honey bees which were exposed to simulated mobile phone base station radiation for 1 h up to 6 days. This (unpublished) study should yet be replicated before any conclusion can be drawn. In another investigation we found less sister chromatid exchanges (SCEs) in blood cells of workers exposed to RF radiation compared to the frequency in a corresponding control population, and also found that their white blood cells after exposure to the chemical mutagen methyl methane sulfonate (MMS) showed less DNA damage than cells of non-exposed persons with the same MMS exposure (Maes et al. 2006). Similar effects were also found in cells that were exposed to low-level 50 Hz magnetic fields as lower micronucleus frequencies were found compared to unexposed control cells. Increased genetic damage was however found at higher exposure levels (Maes et al. 2016).

5.3.3 Brain and Neurological Disorders

Because mobile phones are used near the head, there have been concerns they could affect the brain (other than by inducing cancer). There is some evidence that RF exposure might have a subtle impact on brain activity, sleep, learning, memory or behaviour, but there is yet no evidence for health relevance. Some studies have for example shown small effects of electromagnetic fields (type GSM radiation) on the EEG (electroencephalogram) profiles and brain activity (van Rongen et al. 2009), but the results are not consistent and subject to interpretation. The effects on the EEG yet seem to be the most (only) reproducible effect of GSM signals, but such small changes in the electrical activity have probably no functional significance. In children, studies show the same result. Overall the studies indicate that there is no robust effect, regardless of the RF signal. Thermoregulation of children would be unaffected. Studies on cognitive effects give no solid evidence of adverse effects following exposure to RF-electromagnetic fields. Overall, there is a lack of evidence that mobile phone RF fields affects cognitive functions in humans but given the available data further research on this issue may nevertheless be recommended.

The blood – brain barrier (BBB) guarantees that certain molecules such as proteins of the blood do not penetrate the brain. If that were the case, this would affect the functioning of the brain. Research on the effects of RF-fields on the blood – brain barrier is therefore important. In the past a team of Swedish researchers reported in a series of papers that the radiation from a cell phone disrupts the BBB (e.g., Salford et al. 2003). These studies have often been criticized in the scientific community due to a number of important shortcomings. Since then, nearly 100 studies were published on the effects of RF-radiation on the BBB. Various critical assessments of these studies invariably came to the conclusion that RF electromagnetic fields are indeed capable of disrupting the BBB, but only when it comes to thermal exposures (Stam 2010; Elder 2010; Perrin et al. 2010). The mobile phone or GSM antennas are therefore not concerned.

Claims that exposure to RF-radiation increases the risk of neurological disorders such as Alzheimer's disease is, so far, totally unjustified as suggestive data comes from epidemiological studies related to exposure to extremely low frequency (ELF) magnetic fields, which are different from RF–fields. The few studies that were conducted on the subject rather point to a 'protective effect' of the RF-fields (Arendash et al. 2012; Mortazavi et al. 2013; Jeong et al. 2015). This is however also not established either. To note, the alleged association between ELF-fields and Alzheimer's disease is also very speculative and probably not true (SCENIHR 2015; Maes et al. 2016).

5.3.4 Reproduction and Development

Numerous large and well-conducted studies have investigated potential effects of RF fields on the development of animals and clearly show that RF fields can cause birth defects when the exposure is well above safety guidelines and therefore high enough to significantly raise temperatures in tissue. No consistent evidence of effects has been found yet at exposure levels that do not cause relevant heating of tissues. The new SCENIHR report (2015) which took the most recent human and animal data into account therefore still concluded that there were no adverse effects on reproduction and development from RF fields at non-thermal exposure levels.

No significant effects were seen from almost continuous, lifetime exposure of mice over four generations. Effects on foetuses from mothers' mobile phone use during pregnancy are judged not plausible due to the low level of exposure. The available data provide no clear evidence of consistent adverse effects on human semen quality. Studies on male fertility are of poor quality and provide little evidence (SCENIHR 2015).

There are still no substantiated indications of any other health effects.

5.3.5 *Non-specific Health Symptoms*

Some people attribute non-specific health symptoms such as headache, fatigue and dizziness to RF (and also ELF) fields. Such complaints have raised concern that certain individuals may be critically more sensitive than others to electromagnetic fields, a phenomenon that has been coined as "electromagnetic hypersensitivity" and which is nowadays rather designated as "Idiopathic Environmental Intolerance attributed to electromagnetic fields" (IEI-EMF; Hansson Mild et al. 2006).

Many studies were already conducted in the past and were followed by more recent ones. The overall conclusion reached by expert groups and the majority of investigators remain unchanged. The conclusion is that there is no evidence that RF-exposure from mobile phones (and other EMFs) may be causally linked to these symptoms. Instead, evidence exists in favour of a "nocebo" effect which means that effects may be caused by the pure belief that something is harmful rather than by the suspected cause itself (Hahn 1997; Spiegel 1997; Porsius et al. 2015, 2016; De Ridder and Braeckman 2015). One of the findings from a large investigation on biological effects from non-ionizing radiations which was performed in the Netherlands was that the actual distance of a self-declared 'hypersensitive' individual to a GSM antenna (or high-voltage power line) is not at all associated with non-specific symptoms but the perception of closeness is (de Jongh et al. 2015; Porsius et al. 2016).

It may be interesting to note that some efforts were and are presently undertaken to use commercially available (largely immunological) tests to identify electrosensitive subjects. According to an unpublished preliminary report some immunological parameters were considerably disturbed in RF-exposed cells from a self-declared electromagnetic hypersensitive subject with IEI-EMF and some considered that, provided this could be confirmed by other investigations, this could be seen as a proof of the existence of electro(magnetic) hypersensitivity in people and a way to diagnose such condition. However, if many external stress situations, including but not exclusively electromagnetic fields, are capable of stimulating the immune system, especially in cells from sick subjects or from cells with a deficiency as suggested before, this cannot be considered proof of electromagnetic hypersensitivity (IEI-EMF) and cannot be a diagnostic tool for it. It just shows that cells that present some deficiency, compared to 'normal' cells do react more to an external stimulus, including, but not exclusively, to (low-level) electromagnetic fields. It yet may prove that, at least some of the IEI-EMF-subjects do have real symptoms and hence a real health problem (something which has always been recognized by the WHO 2005), but it does not prove that hypersensititivity to electromagnetic fields is on the origin of the problem.

Belpomme et al. (2015) recently also claimed having found reliable disease biomarkers characterizing and identifying electrosensitivity and multiple chemical sensitivity. They investigated a great number of subjects with self-reported symptoms of IEI-EMF and multiple chemical sensitivity (MCS) and found, amongst others, increased histaminemia and increased Hsp27 and/or Hsp70 in 30–40% of the patients. Their findings may be very interesting but again, they do not prove a causality between EMF and the hypersensitivity. The fact that patients with MCS present the same symptoms indicate a common pathological mechanism but not a causal relationship between EMF and the symptoms.

5.4 Review from Recent *In Vitro* and *In Vivo* Experimental Studies (2014–2015)

Despite the rather reassuring conclusions reached by, e.g., international expert groups, there yet remains a lot of discussion and societal concern with respect to the potential adverse health effects of radiation from mobile communication devices and other electromagnetic fields (EMF)- emitting technologies. This is because there are numerous scientific publications showing that the electromagnetic fields they produce are harmful, and explains contradictory opinions as for example addressed by Sage et al. (2016). Others yet again replied to these (e.g., Leitgeb 2016) keeping the debate lively.

It is clear anyway that many alarming publications continue to be published. This can be illustrated by a literature review of the newly published scientific papers that was performed on *in vitro* and *in vivo* investigations of biological effects of RF-radiation (approximately during the years 2014–2015). For the review common or specialized search-engines were used (Web of Science; PubMed; EMF-Portal; etc.). All papers that were found were taken into account. This means that we did not *a priory* rejected papers for one or another reason (for example bad quality). This is because laymen, engaged citizens or members of action groups against mobile phone base stations and other radiation sources in the environment also did not make this selection and we wanted a 'view' of the literature as they can and do see it. We concluded the literature search after having found and read a total of approximately 200 papers. Amongst them there were 31 review papers (Table 5.1) and 173 research papers (Table 5.2). The tables give a short description of what was done, what EMF-fields were used and what results were reported. They also give the country of origin of the study and the 'official' impact factor of the publication according to Thomas Reuters Citation Reports. This only gives a very rough indication of the quality of the papers or the status of the journals they were published in, but may nevertheless be of some help in the evaluation of the investigations. Table 5.2 shows that no less that 148 papers did report (adverse) effects. This is almost 6 times more than the papers reporting no effect. It is therefore not difficult to understand that public perception is usually rather unfavourable as action groups and 'engaged citizens' are aware of the huge amount of alarming papers and emphasize the importance of the reported negative data. However, they do not make a weight of evidence evaluation, largely ignore above formulated problems and do not know which studies are well conducted and reported in peer reviewed high quality journals and which are poor quality studies and poor quality journals. From the 200 papers (Tables 5.1 and 5.2) many were obviously published in so called 'predatory journals' which provide easy and fast publication at some cost, have questionable editorial policies and poor quality control (Clark and Smith 2015; Devnani and Gupta 2015; Gasparyan et al. 2015). These journals often publish poor quality research from authors in low and middle income countries who do not speak and write English sufficiently well. Predatory journals also have attractive names resembling the names of very reputed journals and enthuse fake impact factors. Articles that are published in predatory journals often are of such a poor quality that they actually should be totally disregarded, but often they are not. The public does not make the difference between such articles and those published in peer reviewed high quality scientific journals. This does not contribute to a correct perception of risk. This does not mean that papers in other more reliable journals are 'per definition' of a much higher quality. Increasingly interdisciplinary and multicentre research, as is typical for research on bio-effects from non-ionizing radiations, puts additional pressure on the filtering system and reveals gaps in the qualifications of referees, who are able to adequately analyse some, but not all parts of the submissions (Gasparyan et al. 2015; Dye 2007).

When this is sorted out the literature review gives a completely different picture compared to what one may have believed at first sight. For example, considering only papers published in the journals *Bioelectromagnetics*, *Radiation Research* and *International Journal of Radiation Biology* which may be considered specialized peer reviewed journals in this field of research, we now find only about 50% of papers reporting adverse effects, compared to the more than 85% overall. From these 50% many reported only minor effects, or if substantial, the effects were seen at rather high exposure levels that are usually not found in daily life. Minor effects can for example be the above mentioned induced meristem production in flax (*Linus usitatissimum L.*) that reflect a mild stress which is also found when cells/organisms are exposed to other abiotic treatments such as a mild cold shock or simply touching (Tafforeau et al. 2004; Verschaeve 2013). Several other examples can be found in the present literature review (Table 5.2).

5.5 Conclusion

There are many publications with too many shortcomings to prevent them from being credible and convincing. Overall, critical evaluations of the scientific data show that these are rather reassuring. It is important that this is known. However, this does not mean that exposure to radiation from wireless communication systems (and some other non-ionizing radiations) are cleared from any suspicion, and that no protective measures need to be taken by authorities in order to *reasonably* apply the precautionary principle. Indeed, no evidence of harm does not equal evidence of no harm, and we may not expect totally consistent results from scientific research when such a complex matter is concerned. Bradford Hill (1965) said that "*whether chance is the explanation or whether a true hazard has been revealed may*

Table 5.1 An overview of recent review papers on biological effects of radiofrequency fields from wireless communication devices

Author(s)	Impact factor[a]	Adverse effects Yes/No	Short summary
Dahal KP (2013) Country: Nepal Mobile communication and its adverse effects Himalayan Phys. 4, 51–59	No impact factor	Y	This publication gives an overview of harmful health effects of mobile communication systems. It describes both thermal and non-thermal effects; in particular effects on genes, cell biological features, electromagnetic hypersensitivity, sleep problems, arthritis, schizophrenia and various forms of cancer. Also effects on plants, birds and other animals are discussed. According to the author all these effects are demonstrated but he gives no references nor justification for this assertion. This article is therefore useless and is only mentioned for completeness
Foster K, Moulder J. (2013) Country: USA Wi-Fi and health: Review of current status of research Health Phys. 105, 561–575	IF = 1.07	N	In a literature review on Wi-Fi and health a series of high quality studies were highlighted which give a good idea of our exposure to "Wi-fi radiation", but also of the lack of sufficient good studies on health-related aspects. The available literature and present knowledge on possible mechanisms of RF radiation yet allow to conclude that biological effects in Wi-Fi users are not expected
Markov MS. (2014) Country: USA Electromagnetic fields and life J. Electr. Electron. Syst. 3, 1000119	No impact factor	N/?	This paper can be seen as a review paper as well as a statement paper. It discuss in a very concise way the role of magnetic/electromagnetic fields in the origin and evolution of life, benefits and clinical applications of EMF and magnetotherapy. With respect to hazards (for example of Wifi communications) a large paragraph actually gives no useful information in terms of what is known or unknown
Cao Y., Tong J. (2014) Country: China Adaptive response in animals exposed to non-ionizing radiofrequency fields: some underlying mechanisms Int. J. Environ. Res. Pub. Health 11, 4441–4448	IF=1.998	?/N	This is a short review paper which provides another view on many 'positive' results that at first appear to be alarming. This paper deals with adaptive responses in animals exposed to radiofrequency fields. The results of different studies indicate a significant increase in survival, decrease in genetic and oxidative damage and alterations in several cellular process. The conclusion is that "*exposure of animals (and human cells) to non-ionizing radiofrequency fields give an adaptive response and may generate some 'mild stress' which may at the end activate cell defence mechanisms*". Nothing to worry then; instead of being proof of 'adverse effect' it is rather the opposite

(continued)

Table 5.1 (continued)

Author(s)	Impact factor[a]	Adverse effects Yes/No	Short summary
Mohamed MA, Abdelrazek MM, Zewita MS. (2014) Country: Egypt A survey of health effects of electromagnetic fields Int. J. Comp. Sci. Issues 11, 167–171	No impact factor	?	This is a brief review of studies which covered potential health effects of both ELF and RF EMF. With respect to RF EMF possible effects on cognition, nonspecific symptoms (EHS, headaches), hormones, genes, sleep, vision and hearing, the endocrine system, the brain, and reproduction were considered. The authors noted the limited knowledge about the long-term effects of EMF but also that existing exposure guidelines are well below the established threshold for harm. As such this paper has no added value compared to other more exhaustive reviews
Dinđić B, Radovanović Z, Pešić S, Zlatanović D, Sokolović D, Đinđić N, Milenković T. (2013) Country: Serbia Cytogenetic damages induced by chronic exposure to non-ionizing radiofrequency fields Acta Med. Medianae 52, 48–52	No impact factor	Y	This is a short review on RF EMF-induced cytogenetic damage following exposure to microwaves. The authors noted increased frequencies of DNA damage and oxidative stress, production of reactive oxygen species and prolonged disruption of DNA repair mechanisms as possible mechanisms for microwave induced cytogenetic damages, even at low-level electromagnetic fields. The authors therefore recommend keeping mobile phones as far as possible from the body and to take antioxidant-rich food or drugs to prevent genotoxic effects that could be caused by mobile phone use. The paper is limited in this sense that it does not provide an extensive literature review and hence that only selected papers were taken into consideration. It is not clear what criteria were used for this selection
Beinart R, Nazarian S. (2013) Country: USA & Israël Effects of external electrical and magnetic fields on pacemakers and defibrillators Circulation 128, 2799–809	IF=15.202	N	In an informative article focused mainly on medical personnel, a summary was given of the effects of external electric and magnetic fields on cardiovascular implantable electronic devices such as pacemakers and implantable defibrillators. The article did not only focus on ELF-fields but also on RF-fields including those from cellular phones. The authors concluded that "*… no special precautions are required when household appliances are used*". Environmental and industrial sources of electromagnetic incompatibility (EMI) are relatively safe if the exposure time is limited and distance to cardiovascular implantable electronic devices (CIEDs) is maximized. The risk of EMI-induced events is highest within the hospital environment. The physician awareness of the possible interactions and methods to minimize exposure remains justified. This is an interesting paper but it is not really relevant to 'normal' RF exposures of the general population

(continued)

Table 5.1 (continued)

Author(s)	Impact factor[a]	Adverse effects Yes/No	Short summary
Leszczynski D. (2014) Country: Finland Radiation proteomics: A brief overview Proteomics 14, 481–488	IF=4.132	?	This is a short review paper which essentially has one message: Because low dose (ionizing and non-ionizing) radiation effects are difficult to establish and to replicate, proteomics, supplemented with other "omics" techniques, might be the best way forward to find out the target molecules of radiation, the biomarkers of radiation exposure and the physiological and health significance of the acute and delayed biological effects caused by exposures to high- and low-dose radiation. However, the currently available database of radiation effects on proteomes is far too small to be useful in the formulation of new hypotheses concerning health consequences of radiation exposures. The only way forward is the high-throughput screening using transcriptomics, proteomics, and metabolomics but these are expensive and demanding methods, and more funding is needed
Vijayalaxmi, Cao Y, Scarfi MR. (2014) Country: USA & Italy Adaptive response in mammalian cells exposed to non-ionizing radiofrequency fields: A review and gaps in knowledge Mutat. Res. 760, 36–45	IF=8.202	?/N	In a literature review on RF EMF and adaptive response (AR), defined by the authors as "*a phenomenon in which cells which were pre-exposed to extremely low and non-toxic doses of a genotoxic agent became resistant to the damage induced by subsequent exposure to a higher and toxic dose of the same, similar (in action) or another genotoxic agent*," they concluded that RF-EMF-induced AR poses opportunities for beneficial uses. This paper is in accordance with others on the same subject and reaches the same conclusions. Follow up is recommended
Jayasanka SMDH, Asaeda T. (2013) Country: Japan The significance of microwaves in the environment and its effect on plants Environ. Rev. 22, 220–228	IF=2.649	Y	This literature review is on effects of RF EMF exposure on plants. The authors concluded that though the number of studies is limited, most findings indicate that there are effects on plants at exposure levels below the US Federal Communications Commission (FCC) and ICNIRP guidelines. However, any such effects vary by the type of plant and the characteristics of the radiation

(continued)

Table 5.1 (continued)

Author(s)	Impact factor[a]	Adverse effects Yes/No	Short summary
Gherardini L. , Ciuti G., Tognarelli S., Cinti C. (2014) Country: Italy Searching for the Perfect Wave: The effect of radiofrequency electromagnetic fields on cells Int. J. Molec. Sci. 15, 5366–5387	IF=2.464	Y/N/?	In this review *in vitro* and *in vivo* studies and epidemiological investigations were cited that the authors consider to be of particular importance. This is thus not a full literature review but a review of papers that the authors consider 'decisive'. Research is cited that is alarming, in particular with relations to the brain which appears to be particularly sensitive to electromagnetic (RF) fields. The authors then give examples of research that immediately contradict these alarming messages and underline for example the studies on an adaptive response and hence a beneficial impact of RF. The same applies to studies regarding RF-induced ROS (Reactive oxygen species) and associated inflammation of cortical and subcortical tissues in the brains of Alzheimer disease patients. Then, reference is made to studies which show that pulsed RF fields recover deposition of β-amyloid and cognitive functions in mice, which again is favourable. A similar discussion follows regarding Parkinson's disease. The authors end up with indicating that specific tuning of experimental and clinical RF exposure might lead to beneficial health effects and that studies should be done in compliance with good standard. This review has some merits but brings no new information. Yet, emphasis on the complexity of the various often conflicting data is useful to bring results from individual research into perspective
Yakymenko I, Sidorik E, Henshel D, Kyrylenko S. (2014) Country: Ukraine Low intensity radiofrequency radiation: a new oxidant for living cells Oxid. Antioxid. Med. Sci. 3, 1–3	No impact factor	Y	A review of 80 papers on RF EMF and oxidative stress in living cells was reported in a brief editorial. According to that 76 (92.5%) of the studies reported the detection of significant oxidative effects. A non-phagocyte NADH oxidase, a known enzymatic source of ROS, was in particular shown to be significantly activated just after a few minutes of exposure to low intensity RFR. These effects have been observed at RF EMF levels as low as 0.1 $\mu W/cm^2$ with SAR levels of 0.3 μW/kg. In addition, different antioxidants have been shown to reverse the oxidative stress caused by RF EMF. The authors urge a precautionary approach

(continued)

Table 5.1 (continued)

Author(s)	Impact factor[a]	Adverse effects Yes/No	Short summary
Jalota C. (2014) Country: India Bluetooth technology Int. J. Computer Sci. Management Res. 3, 3858–3864	No impact factor	Y	In a generally positive review of Bluetooth technology, an author from India included a section on health hazards. The author mentions biological effects such as weight gain, brain cancer, birth defects, miscarriage, hair loss, Alzheimer's disease, neck pain, and headache. But apart from mentioning these there are no data presented nor references given. Therefore this paper has no added value with respect to the health issue
Shckorbatov Y. (2014) Country: Ukraine The main approaches of studying the mechanisms of action of artificial electromagnetic fields on cell J. Electr. Electron. Syst. 3, 1000123	No impact factor	Y/?	The effects of electromagnetic fields (EMFs) on health were discussed in this paper, especially effects of EMFs on cell membranes, free radical concentration in cells, and intracellular regulatory systems, to understand the mechanism of the biological action of EMFs. According to the author, the analysis of experimental results proves the multilateral nature of the impact of EMFs on cell. In his opinion the primary reaction to electromagnetic fields may be connected with the level of gene activity regulation. Here RF and ELF fields were both discussed. This paper does not give clear explanations or hypotheses other than that effects of EMF 'may be' related to gene activity regulation and that the investigations of molecular mechanisms can be important to give new information on mechanisms of EMF action on cells. The paper has some merits but clearly does not add much to the discussion
Singh AP, Kaur R. (2014) Country: India Electromagnetic fields: Biological implications on various life forms Int. J. Bioassays 3, 2030–2040	IF=2.94	Y	This is a review on the "biological implication of electromagnetic fields on all living forms…". Although the connection between biological effects and human health effects is not clear, the authors say, every individual is altered in some way. They recommend the use of shielding of all sources of EMF. This review did clearly not take into account all scientific papers. As a matter of fact it focused only on selected alarming papers ignoring many other reports and is therefore rather misleading than scientifically correct

(continued)

Table 5.1 (continued)

Author(s)	Impact factor[a]	Adverse effects Yes/No	Short summary
Mortazavi SAR, Tavakkoli-Golpayegani A, Haghani M, Mortazavi SMJ. (2014) Country: Iran Looking at the other side of the coin: the search for possible biopositive cognitive effects of the exposure to 900 MHz GSM mobile phone radiofrequency radiation J. Environ. Hlth. Sci. Engineer 12(75)	IF=1.23	?	In a review of possible positive health effects of exposure to low levels of RF EMF, the authors cite the following effects: adaptive response, in which pre-exposure to RF EMF protects cells against the effects of subsequent exposures to other toxic agents; improved visual reaction time in humans; induced cognitive benefits; and a possible protective effect against symptoms of Alzheimer's disease. This paper thus essentially provide evidence of health benefits in relation to mobile phone radiation exposure. It should be stressed that, if there are no strong arguments in favour of deleterious effects from exposure to mobile phone radiation, there are at present also no proofs of beneficial effects
Janković SM, Milošev MZ, Novaković MLJ. (2014) Country: Serbia The effects of microwave radiation on microbial cultures Hospital Pharmacol. 1, 102–108	No impact factor	Y	This is a review of the scientific literature on the effects of microwave radiation on microbial cultures. According to this paper it is clear that microwaves produce significant effects on the growth of microbial cultures, which vary from the killing of microorganisms to enhancement of their growth. The nature and extent of the effect depend on the frequency of microwaves and the total energy absorbed by the microorganisms. Low energy, low frequency microwaves enhance the growth of microorganisms, whereas high energy, high frequency microwaves destroy the microorganisms. This paper is interesting but one should go to the original papers that it describes to be sure of the validity of the claims. It is not sure that all so-called non-thermal effects were really non-thermal for example. Thermal effects are obvious as microwaves are sometimes also used as pest-killers and hence high energy effects are not surprising and do not give direct information with respect to low-level radiation effects on human health. The paper is yet interesting from a scientific point of view

(continued)

Table 5.1 (continued)

Author(s)	Impact factor[a]	Adverse effects Yes/No	Short summary
Liu K, Li Y, Zhang G, Liu J, Cao J, Ao L, Zhang S. (2014) Country: China Association between mobile phone use and semen quality: a systemic review and meta-analysis Andrology 2, 491–501	IF = 2.298	Y	Relevant studies published up to May 2013 were identified from five major international and Chinese literature databases. The review of these studies (on humans, animals and *in vitro* studies) showed that results of most of the *in vivo* and *in vitro* laboratory studies indicated that mobile phone use or radiofrequency exposure had negative effects on the various semen parameters studied (e.g., sperm motility and viability). However, meta-analysis indicated that mobile phone use had no adverse effects on semen parameters in human studies. Animal studies showed that RF exposure has harmful effects on sperm concentration and motility. The data suggest potential harmful effects of mobile phone use on semen parameters. Interesting review of carefully selected papers (with well-defined selection criteria). Many evaluated papers are actually from human epidemiological studies. For the *in vitro* and animal studies one should consult the original papers to estimate the exact exposure and experimental conditions. It is clear that there were many differences from one study to another in this respect. For this reason the real relevance for mobile phone users is unclear
Verschaeve L. (2014) Country: Belgium Environmental impact of radiofrequency fields from mobile phone base stations Crit. Rev. Environ. Sci. Technol. 44, 1313–1369	IF = 3.383	N	This review examines the potential environmental impact of radiofrequency (RF) fields emitted by mobile phone base station antennas and other sources of RF radiation. Overall, many alarming investigations were found but most are characterised by severe methodological shortcomings. For this reason these studies do not provide strong evidence that observed biological effects are associated with exposure to the electromagnetic fields. So far, the studies do not prove that environmental exposures to mobile phone base station radiation (and other environmental RF exposures) are harmful to wildlife

(continued)

Table 5.1 (continued)

Author(s)	Impact factor[a]	Adverse effects Yes/No	Short summary
Balmori A. (2014) Country: Spain Electrosmog and species conservation Sci. Total Environ. 496C, 314–316	IF = 4.099	Y	In his continued efforts to bring attention to the potential effects of EMF on wildlife this author has published a new concise literature review of RF EMF effects on animal behaviour, navigation, reproduction, and habitat loss, plus data on effects on plants. The researcher asserts that "*there are reasonable grounds for believing that microwave radiation constitutes an environmental and health hazard*," and calls for more research and use of precautionary measures to protect wildlife. The author only refers to alarming publications ignoring all others without giving any rationale for the selections he made
MINOR UPDATES IN WHO FACT SHEET ON CELL PHONES AND HEALTH (2014) Country: International http://www.who.int/mediacentre/factsheets/fs193/en/	No impact factor	N	The WHO has updated its Fact Sheet on cell phones and health. As before, it notes the IARC classification of RF EMF from mobile phones as a possible human carcinogen, but states that, "*To date, no adverse health effects have been established as being caused by mobile phone use*". WHO yet remains vigilant with respect to mobile phones and brain cancer
ARPANSA Fact Sheet (2015) Country: Australia http://www.arpansa.gov.au/radiationprotection/FactSheets/is_wifi.cfm	No impact factor	N	A newly published Fact Sheet on Wi-Fi and health from the Australian Radiation Protection and Nuclear Safety Agency (ARPANSA) concludes: "*There is no established scientific evidence showing that the low exposure to radiofrequency electromagnetic energy from Wi-Fi adversely affects the health of children or the general population.*" It explains that public RF EMF exposures are limited by the standards set by ARPANSA, and that Wi-Fi exposures tend to be much lower than the ARPANSA limits. On the basis of existing scientific knowledge, "*ARPANSA sees no reason why Wi-Fi should not continue to be used in schools and in other places*," the Fact Sheet states

(continued)

Table 5.1 (continued)

Author(s)	Impact factor[a]	Adverse effects Yes/No	Short summary
Vijayalaxmi, Scarfi MR. (2014) Country: USA and Italy International and national expert group evaluations: Biological/health effects of radiofrequency fields Int J Environ Res Public Health 11, 9376–408	IF=1.993	N/?	This paper is a compilation of the conclusions, on the biological effects of RF exposures, from various national and international expert groups, based on their analyses. It is almost identical to a previous book chapter (Verschaeve L., 2012). Conclusions were also the same, namely that in general, the expert groups did not consider that there is a demonstrated health risk from RF exposure from mobile telephones and other wireless communication devices. Because of uncertainties they yet suggested a reduction in exposure levels, a precautionary approach, and further research
Balmori A. (2015) Country: Spain Anthropogenic radiofrequency electromagnetic fields as an emerging threat to wildlife orientation Sci.Total Environ. 518–519, 58–60	IF=4.099	Y	This short paper reviewed adverse effect of RF EMF on wildlife orientation. According to this, radio frequency noise interferes with the primary process of magnetoreception, and existing guidelines do not adequately protect wildlife. No attention is however paid to studies with negative findings
Redlarski G, Lewczuk B, Gak A, Koncicki A, Krawczuk M, Piechocki J, Jakubiuk K, Tojza P, Jaworski J, Ambroziak D, et al. (2015) Country: Poland The influence of electromagnetic pollution on living organisms: Historical trends and forecasting changes BioMed. Res. Int. 2015:234098	IF=2.706	N/?	Polish researchers conducted a review of the literature on effects of EMF exposure on living organisms, giving particular focus to 'lower' organisms (e.g., bacteria, *E. coli*, nematodes, *C. elegans*, land snails, fruit flies, etc.). They concluded that despite the extensiveness of the available research results, no clear answers have been given yet to the question of whether electromagnetic pollution has a bad influence on living organisms

(continued)

Table 5.1 (continued)

Author(s)	Impact factor[a]	Adverse effects Yes/No	Short summary
Ministry of Health (2015) Interagency Committee on the Health Effects of Non-ionising Fields: Report to Ministers 2015. Wellington: Ministry of Health Country: New Zealand http://www.health.govt.nz/system/files/documents/publications/interagency-committee-on-health-effects-on-nonionising-fields-may15.pdf	No impact factor	N	New Zealand's Interagency Committee on Non-Ionizing Fields has published a 56-page report that "highlights key findings from comprehensive reviews under-taken in recent years by national and international health and scientific bodies. The conclusion was that, despite much new research results since their latest report in 2004 and since the IARC classification in 2011, none of this causes the Committee to consider that current policies and recommendations should be reviewed. It is interesting to note from this new report that almost all of the epidemiological data that went into the IARC review were based on GSM (2G) or older-generation cell phones, which typically operate at powers 50–100 times greater than 3 or 4G phones, and so produce exposures to the head that are correspondingly higher. The committee also commented on the Bioinitiative report (http://www.bioinitiative.org/) and estimated that this report has weaknesses that undermine its credibility and conclusions and does not place any weight on its findings or recommendations. Finally, the committee points to several reviews and meta-analyses published since the IARC assessment which consider that more recent research weighs against there being a cause and effect relationship
Swedish Radiation Safety Authority (Strålsäkerhetsmyndigheten, SSM) (2015) Country: Sweden Tenth report from SSM's Scientific Council on Electromagnetic Fields http://www.stralsakerhetsmyndigheten.se/Global/Publikationer/Rapport/Stralskydd/2015/SSM-Rapport-2015-19.pdf	No impact factor	N	This new report includes publications from October 2013 through September 2014. It covers static, low frequency (LF), intermediate frequency (IF), and RF electro-magnetic fields. As in the past, the Council examined peer-reviewed, English-language publications that were released since the previous report. Studies were weighted based on a quality assessment, and some studies were excluded due to irrelevance or poor quality. With regard to mobile phones and tumours of the head region, the Council did not find evidence to change their previous conclusion that evidence is not convincing in linking mobile phone use to brain tumours. However, they note, uncertainty remains for use longer than 15 years and consider that it is also too early to draw firm conclusions regarding risk of brain tumours in children and adolescents.

(continued)

Table 5.1 (continued)

Author(s)	Impact factor[a]	Adverse effects Yes/No	Short summary
			However, based on the available literature to date they do not see any indication of an increased risk. The observed effects on mobile phone exposure and sleep EEG in human volunteer studies are also considered weak and no behavioural or other health effects are expected from the data. They also consider that studies so far have not provided evidence that exposure to electromagnetic fields is a causal factor in supposed electromagnetic hypersensitivity. They note that several new studies have found an association between mobile phone use and symptoms in adolescents, but that the cause is unclear and may not be related to RF EMF. They also see no reason to believe that there might be any health risks for the general public related to exposure from radiofrequency electromagnetic fields from far-field sources, such as base stations and radio and TV transmitters, or wireless local data networks in schools or at home
ZonMw study report (2015) Country: The Netherlands http://www.zonmw.nl/en/programmes/electromagnetic-fields-and-health-research/evaluation/	No impact factor	N	A large research project was conducted in the Netherlands (2008–2014) as a result of increasing societal concern about the potential adverse health effects of mobile communication and other EMF-emitting technologies. A total of 60 research projects in four thematic areas were funded by the project: (1) *In vitro*/*in vivo* and molecular research; (2) Epidemiology; (3) Sociology (role of risk perception and risk communication); (4) Technology (models and monitoring equipment's). The studies were on many different aspects of EMF research including static fields, ELF-magnetic fields and the mobile phone technology. A large part of the investigations was actually devoted to mobile phone radiation. The majority of these studies have found no significant effect on any of the parameters studied (*in vitro*/*in vivo*/epidemiology). This does not make EMF an attractive field for a research career. Consequently many of the investigators decided not to continue research on EMF effects. As one principal investigator stated: "*if you do not find any positive effects, you essentially have two choices: either continue looking to*

(continued)

Table 5.1 (continued)

Author(s)	Impact factor[a]	Adverse effects Yes/No	Short summary
			see if you can still find anything, anywhere, or you draw the conclusion there is nothing there. I fall into that second category: I'm done with it." A study on risk communication in relation to EMF has found that perceived risk from exposure to EMF is associated with lack of knowledge about EMF and a lack of trust in the role of the government. These findings may be of use in development of better communication protocols
Eghlidospour M, Mortazavi S M J, Yousefi F, Mortazavi SAR. (2015) Country: Iran New horizons in enhancing the proliferation and differentiation of neural stem cells using stimulatory effects of the short time exposure to radiofrequency radiation J Biomed Phys Eng. 5, 95–104	No impact factor	Y/N	Here studies were reviewed on the potential effects of RF EMF on stem cells and their proliferation and differentiation. The authors concluded from their review that the stimulatory or inhibitory effects of RF radiation on the proliferation and differentiation of stem cells depend on various factors such as the biological systems, experiment conditions, the frequency and intensity of RF and the duration of exposure
Dasdag S, Akdag MZ (2015) Country: Turkey The link between radiofrequencies emitted from wireless technologies and oxidative stress J Chem Neuroanat. 75, 85–93	IF=1.500	Y/?	This is a review of *in vitro* and *in vivo* studies of RF EMF on oxidative stress. An overwhelming number of the studies reviewed here indicated that mobile phones or similar wireless equipment increase the oxidative stress. The authors yet point to the fact that there is a need of human studies where maximal attention should be paid to methodology, exposure parameters and apparatus. Many studies so far are insufficiently precise and therefore they cannot be very convincing
Pall ML (2015) Country: USA Microwave frequency electromagnetic fields (EMFs) produce widespread neuropsychiatric effects including depression J Chem Neuroanat.75, 43–51	IF=1.500	Y	This literature review provided evidence that various non-thermal microwave exposures produce diverse neuropsychiatric effects

[a]Impact factor as from 2014 Thomson Reuters, Journal Citation Reports

Table 5.2 An overview of in vitro and in vivo (animal) studies on biological effects from radiofrequency fields

	Country	Organism/cells	Author & Journal	IF[a]	Exposure	Effect	Conclusion from author	Remarks
Bacteria								
Genome changes								
	Iraq	*E. Coli* & *S. aureus*	Latif et al. (2013) Iraqi J. Biotech., 12, 63–74	No impact factor	900 MHz/cells between two mobile phones. Exposure 15–45 min	Y	Genomic changes found	Questionable procedures
Antibiotic effect	Nigeria	Different bacteria in soil	Adebayo et al. (2014) J. Appl. Sci. Environ. Manag., 18, 669–674	No impact factor	Base station antenna. Distance from antenna at 0–100 ft	Y	Effect on bacterial selection possibly contributing to antibiotic resistance	100 ft. =~30 m; exposure should be quasi nihil (well below radiation beam). Very questionable!
	Iran	Klebsiella pneumonia	Taheri et al. (2015) J. Biomed. Phys. Eng., 5, 115–120	No impact factor	Exposure from Wifi router (2400 MHz); 3, 4.5 and 8 h	Y	Non-linear response and increased sensitivity to antibiotics after 4.5 h of exposure. This fell after 8 h	Questionable procedures
Fungi, Yeast								
	Iran	*Candida alba* (yeast)	Shahin-Jafari et al. (2015) Saudi J. Biol. Sci., 23, 426–433	IF = 1.257	960–851 MHz. Maximum output power of 900 mW; different exposure times of 10, 70, 210, 350 and 490 h	N	No effects on alpha-INT gene sequence	Exposure set-up insufficiently described

Non mammalian cells								
Immunology	India	Chicken blood cells	Das et al. (2014) Adv. Anim. Vet. Sci., 2, 31–36	No impact factor	850–1200 MHz Exposure from a standard mobile phone (SAR 1.7 W/Kg) for 5 , 15 , 25, 40 and 60 min	Y	Increased cellular stress	Mobile phone in direct contact with the plates with cells
Oxidative damage, fertility and other parameters	Greece	Nematoda (*C. elegans*)	Fasseas et al. (2015) Int. J. Radiat. Biol., 91, 286–293	IF = 1.687	Different E-field levels below ICNIRP guidelines. Exposure for 30 min, 1 h, or 3 h (mobile phone), or 30 min, 1 h, 3 h, 6 h, or 24 h (DECT base or Wi-Fi router)	N	No effect on all investigated parameters (changes in growth, fertility, lifespan, chemotaxis, short-term memory, ROS (Reactive Oxygen Species) production and apoptosis)	
Mammalian Cells								
Immunology	Italy	PC12 rat tumor cells	Valbonesi et al. (2014) Int. J. Radiat. Biol., 90, 382–391	IF = 1.687	GSM 1800 MHz – 217 Hz. Exposure for 4, 16 or 24 h to 1.8 GHz continuous wave signal or to two different GSM modulation schemes, GSM-217 Hz and GSM-Talk (SAR was 2 W/kg)	Y/N	Effect on HSP70 expression possible and possible role of signal modulation	

(continued)

Table 5.2 (continued)

	Country	Organism/cells	Author & Journal	IF[a]	Exposure	Effect	Conclusion from author	Remarks
	China	N9 microglia cells	He et al. (2014) J. Neuroinflamm., 11, 49	IF = 5.408	2450 MHz; exposure for 20 min to pulsed microwaves at an average specific absorption rate. of 6 W/kg	Y	Changes in inflammatory response and phagocytosis. Curcumin is protective	OK but high (thermal) exposure levels
	Korea	Adipose derived stem cells	Lee et al. (2014) J. Craniofac. Surg., 25, 1902–1907	IF = 0.678	Exposure via a smartphone connected to a Wi-Fi network, which was uploading files at a speed of 4.8 Mbps for 10 h a day, for a total of 5 days	N	No effects on apoptosis, cell proliferation etc.	
	China	Embryonic neural stem cells	Chen et al. (2014) Sci. Reports, 4: 5103	IF = 5.578	1800 MHz; 1800 MHz with intermittent exposure (5 min field on, 10 min field off) in a GSM Talk-signal mode. Average SAR was 1, 2, or 4 W/kg	Y/N	No influence on apoptosis, proliferation, cell cycle or mRNA expressions of related genes. Also no alteration of the ratio of eNSC differentiated neurons and astrocytes. However, neurite outgrowth of eNSC differentiated neurons was inhibited after 4 W/kg RF-EMF exposure for 3 days	Effects only at high exposure level (4 W/kg ≅ thermal effect?)

	Russia	RAW264.7 mouse macrophages	Glushkova et al. (2015) Int. J. Radiat. Biol., 91, 321–328	IF = 1.687	Sweep-type generator providing swinging frequencies in the range 8.15–18 GHz at an output power of 2 mW for 1 h	Y	Effects on cell signalling and cytokine production at low intensities	OK but might be mild stress without consequences. No mobile phone frequencies
ROS – oxidative damage	Croatia	V79 cells	Marjanovic et al. (2015) Electromagn. Biol. Med., 34, 381–386	IF = 1.194	1800 MHz; exposure during 10, 30, and 60 min at a SAR of 1.6 W/kg	Y/N	ROS levels were altered after 10 min but not after 30 min indicating activation of an antioxidant defence mechanism. Other endpoints (metabolic activity and total protein content) did not change	Mild effect which can be related to an adaptive response
	China	GC-2 mouse spermatocyte cells	Liu et al. (2014) Toxicol. Lett., 228, 216–224	IF = 3.262	1800 MHz signals in GSM-Talk mode at a SAR of 1 w/kg, 2 w/kg or 4 w/kg for 24 h	Y	Effect on intracellular ROS-levels at 4 W/kg but no effect on apoptosis	Mild effect which can be related to an adaptive response. Effect at 4 W/kg ≅ thermal effect (?)
BBB (in vitro)	China	ECV304 cell line and primary rat cerebral astrocytes	Wang et al. (2015) Mol. Neurobiol., 52, 478–491	IF = 5.137	2856 MHz; 50 mW/cm^2, 5 min	Y	Effect on BBB via activation of vascular endothelial growth factor	

(continued)

Table 5.2 (continued)

	Country	Organism/cells	Author & Journal	IF[a]	Exposure	Effect	Conclusion from author	Remarks
Genetic effects & gene expression	China	Bone marrow mesenchymal stem cells (from C57BL/6 mice)	Wang et al. (2015) PLoS ONE 10(2): e0117550. doi:10.1371/journal. pone.0117550	IF = 3.234	2856 MHz (4 W/kg); 50 pulses per second, with a pulse width of 500 ns. Peak field power densities were set to 200 W/cm^2	Y/N	No effect on apoptosis, cell differentiation and viability but influence on mRNA expression levels of osteopontin and osteocalcin	
	Japan	W138VA13 2RA human fibroblasts	Mizuno et al. (2015) Int. J. Environ. Res. Publ. Hlth., 12, 3853–3863	IF = 2.063	Wireless power transfer (WPT) performed using a helical coil-based exposure system designed to transfer power with 85.4% efficiency at a 12.5-MHz resonant frequency	N	No effects on cell growth, DNA breaks and gene mutations	OK, but no mobile phone frequency
	China	Sprague Dawley rat neurons	Zuo et al. (2015) J. Neuroinflamm., 12:105	IF = 5.408	1800 MHz; 24 h intermittent exposure at an absorption rate of 2 and 4 W/kg	Y	RF-radiation could not directly induce DNA damage. However, the 4 W/kg combined with LPS (40 μg/ml) showed some effects (e.g., karyopyknosis, presence of lysosomes and autophagosome, and increasing expression of LC3-II and Beclin1). The ROS values significantly increased in the 4 W/kg exposure and 4 W/kg combined with LPS	OK but Effects at 4 W/kg ≅ thermal effect level

	Australia	Long bones of Sprague Dawley rats	Kumar et al. (2015) Int. J. Radiat. Biol., 91, 664–672	IF = 1.687	900, 1800 MHz; SAR levels of 2 and 10 W/kg or 2.5 and 12.4 W/kg resp	N	No genotoxicity in lymphoblasts and no change in cell proliferation	Moderate to high exposure levels
	Russia	Mouse whole blood cells	Gapeyev & Lukyanova (2015) Biofizika, 60, 889–897	No impact factor	42.2 GHz, 0.1 mW/cm^2, 20-min exposure, modulation frequencies of 1 and 16 Hz	Y	Protective effect of pulse-modulated electromagnetic radiation (not continuous) against DNA damage induced by X-rays, MMC and H_2O_2. Interpreted as adaptive response due to induction of nanomolecular concentrations of reactive oxygen species	In Russian, only abstract available
Neurological effects	Germany	Mouse retinal cells	Ahlers & Ammermüller (2014) Bioelectromagnetics, 35, 16–29	IF = 1.705	900, 1800, 2200 MHz; SAR = 0 (sham), 0.02, 0.2, 2, and 20 W/kg	N	No acute effects on retinal ganglion cell responses under constant temperature conditions	
	China	Hippocampal neurons of rats and in PC12 cells	Wang et al. (2015) Pathobiol., 82, 181–194	IF = 2.48	2856 MHz, 30 mW/cm^2 for 5 min	Y	Microwave radiation may induce neuronal injury via changes in the molecular organization of postsynaptic density and modulation of the biochemical cascade that potentiates synaptic transmission	

(continued)

Table 5.2 (continued)

	Country	Organism/cells	Author & Journal	IF[a]	Exposure	Effect	Conclusion from author	Remarks
	Italy	Rat PC12 cells	Valbonesi et al. (2015) Int J. Radiat Biol., 92, 1–10	IF = 1.687	24 h to 1.8 GHz 217-GSM signal. SAR = 2 W/kg		Effects on AChE activity in vitro	
HUMAN CELLS	Ukraine	Human sperm cells	Gorpinchenko et al. (2014) Central Eur. J. Urol., 67, 65–71	No impact factor	900–1800 MHz; mobile phone in combined standby/talk mode placed at 5 cm from the sperm samples	Y	DNA fragmentation, lower mobility etc.	Highly questionable experiment and unlikely conclusions
Adaptive response	Italy	Human lymphocytes	Sannino et al. (2014) J. Rad. Res., 55, 210–217	IF = 1.797	1950 MHz; 20 h exposure to 1950 MHz at an average specific absorption rate of 0.3 W/kg. Then exposure to 1.0 or 1.5 Gy X-irradiation	**Y**	Adaptive response with respect to X-rays (study of micronucleus induction)	
Oxidative damage, apoptosis, Ca, etc.	Korea	Mouse/rat/human neuronal cells	Kang et al. (2014) J. Rad. Res., 55, 265–276	IF = 1.797	837 & 1950 MHz; combined (CDMA at 2 W/kg plus WCDMA at 2 W/kg) signals for 2 h	N	No effect on intracellular ROS levels. Also no influence on effects of the ROS-inducers menadione and hydrogen peroxide	

	China	Human red blood cells	Yong et al. (2013) Engineering, 5, 450–454	No impact factor	900 MHz; maximum output power approximately 10 mW/cm^2	Y	Effects on cell morphology etc. No effects of ultra-low density (15 μW/cm^2, 31 μW/cm^2) electromagnetic wave radiation at radiation time less than 30 min. Significant reactions with power densities of 1 mW/cm^2 and 5 mW/cm^2	No conclusions with respect to human health
	Turkey	Human embryonic kidney cells (HEK293T cells)	Sefidbakht et al. (2014) Photochem. Photobiol. Sci., 13, 1082–1092	IF = 2.267	940 MHz; SAR of 0.09 W/kg for 15, 30, 45, 60, and 90 min	**Y**	Mild and rather protective effects = adaptive response	Interesting but no conclusion so far
	Turkey	MCF-7 breast cancer cells	Çig et al. (2015) Biochim. Biophys. Acta -Biomembranes, 1848, 2756–2765	IF = 3.836	1 h exposures to 900, 1800, 2450 MHz; power densities of 1.2, 12 or 120 μW/cm^2	Y	Effects on all investigated parameters (calcium signalling, cytosolic ROS production, cell viability, apoptosis, plus caspase-3 and -9 values) but only close to RF-source	Conclusions made can be questioned
	Iran	Peripheral blood samples	Kazemi et al. (2015) J. Biomed. Phys. Eng., 5(3)	No impact factor	2 h exposure to 900 MHz; 2 W power output	Y	Increased ROS production in cell populations rich in monocytes but not in lymphocytes	Authors concluded that the observed effects can be beneficial

(continued)

Table 5.2 (continued)

	Country	Organism/cells	Author & Journal	IF[a]	Exposure	Effect	Conclusion from author	Remarks
Neurological effects	China	eNCSC (embryonic neural stem cells)	Chen et al. (2014) Sci. Rep., 4, 5103	IF = 5.578	1800 MHz; SAR = 1, 2 and 4 W/kg; for 1, 2, and 3 days	Y	No influence on eNSC apoptosis, proliferation, cell cycle or mRNA expressions of related genes. Also no alteration of the ratio of eNSC differentiated neurons and astrocytes. However, effects on neurite outgrowth of eNSCs and mRNAand protein expression at 4 W/kg	Effects found only at threshold level for thermal effects
	China	Glioblastoma cell lines U251-MG and U87-MG)	Liu et al. (2015) BMC Public Hlth., 15, 764	IF = 2.264	12, 24, and 48 h. 1950-MHz-exposure; maximum SAR = 5.0 W/kg)	N	No significant cellular effects (apoptosis, cell growth, gene expression etc.) or tumorigenesis	
Apoptosis and synergy	Turkey	Burkitt lymphoma cells	Canseven et al. (2015) Electromagn. Biol. Med., 34, 322–326	IF = 1.194	24 h exposure to 1800 MHz electromagnetic fields; SAR level of 0.350 W/kg. With and without pre-treatment by 5 μM gemcitabine 24 before exposure	Y	Increased apoptosis and decreased viability. Synergy with gemcitabine	

Chemotaxis and phagocytosis	Japan	HL-60 cells	Koyama et al. (2015) J. Radiat. Res., 56, 30–36	IF = 1.797	2450 MHz; 2 and 10 W/kg for 4 and 24 h	N	No effects on apoptosis and chemotaxis	
Hsp70 chaperone system	Switserland	E. coli system	Beyer et al. (2014) Bioelectromagnetics, 35, 470–478	IF = 1.705	Exposure to sinusoidal signals of 100, 1000 and 1900 MHz at various field strengths up to 5.0 kV/m and to GSM signals at 0.3 kV/m in the protein solution	N	No effect on GrpE system	
Enzymes	Italy	LPO and HRP enzymes	Barteri et al. (2015) Electromagn. Biol. Med., 21, 1–7	IF = 1.194	900 MHz; (0.04 mW/kg; 30 min)	Y	Conformational changes and effect on free radical formation and stability	
Hemoglobine	Italy	Human hemoglobin	Calabrò et al. (2015) Spectroscop. Lett., 48, 74–747	IF = 0.852	900 MHz, 4 h exposure at an average H-field intensity of 42 mA/m	Y	Structural changes in hemoglobine in water	Significance in terms of human health unclear
Plants								
	India	Pea and Fenugreek seeds	Sharma and Parihar (2014) Curr. World Environ., 9(1)	No impact factor	850–1850 MHz and 900-1900 MHz; exposure to the emissions of a Nokia 2690 phone and a Samsung GT B7722 phone for 0.5, 1, 2, 4, or 8 h	Y	Changes in morphology and biochemistry of the plants were observed	Not convincing due to many shortcomings

(continued)

Table 5.2 (continued)

	Country	Organism/cells	Author & Journal	IF[a]	Exposure	Effect	Conclusion from author	Remarks
	Japan	Spinach	Miyasaka et al. (2014) Environ. Control Biol., 52, 29–36	No impact factor	Exposure to 2450 MHz fields for 1, 3, and 5 weeks. The microwave output was high (around 20 W output). Pots were exposed at levels between 0.16 and 8.93 mW/cm^2	Y	Effects after 5 weeks exposure with regard to dry matter, fresh matter of above-ground part, and the third leaves area	Not directly related to mobile phone usage. High exposure levels
	Soudan	Green mint	Kafi et al. (2014) Adv. Electromagn., 3(1)	No impact factor	Pots 10, 20, 30, and 50 m from a rooftop antenna (800–1000 MHz). Power densities of 10.5, 8.2, 5.3, and 0.5 μW/cm^2 respectively	Y	Stressing effects on the plants (growth etc.) in all cases	Not directly related to mobile phones; may be mild stress without consequence
	Japan	Myriophyllum aquaticus	Senavirathna et al. (2014) Plant Signal Behav., 9, e28590	No impact factor	2000 MHz; maximum of 1.42 W/m^2 for 1 h	Y	Effects on growth dynamics	
	Iran	Maize	Zare et al. (2014) Int. J. Farming Alli. Sci., 3, 365–367	No impact factor	940 MHz; 5 h/day up to 25 days	Y	Effects on growth	Insufficient info on experimental set-up. Questionable with respect to relevance to humans. Paper not well written

	Romania	Aromatic plants	Soran et al. (2014) J. Plant Physiol., 171, 1436–1443	IF = 2.557	860–910 MHz (GSM; 100 W/m^2) and 2412–2480 MHz (WLAN; 70 mW/m^2).	Y	Physiological changes; thinner cell walls, smaller chloroplasts and mitochondria, and enhanced emissions of volatile compounds. Essential oil content increased for GSM but decreased for WLAN exposures	Significance unclear
	Australia	Soybean	Halgamuge et al. (2015) Bioelectromagnetics, 36, 87–95	IF = 1.705	2 h exposure to GSM mobile phone pulsed radiation or continuous wave (CW) radiation at 900 MHz with amplitudes of 5.7 and 41 V/m	Y	Effects on growth depending on radiation intensity and amplitude modulation	Significance for humans unclear
	India	Seeds of Zea mays	Kumar et al. (2015) Protoplasma 253, 1043–1049	IF = 2.651	1800 MHz; SAR of 1.69 for ½, 1, 2, and 4 h	Y	Significant growth and biochemical alterations at 4 h exposure	
Insects								
Behaviour	Belgium	Ants	Cammaerts et al. (2014) J. Insect Behav., 27, 514–526	IF = 1.143	940 MHz at 1.5 V/m, during 10 min	Y	Behavioural effects (locomotion, etc.)	OK but no link with human health
	India	Male and female Wistar albino rats	Kameshwaran et al. (2014) Acta Biomed. Sci., 1, 40–44	No impact factor	900 & 1800 MHz mobile phone in silent mode, 5 min calling/5 min interval for 1 h/day for 2 months	Y	Behavioural effects. Tecoma stans protects	Insufficient info on experimental set-up

(continued)

Table 5.2 (continued)

	Country	Organism/cells	Author & Journal	IF[a]	Exposure	Effect	Conclusion from author	Remarks
	Lebanon	Honeybees	Halabi et al. (2014) 17th IEEE Mediterranean Electrotechnical Conference (MELECON), 408–14	No impact factor	900, 1800, 2100 MHz; in the main lobe of the antenna at a distance of 100 m (4.091 mW/m^2), and another at a distance of 500 m (0.1636 mW/m^2); and, in the back lobe of the antenna at a distance of 100 m. 1 month observation	Y	Effects on viability, egg production, etc. at 100 m (not 500 m)	Questionable, but insufficient info available
	Russia	Drosophila melanogaster	Fedorov, Weisman (2015) Biofizika, 60, 1009–1017	No impact factor	0.1–2.2 THz; 30 min exposure of stressed flies	Y	No effect of individual lifespan of F1 progeny but sexual dimorphism was detected and survival curves of males but not females (from mature and immature oocytes at the time of irradiation) were different from that of controls	Article in Russian. Only abstract available. No details available

	India	Bees	Dalio (2015) J. Res. Agricult. Animal Sci., 2, 6–10	No impact factor	Two cell phones were kept in working condition (call mode) in each test colony for 15 min for 20 days at the time when foraging activity of bees was at peak (12:30 pm)	Y	Increased aggression, decreased number of bees, etc.	Too many shortcomings to be taken seriously
Apoptosis	Greece	Drosophila melanogaster	Sagioglou et al. (2015) Electromagn. Biol. Med., 21, 1–14	IF = 1.194	Continuous Wave (CW) or Frequency Modulation (FM) emission modes, at power output values of 10 dB, 0, −10 dB; under carrier frequencies 100, 395, 682, 900 MHz. Newly emerged flies were exposed either acutely (6 min or 60 min on the 6th day), or repeatedly (6 min or 60 min daily for the first 6 days of their life)	Y	Effect on mean ovicity in 2nd generation with possible radiofrequency sensitivity difference between males and females. All exposure protocols resulted in an increase of apoptotic cell death (ACD) observed in egg chambers, even at very low electric field strengths. FM waves seem to have a stronger effect in ACD than continuous waves	Near field exposure (does not correspond to exposure from base station antenna)
	Greece	Drosophila melanogaster	Geronikolou et al. (2014) Plos ONE, 9(11): e112139	IF = 3.234	900 MHz or 1880 MHz for 20 min/day for 5 days. Field characteristics ‘as in typical use’	Y	Possible radiofrequency sensitivity difference in insects which may be due to the distance from the antenna or to unexplored intimate factors	

(continued)

Table 5.2 (continued)

	Country	Organism/cells	Author & Journal	IF[a]	Exposure	Effect	Conclusion from author	Remarks
Birds								
Reproduction/ teratology	Turkey	Chicken embryo	Umur et al. (2013) Turk Neurosurg., 23, 742–752	IF = 0.576	Eggs incubated for 30, 48, 72 h. and exposed to RF-fields by placing them at a distance <25 cm from a mobile phone (SAR Value: 0.77 W/kg)	Y	Developmental delay in chick embryos in early period	Highly questionable
	Poland	Chicken embryo	Pawlak et al. (2014) Int. J. Occup. Med. Environ. Hlth., 27, 114–122	IF = 1.365	1800 MHz; chick eggs exposed ten times over a 24 h period at power density of 0.090–0.110 W/m^2 for 4 min	Y	Effects on enzyme levels and embryogenesis process	
	India	Chicken embryo	D'Silva et al. (2014) Adv. Anatomy, 798425, 9 pages	No impact factor	900, 1800 MHz (SAR = 2.0 W/kg). For exposure, a cell phone was rung from another cell phone for a duration of 3 min each, every half an hour, with the first exposure delivered at 12th hour of incubation	Y	Structural changes in lens epithelial cells and other lens defects and DNA damage	Many shortcomings

Magnetoreception	Germany	Domestic chickens and European robins	Wiltschko et al. (2014) J. R. Soc. Interface, 12, 20141103	IF = 3.917	7.0 MHz field of 480 nT in autumn, in the following spring a 1.315 MHz field of 15 nT was used	Y	Perturbation of magneto-reception but not permanent	
Bird nest count	West Bengal	*Erithacus rubecula* (Turdidae)	Bhattachyara and Roy (2014) J. Innov. Res. Sci., 3, 32–36	No impact factor	Mobile phone base stations	Y	No bird nests close to antennas	Highly questionable
Development & mortality	India	Chick embryos	Dharmik (2015) Global J. Multidiscipl. Res., 4, 100–111	No impact factor	900–1900 MHz; mobile phone mounted on top of the incubator in which the eggs were housed. Exposure up to 21 days, in silent mode, rang up to four times for 15 min from another mobile phone with time interval of one hour. After 7 or 14 days, embryos were extracted	Y	Developmental effects, increased mortality, enhanced growth	Highly questionable. Standard egg incubator with metal parts everywhere
	China	Chick embryos	Ye et al. (2016) Anat. Histol. Embryol., 45, 197–208	IF = 0.672	900 MHz mobile phone; 3 h per day during incubation to 900-MHz EMF (SAR ~1.07 W/kg) intermittently during incubation	Y	Myocardial changes, DNA damage and mortality	Highly questionable

(continued)

Table 5.2 (continued)

	Country	Organism/cells	Author & Journal	IF[a]	Exposure	Effect	Conclusion from author	Remarks
Rat								
Circadian rythm & melatonin	China	Adult male Sprague-Dawley	Qin et al. (2014) Chronobiol. Int., 31, 123–133	IF = 3.343	1800 MHz; exposure during 32 days for 2 h/day at a power density of 205 μw/cm^2, 0.04 W/kg	Y	Alterations in the circadian rhythm of reproductive functional markers	
	China	Adult male Sprague-Dawley	Cao et al. (2015) Int. J. Environ. Res. Public Health, 12, 2071–2087	IF = 2.063	1800 MHz; SAR = 0.05653 W/kg [whole body]. Exposure for 2 h/day and for 32 consecutive days	Y	Effect on daily antioxidant levels and circadian rhythm	
	Korea	Adult male Sprague-Dawley (diabetic and non-diabetic)	Kim et al. (2015) Int. J. Rad. Biol., 91, 898–907	IF = 1.687	915 MHz; whole body SAR = 4 W/kg, 5 days/weeks, for 2 wks, with exposures occurring either during the daytime (diurnal) or night-time hours	Y	Nocturnal RFID exposure can cause reductions in the levels of urinary melatonin and 6-OHMS, possibly due to decreased melatonin biosynthesis via suppression of AANAT gene transcription in the rat pineal gland	
Genetic effects and oxidative damage	Turkey	Exfoliated bladder cells of non-diabetic and diabetic rats	Gurbuz et al. (2015) Endocr. Regul., 49, 126–130	No impact factor	2100 MHz; whole body average SAR = 0.24 W/kg with an ERMS field of 17.5 V/m in non-thermal levels	N	No induced micronuclei in the cells of the rats; also those with the chronic disorder	

	Brazil	Young rats	Furtado-Filho et al. (2015) Int. J. Radiat. Biol., 91, 891–897	IF = 1.687	950 MHz; 30 min/day for 27 days (21 days during the neonatal period and 6 days after birth)	Y/N	No evidence of genotoxicity or oxidative stress in the left- or right cerebral cortex. Some changes were found in protein and glucose levels	
	Turkey	Male Wistar albino rats	Tök et al. (2014) Indian J. Ophthalmol., 62, 12–15	IF = 0.900	2.45-GHz Wi-Fi signals (0.1 W/kg whole body SAR) for 1 h/day over 30 days or to Wi-Fi in the same manner but with addition of daily injections of melatonin	Y/N	Poor oxidative toxic effects. Melatonin has a protective effect	
	Iran	Adult male Sprague-Dawley rats	Jelodar et al. (2014) Zahedan J. Res. Med. Sci., 16, 46–50	No impact factor	900 MHz; 45 days treatment with vitamin C (200 mg/kg of body weight/day orally), exposure to a 900-MHz base transceiver station (BTS) antenna for 4 h/day, and vitamin C plus exposure	Y	Lower antioxidant enzyme activity and increased malondialdehyde levels. Vit C protects	Questionable experimental set up
	Turkey	Pregnant rats	Çetin et al. (2014) J. Matern. Fetal Neonatal Med., 27, 1915–1921	IF = 1.367	900 MHz and 1800 MHz; exposure for 60 min/d during pregnancy and neonatal development	Y	Mobile phones can induce oxidative damage in brain and liver of developing rats	

(continued)

Table 5.2 (continued)

	Country	Organism/cells	Author & Journal	IF[a]	Exposure	Effect	Conclusion from author	Remarks
	Iran	Male Sprague-Dawley rats	Akbari et al. (2014) Toxicol. Mech. Meth., 24, 347–352	IF = 1.517	900 MHz; 45 days treatment with vitamin C (200 mg/kg of body weight/day orally), exposure to a 900-MHz base transceiver station (BTS) antenna for 4 h/day, and vitamin C plus exposure	Y	Decreased antioxidant enzymes activity and increased MDA compared with the controls. VitC had a protective effect	Apparently the same work as the one reported in Jelodar et al. (2013) Zahedan J. Res. Med. Sci. 16, 46–50)
	Turkey	Wistar rats	Gürler et al. (2014) Int. J. Radiat. Biol., 90, 892–896	IF = 1.687	2450 MHz; 1 h/day over 30 days to 1.04-mW/cm^2	Y	2.45 GHz radiation increases the DNA damage in both brain tissues and plasma of the rats whereas it increases protein oxidation only in plasma. The use of garlic decreases these effects	
	Egypt	Male albino rats	Abd el Rahman et al. (2014) J. Am. Sci., 10, 170–176	No impact factor	950 MHz; exposure at a SAR of 1 W/kg for 2 h/day, three times per week, for 2 months	Y	Oxidative stress which is annihilated by Vit E or silymarine given during 2 months before exposure	

	Turkey	Wistar rats	Ghazizadeh et al. (2014) Metab. Brain Dis., 29, 787–799	IF = 2.638	2450 MHz; Wistar rats in whom epileptic symptoms had been induced by pentylentetrazol (PTZ) were exposed to the Wi-Fi signal for 1 h after induction of epilepsy, either with or without incubation with capsazepine (CZP)	Y	Increased Ca^{++} concentration. CZP protects	Difficult to evaluate due to incompletely provided information
	Morocco	Male and female Wistar rats	Tazrout et al. (2014) Ind. J. Res., 3, 1–7	No impact factor	930 MHz exposure, 1 or 2 h/day during 3 months. Cage average power density ~ 100 μW/cm^2 (19.4 V/m)	Y	Biochemical changes indicating oxidative stress in plasma and some organs. Females more affected than males	Apparently well-done experiments. Not sure that exposure system is appropriate but this should be carefully controlled by an experienced engineer
	Serbia	Male wistar rats	Djordjevic et al. (2015) Bratisl. Lek. Listy, 116, 96–100	IF = 0.439	900 MHz; 100–300 μT, 54–160 V/m) for 4 h/d for 20, 40, or 60 days	Y	Biochemical changes indicating oxidative stress. Melatonin has an antioxidant effect in the liver	
Biochemstry & Histopathology	Turkey	rats	Aydogan et al. (2015) Am. J. Otolaryngol., 36, 39–46	IF = 0.984	2100 MHz; simulating a third generation mobile phone for 6 h/day, 5 days/week, for 10 or 40 days	Y	Histopathological changes regarding salivary gland damage increased with exposure time	

(continued)

Table 5.2 (continued)

	Country	Organism/cells	Author & Journal	IF[a]	Exposure	Effect	Conclusion from author	Remarks
	Turkey	Male wistar rats	Cam et al. (2014) Cell Biochem. Biophys., 70, 643–649	IF = 1.680	900 MHz; 20 min/day over 3 weeks. The estimated SAR was 1.35 W/kg	Y	Increased hydroxyproline levels in skin	Relevance to humans can be questioned
	Slovak Republic	Male wistar rats	Holovska et al. (2015) J. Toxicol. Environ. Health A, 78, 353–356	IF = 2.351	Exposure to a 2.45 GHz, 2.8 mW/cm^2 electromagnetic field for 3 h/day for 3 weeks on the livers of mature Wistar rats	Y	Ultrastructural changes in the liver	
	Spain	Female Sprague-Dawley rats	Misa-Agustiño et al. (2015) Exp. Biol. Med., 240, 1123–1135	IF = 2.165	2450 MHz; Rats were individually exposed once or repeatedly (10 times in 2 weeks) for 30 min in a commercial chamber at different non-thermal specific absorption rates	Y	Changes in the rat thyroid morphology and expression of HSP-90	
	Turkey	Male Wistar Albino rats	Eris et al. (2015) Bratisl. Lek Listy, 116, 101–103	IF = 0.439	900 MHz; 45-min exposure (15.1 V/m field strength in the head region)	Y	Increased 5-HT serotonin levels with possible implications on learning behaviour and cognition	Subtle changes, speculative conclusions

	Poland	Wistar rats	Bodera et al. (2015) Int.J. Occup. Med. Environ. Hlth., 28, 751–759	IF = 1.365	1800 MHz for 15 min, five times (day after day). Exposure to the far-field range of an antenna (effective electric field 20 V/m and effective magnetic field value 0.05 A/m)	Y	Repeated EMF exposure applied together with the opioid drug tramadol slightly enhanced lipid peroxidation level in brain, blood, and kidneys	
	Turkey	Wistar albino female rats	Yüksel et al. (2015) Endocrine 53, 352–362	IF = 3.878	900, 1800 and 2450 MHz; 60 min/day and 5 days/week (total of 52 weeks -from four generations)	Y	Effects on oxidative stress and plasma hormone levels in pregnant rats and their offspring	
Radioprotection & genetic effects	Egypt	Male Sprague-Dawley rats	Mohamed et al. (2014) Comp. Clin. Pathol., 23, 1719–1726	No impact factor	A digital mobile phone at GSM 900-MHz frequency was used (SAR 1.28 W/kg). The mobile phones used in the study were in the standby position and called intermittently (six times a day for 1 h with 3-h intervals)	Y	Oxidative stress and plasma hormone levels in pregnant rats and their offspring	Many shortcomings and questions

(continued)

Table 5.2 (continued)

	Country	Organism/cells	Author & Journal	IF[a]	Exposure	Effect	Conclusion from author	Remarks
	Turkey	Pregnant rats	Odaci et al. (2015) Biotech. Histochem., 90, 93–101	IF = 1.444	900 MHz EMF, 1 h/day, on days 13–21 of pregnancy. Exposure to 13.64 V/m (0.49 W/m^2); SAR = 0.024 W/kg	Y	Pathological changes in kidney attributed to oxidative stress	
Reproduction/ teratology	Iran	Male Sprague–Dawley rats	Sepehrimanesh et al. (2014) Int. J. Biometeorol., 58, 1657–1663	IF = 3.246	900 MHz continuous radiation (86 mW/cm^2, SAR estimated at 0.19–1.22 W/kg) via a dipole antenna for 30 days and 1, 2 or 4 h per day	Y	Perturbation of reproductive hormone levels	OK but yet questions on experimental set-up
	Croatia	Adult male Wistar rats	Trosic et al. (2013) Arh. Hig. Rada Toksikol., 64, 513–519	IF = 0.932	915 MHz RF field, power density of 2.4 W/m^2 and strength of 30 V m-1. Exposure 1 h per day during 2 weeks. SAR was calculated as 0.6 W kg^{-1}	N	No effects on testicular functions	

	Iran	Adult male Wistar rats	Ghanbari et al. (2013) Int. J. Fertil. Steril., 7, 21–28	No impact factor	915-MHz GSM phone signal at an average power density of 1.6 mW/cm^2 for 14 days or 21 days. Other rats were exposed to base station type RF EMF at 950 MHz for 14 days	Y	Viability, motility, and total antioxidant capacity in all exposure groups decreased significantly compared to the control group. More effect at longer exposure times	
	India	6 weeks old male Swiss albino rats	Kumar & Shukla (2014) J. Intern. Acad. Res. Multidiscipl., 2, 518–526	No impact factor	CDMA mobile phone; 6 h (3+3)/day for 5 months	Y	Effects on sperm morphology and others	Exposure was to mobile phone radiation but no details were given. It is not possible to estimate the validity of this study
	Iran	Pregnant Wistar rats	Razavinasab et al. 2016. Toxicol Ind Health 32, 968–979	IF = 1.859	900 MHz; 6 h/day from the start of pregnancy until birth	Y	Effects in offspring; i.e. adverse effects on cognition	
	Slovak Rep.	Adult male Wistar rats	Almasiova et al. (2014) J. Toxicol. Environ. Hlth., A, 77, 747–750	IF = 2.351	2450 MHz; mean power density 2.8 mW/cm^2. Exposure 3-h daily for 3 weeks	Y	Degenerative changes in testes	
	Turkey	Rats	Ulubay et al. (2015) Int. J. Radiat. Biol., 91, 35–41	IF = 1.687	900 MHz; power flux density 1 mW/cm^2; SAR 2 W/kg	Y	Effects on the kidney of offspring's and protective effect from melatonin and mega-3	Shortcomings in the study

(continued)

Table 5.2 (continued)

	Country	Organism/cells	Author & Journal	IF[a]	Exposure	Effect	Conclusion from author	Remarks
	Turkey	Rats	Türedi et al. (2015) Electromagn. Biol. Med., 34, 390–397	IF = 1.194	Pregnant experimental group rats were exposed to 900 MHz, 13.77 V/m (0.50 W/m^2); 1 h/d, on days 13–21 of pregnancy	Y	Oxidative stress and histopathological changes in heart tissue of male rat pups	Shortcomings in the study
	Iran	Male Sprague–Dawley rats	Sepehrimanesh et al. (2014) Electrophoresis, 35, 3331–3338	IF = 3.028	900 MHz radiation for 1, 2, or 4 h/day for 30 consecutive days, simulating a range of possible human cell phone use	Y	Non-thermal stress in testes (Significant effects on proteome, particularly in protein species in the rodent testicular tissue)	
	India	Male Wistar rats	Kumar et al. (2014) Ind. J. Exp. Biol., 52, 890–897	IF = 0.835	1910.5 MHz; SAR = 0.28 and 0.0226 W/kg. Exposure during 60 days (2 h/day, 6d/wk)	Y	Significant decrease in spermatozoa, DNA damage and other effects	Methodology may be questioned
	Turkey	Male rats	Odaci et al. (2015) Int. J. Radiat. Biol., 91, 547–554	IF = 1.687	900 MHz; 1 hour/day for 30 days. Electric field strength: 10 V/m; mean value power flux density: 0.265 W/m^2. Average SAR: 0.025 W/kg	Y	Biochemical and histological changes in the testes	

	Turkey	Sprague Dawley rats	Topal et al. (2015) Turk J. Med. Sci., 45, 291–297	IF = 0.497	900 MHz; 1 h daily during days 13–21 of pregnancy (0.54 W/m^2). Whole-body SAR was calculated as 0.027 W/kg	Y	Oxidative stress and pathological alterations in the liver in the postnatal period	Experimental set-up unclear
	Turkey	Male Albino rats	Karaman et al. 2014 Arch. Ital. Urol. Androl., 86, 274–277	No impact factor	Cell phones were suspended 1 cm above the cover of the animal's cages. Phones with SAR = (1.52 W/kg were used. They remained on talk mode for 8 h and standby mode for 16 h	Y	Histological changes in testes	Very questionable experimental set-up and results
	Turkey	Pregnant Sprague-Dawley rats	Bedir et al. (2015) Ren. Fail., 37, 305–309	IF = 0.944	900 MHz; exposure during the first 20 days of pregnancy	Y	Mild tubular effects in kidney of prenatally exposed animals	Experimental set-up questionable
	Turkey	Pregnant Sprague-Dawley rats	Hanci et al. (1015) Biotech. Histochem., 90, 535–543	IF = 1.444	900 MHz; 1 h application on the spleen and thymus of 21-day-old male rat pups 13.77 V/m; Whole body SAR calculated as 0.025 W/kg	Y	Pathological and enzymatic changes in spleen and thymus	

(continued)

Table 5.2 (continued)

	Country	Organism/cells	Author & Journal	IF[a]	Exposure	Effect	Conclusion from author	Remarks
	Turkey	Wistar-Hannover albino rats	Bakacak et al. (2015) Kaohsiung J. Med. Sci., 31, 287–292	IF = 0.803	900 MHz; exposure of the abdominal region for 15 min/d for 15 days. SAR varied from 0.018 W/kg to 4 W/kg for the entire body	Y	Significant decrease in the number of ovarian follicles in rats exposed to an EMF	
	Egypt	Pregnant Sprague Dawley Albino rats	Ismail et al. (2015) Biochem. Physiol., 4:2	No impact factor	900, 1800, 2100 MHz; Animals were exposed to 15 missed calls per day during the pregnancy period (21 day). Therefore a mobile phone was placed at 15 cm distance of the cage of rats. New-born rats were also subjected to the same condition of exposition to EMF. Brain samples of newborn rats were taken on 1, 7, 14, 21 and 28 days after delivery	Y	Significant lower serotonin and melatonin levels in the brain	Serious questions related to experimental set up

	Turkey	Pregnant Sprague Dawley rats	Sehitoglu et al. (2015) Arch. Esp. Urol., 68, 62–68	IF = 0.309	Pregnant rats exposed to a cell phone set to talking mode for 24 h a day for 20 days	Y	Reductions in serum total testosterone levels and in the size and weight of the testes as well as a modest increase in apoptosis compared to controls	
	China	Male Sprague-Dawley rats	Liu et al. (2015) Reproduct. Hlth., 12:65	IF = 1.883	2 h/day exposure to 900 MHz; SAR = 0.66 ± 0.01 W/kg	Y	The exposure increases the ROS level and decreases total antioxidant capacity in rat sperm. Also effects on gene expression levels	
	Iran	Male Wistar rats	Shokri et al. (2015) Cell J., 17, 322–331	IF = 1.105	2450 MHz; 1 or 7 h/day for 2 months. Exposure via two Wi-Fi antennas, 8.5 dBi	Y	Decrease in sperm parameters an increased number of apoptotic cells and caspase-3 activity in the seminiferous tubules, and reduced weight of the seminal vesicle	
	Egypt	Male Sprague dawley albino rats	El-Hady et al. (2015) J Biosci. Appl. Res., 1, 97–111	No impact factor	950 MHz; 2 h/day, 3 days/week for 2 months; some animals received also vitamin E or silymarin	Y	The radiation had a negative effect on the testicular structure through the induction of oxidative stress and disruption of the testicular antioxidant status. Vitamin E and silymarin ameliorated the testicular damage	Not sure that exposure system and dosimetry are correct. No information provided on dosimetry

(continued)

Table 5.2 (continued)

	Country	Organism/cells	Author & Journal	IF[a]	Exposure	Effect	Conclusion from author	Remarks
	Turkey	Male Sprague Dawley rats	Saygin et al. (2016) Environ. Toxicol., 31, 1771–1784	IF = 3.197	2450 MHz; 0.1–45.5 V/m, for 3 h/day for 30 days. SAR =3.21 W/kg. Also co-exposure with gallic acid (GA (30 mg/kg/ daily)	Y	Long term exposure resulted in testicular physiopathology via oxidative damage and inflammation. Gallic acid appears to have a beneficial effect on the pre-pubertal rat testes physiopathology	
	Egypt	Male Wistar rats	Bin-Meferij & El-Kott (2015) Int. J. Clin. Exp. Med., 8, 12, 487–12, 497	IF = 1.277	900 MHz; 1 h/day and 7 days/week. Mobile phone was located in the cage centre, while the distance between mobile phone from the bottom of cage was 3 cm and maximal distance from the cage corners was 28.2 cm	Y	Exposed animals exhibited testicular injury which was prevented by aqueous extract of *Moringa aliefera* leaves	Questions with respect to the exposure system. No dosimetry
Neurological effects	Turkey	Wistar Albino adult male rats	Eser et al. (2013) Turk neurosurg., 23, 707–715	IF = 0.576	900, 1800 , 2450 MHz; SAR = 1.04 W/kg; exposure 1 hour per day for a period of 2 months	Y	Serious degenerative changes in the brain	Questions on experimental set-up

	Brazil	Male Wistar rats	de Caires Junior et al. (2014) Neurol. Res., 36, 800–803	IF = 1.439	1800 MHz; 2 V/m during call, 0.1 V/m in standby mode. Intermittent exposure for 3 days (25 s phone calls every 2 min)	Y/N	No effect on central nervous system but behavioural (stress) effects possible	
	China	Male Wistar rats, hippocampal synaptosomes, and differentiated neuronal P12 cells	Qiao et al. (2014) PLoS ONE 9(4): e95503. doi:10.1371/journal.pone.0095503	IF = 3.234	2856 MHz, 2.856-MHz field at 30 mW/cm^2 for 5 min	Y	Effects on cognitive and memory function	Exposure levels much higher than normal human exposure
	India	Rats	Saikhedkar et al. (2014) Neurol. Res., 36, 1072–1079	IF = 1.439	900 MHz; exposure by means of a mobile hand set for 4 h per day for 15 days	Y	Extensive neuronal damage results in alterations in behaviour related to memory and learning	
	Germany	Male and female rats	Schneider & Stangassinger (2014) Behav. Neurosci., 128, 633–637	IF = 2.728	900 and 1966 MHz at 0.4 W/kg. Exposure periods of 4 min	Y	Effects on memory performance	
	Japan	Sprague–Dawley rats	Shirai et al. (2014) Bioelectromagnetics, 35, 497–511	IF = 1.705	2140 MHz; exposure to low and high SAR values (<0.24 W/kg and <0.08 W/kg)	N	No effects on brain function	

(continued)

Table 5.2 (continued)

	Country	Organism/cells	Author & Journal	IF[a]	Exposure	Effect	Conclusion from author	Remarks
	Germany	Female Wistar rats	Klose et al. (2014) Radiat. Res., 182, 435–447	IF = 2.911	900 MHz; exposure by means of a mobile hand set for 4 h per day for 15 days	N	No behavioural and cognitive effects	
	India	Albino Wistar rats	Narayanan et al. (2014) Bratisl. Lek. Listy, 115, 260–266	IF = 0.439	900 MHz; exposure via a GSM mobile phone kept in silent mode (no ring tone and no vibration) for 1 h/day for 4 weeks	Y	Behavioural effects and biochemical changes (oxidative stress) in brain	Effects in 'stand by' mode is highly questionable
	China	Wistar rats	Wang et al. (2015) Int. J. Radiat. Biol., 91, 306–311	IF = 1.687	856 MHz; exposure at a power density of 0 mW/cm^2 and 50 mW/cm^2 microwave for 6 min	Y	Deficiencies in cognition and learning ability and biochemical changes	
	Korea	Male Sprague-Dawley rats	Kim et al. (2015) Int. J. Radiat. Biol., 91, 354–359	IF = 1.687	848.5 MHz; SAR of 2 W/kg for 1–8 h/day, 5 days/week, for 2 weeks	N	No effects on neurogenesis in brain and cell proliferation	
	Turkey	Adult male Wistar Albino rats	Dasdag et al. (2015) Int J Radiat Biol., 91, 306–311	IF = 1.687	900 MHz; estimated SAR = 0.0369 W/kg for 3 h/day, 7 days/week, over 12 months	Y/N	No effects, except one, on microRNA expression	

	India	Swiss albino mice	Salunke et al. (2015) Physiol. Behavior, 140, 32–37	IF = 2.976	2450 MHz (Bluetooth); peak density of 100 mW for 60 min/day for 7, 30, 60, 90, or 120 days	N	No effects on different behavioural parameters	Questions with respect to the experimental set-up
	China	Male Wistar rats	Xiong et al. (2015) Biomed. Environ. Sci., 28, 13–24	IF = 1.26	2856 MHz; 30 mW/cm^2 for 10 min every other day for 3 days	Y	Immuno- cytochemical effects on hippocampus	
	China	Male Wistar rats	Liu et al. (2015) Biomed. Environ. Sci., 28, 72–75	IF = 1.653	2857 MHz; 6 min exposure at SAR = 5, 10, or 50 mW/cm^2	Y	Analyses at At 1, 7, 14, and 28 d and 3, 6, 9, and 12 months after exposure. Pathological changes in sinus node in the 10 and 50 mW/cm^2 exposure groups	Questionable; high exposure levels
	China	Male Wistar rats	Li et al. (2015) Physiol. Behav., 140, 236–246	IF = 2.976	2856 MHz; 30 mW/cm^2 for 10 min every other day for 3 days. 5, 10, 20, or 30 mW/cm^2 for 6 min, 3 times per week, for up to 6 weeks		Effects on learning ability, neural degeneration etc.	
	China	Male Sprague-Dawley rats	Tang et al. (2015) Brain Res., 1601, 92–101	IF = 2.843	900 MHz; 1 mW/cm^2 for 14 or 28 days (3 h per day). SAR varied between 0.016 (whole body) and 2 W/kg (locally in the head)	Y	Alteration of BBB and histological changes in the brain	

(continued)

Table 5.2 (continued)

	Country	Organism/cells	Author & Journal	IF[a]	Exposure	Effect	Conclusion from author	Remarks
	China	Male and female CD1 mice	Zhang et al. (2015) J. Radiat. Res., 56, 261–268	IF = 1.797	9417 MHz; 200-V/m for 12 h/day	Y	Decreased cognitive and learning behaviour following in utero exposure	Experimental set-up unclear
	India	Male Fischer-344 rats	Megha et al. (2015) Cell Biochem. Biophys., 73, 93–100	IF = 1.680	900, 1800 MHz; 2 h/day, 5 days/week, for 30 days, SAR = 5.953×10^{-4} (900 MHz) and 5.835×10^{-4} W/kg (1800 MHz)	Y	Reduced levels of monoamine neurotransmitters	
	Iran	Male Wistar rats	Razavi et al. (2015) Comp. Clin. Pathol., 24, 1271–1276	No impact factor	940-MHz frequency magnetic field for 30 min every day for a period of 30 days, or for 1, 2, and 3 h every day for a period of 30 days	Y	Increased BBB permeability	Dosimetry not well described and hence difficult to evaluate the significance of the findings
	Japan	Rats	Masuda et al. (2015) In vivo, 29, 207–215	IF = 0.974	The cortex tissue was locally exposed for 50 min to 1457 MHz RF at a SAR of 2.0 W/kg in the target area	N	No dynamic changes occurred in cerebral microcirculation even during local cortex exposure	

	India	Male Fisher rats	Desmukh et al. (2015) Int. J. Toxicol., 34, 284–290	IF = 1.290	900–2400 MHz; exposure for a period of 180 days at a SAR of 5.953×10^{-4} W/kg (900 MHz), and 5.835×10^{-4} W/kg (1800 MHz) or 6.672×10^{-4} W/kg (2450 MHz)	Y	Decreased cognitive effects and increased HSP70 levels. DNA damage in brain cells. Effects found in all exposure groups	
	Turkey	Wistar Albino adult male rats	Dasdag et al. (2015) Int. J. Radiat. Biol., 91, 555–561	IF = 1.687	2400 MHz; 0.0369 W/kg for 3 h/day, 7 days/ week, over 12 months	Y	Influence of radiation on the expression of some mRNAs, but not all	
	Japan	Male Sprague-Dawley rats	Masuda et al. (2015) In vivo, 29, 351–357	IF = 0.974	1457 MHz; The cortex tissue was locally exposed for 50 min to 1457 MHz RF at a SAR of 2.0 W/kg in the target area	N	No effect on BBB	
	India	Male Wistar rats	Narayan et al. (2015) Metabol. Brain Dis., 30, 1193–1206	IF = 2.638	900 MHz; SAR-1.15 W/kg with peak power density of 146.60 $\mu W/cm^2$) for 1 h/day, for 28 days	Y	Effects on cognition and learning ability. Also histological changes and decreased longevity	

(continued)

Table 5.2 (continued)

	Country	Organism/cells	Author & Journal	IF[a]	Exposure	Effect	Conclusion from author	Remarks
	Turkey	Male Wistar rats	Sahin et al. (2015) Brain Res., 1624, 232–238	IF = 2.843	900 MHz; 1 h/day for 30 days. EMF-exposure was at a mean EMF intensity of 13.88 V/m inside the jar and 4.42 V/m beneath the jar. The whole-body SAR was calculated at 0.024 W/kg	Y	Evidence that the EMF-exposed animals demonstrated pyramidal neuron loss and histopathological changes in the cornu ammonis	
	Turkey	Sprague Dawley female rats	Odaci et al. (2016) J. Chem. Neuroanat. 75, 105–110	IF = 1.500	900 MHz; exposure during the prenatal period for 1 h during days 13–21 of pregnancy. SAR = 0.01 W/kg	Y	Prenatal exposure to EMF affects the development of Purkinje cells in the female rat cerebellum and this persist after the postnatal period	
	Korea	Male Sprague-Dawley rats	Kim et al. (2015) Electromagn. Engin. Sci., 15, 173–180	No impact factor	915 MHz; 4-W/kg SAR, RFID signal in a reverberation chamber for 8 h/day, 5 day/week, for 2 week	N	RFID exposure induces no significant change in DCX+ neuronal precursor or NeuN+ neuronal cells in DG of rats	

	Nigeria	MaleWistar albino rats	Shehu et al. (2016) Metab. Brain Dis., 31, 355–362	IF = 2.638	900 and 1800 MHz from a mobile phone kept in the rat's cages. Exposure in silent mode, ring tone, vibration and combinations thereof. 10 min calls/day for 4 weeks	Y	4 weeks exposure to electromagnetic radiation, vibration, ringtone or both produced a significant effect on anxiety-like behaviour and oxidative stress in young Wistar rats	Not really surprising and not relevant for human health
Ear, nose & Eye effects	Turkey	Pregnant albino Wistar rats	Seckin et al. (2014) J. Laryngol. Otol., 128, 400–405	IF = 0.672	900, 1800 MHz; 1 h/day from the 12th day in utero through 21 days after delivery. The field at 900 MHz was 9–25 V/m and at 1800 MHz, 9–18 V/m	Y	Cellular damage in cochlea presumably due to thermal effect	If thermal it is of no relevance to 'normal' human exposure
	Turkey	Albino Wistar rats	Özgür et al. (2015) Acta Otolaryngol., 135, 765–770	IF = 1.099	1800 MHz; 24 h/day for 30 days	Y	Histopathological and immunochemical changes suggesting neuronal degeneration of the auditory system	

(continued)

Table 5.2 (continued)

	Country	Organism/cells	Author & Journal	IF[a]	Exposure	Effect	Conclusion from author	Remarks
	Turkey	Albino Wistar rats	Aydogan et aL (2015) Int. Forum Allergy Rhinol., 5, 626–632	IF = 2.082	2100-MHz radiofrequency radiation emitted by a generator, simulating a 3G–mobile phone, 6 h/day, for 10 or 40 days, respectively	Y	Radiofrequency radiation at 2100 MHz damaged the nasal septal mucosa, and disturbed the mucociliary clearance. Ciliary disorganization and ciliary loss in the epithelial cells resulted in deterioration of nasal mucociliary clearance	Shortcomings in experimental set up
	Egypt	Male albino rats	Fathy et al. (2015) Egypt. J. Histology, 38, 102–115	No impact factor	900 MHz; postnatal exposure of rats for 28 days; 30 min daily	Y	Degenerative changes in parotide	No details given and hence no evaluation possible
	Turkey	Wistar albino rats	Aydogan et al. (2015) Am. J. Otolaryngol., 36, 39–46	IF = 0.984	2100 MHz; 6 h/day, 5 days/week, for 10 or 40 days. 16 V/m during the exposure period. SAR = 0.4 W/kg	Y	Numerous histopathological changes regarding salivary gland damage. Increased exposure duration led to increased effects	

Effects on bone								
	Poland	Male Wistar rats	Sieron-Stoltny et al. (2015) Biomed. Res. Int., 896,019, 11 pages	IF = 1.579	900 MHz; generated by a mobile phone (Nokia 5110) placed directly under the animal's cages. SAR stated by the manufacturer was 0.69 W/kg of body weight	Y/N	No effects on different bone parameters but some mechanical changes and lower calcium content	Questionable experimental set-up. Exposure level cannot be estimated from the provided data
Immunology	Egypt	Albino Sprague-Dawley rats	Ragy (2015) Electromagn. Biol. Med., 34, 279–284	IF = 1.194	900 MHz from a mobile phone; 1 h/d for 60 d and idem but left 30d without exposure	**Y**	Impairments in some biochemicals changes and oxidative stress in brain, liver and renal tissue of albino rats. These alterations were corrected by withdrawal ((left without exposure for 30 days)	
	Turkey	12 weeks-old male Wistar albino rats	Yilmaz et al. (2014) Eurp. Rev. Med. Pharmacol. Sci., 18, 992–1000	IF = 1.213	1900 and 2100 MHz; exposed to EMW emitted by mobile phones 7 times a day (mobile phone placed against the rat's head). Each application lasted 5 min and a total of 920 min in 4 weeks		Effect on apoptotic and anti-apoptotic proteins	

(continued)

Table 5.2 (continued)

	Country	Organism/cells	Author & Journal	IF[a]	Exposure	Effect	Conclusion from author	Remarks
	China	Wistar rats and PC12 cells	Zuo et al. (2014) Int. J. Med. Sci., 11, 426–435	IF = 2.003	2856 MHz, 30 mW/cm^2 for 5 (cells) and 15 min (animals)	Y	The mitochondria membrane potential decreased and DNA fragmentation increased, leading to an increase in the apoptotic cell percentage. Increased gene expressions (Bax, Bcl-2, cytochrome c, cleaved caspase-3, and PARP)	High exposure levels
	Greece	Wistar rats and PC12 cells	Hajioun et al. (2014) Health Med., 8, 469–476	No impact factor	900 MHz cell phone; exposure for 10 min, 12 times per day for 1 month	Y	Different effects on thyroid gland. Garlic could compensate for some of the effects	Questions with experimental set-up
	Spain	female Sprague–Dawley rats	Misa-Agustino et al. (2015) Life Sci., 127, 1–11	IF = 2.702	Sub-thermal exposure to 2450 MHz at 0, 1.5, 3.0 or 12.0 W power. Exposure time = 30 min	Y	Several morphological changes in thymus tissue	OK but as a model for diathermy. Unrelated to mobile phones
	Japan	Sprague Dawley rats	Ohtani et al. (2015) J. Radiat. Res., 56, 467–474	IF = 1.797	2140 MHz; Exposures for a total of 9 weeks spanning in utero development, lactation and the juvenile period. Rats were continuously exposed to RF-EMF for 20 h/day, 7 days/week at a SAR of 0.2 W/kg	Y	Some effects on transcription but overall no indication of adverse effects	

Muscles								
	Lebanon	Sprague Dawley rats	Ramadan et al. (2015) Int. J. Advances Life Sci., 7 (1 & 2), 1–9	No impact factor	900 MHz; 6 weeks continuous (24 h/day) whole body exposure (E_{eff}= 25 V/m)	Y	Changes in the distribution of muscles proteins and in the percentage of MHC isoforms	
Xenopus laevis/ frogs	Turkey	*X. laevis* embryo's	Boga et al. (2015) Ecotoxicol. Environ. Saf., 113, 378–390	IF = 2.762	900, 1800 MHz; Average power density = 1.59 W/m^2. The electric field strength within the exposure area was set at30dB and 24.57 V/m to achieve a SAR of 1.0 W/kg. Exposure for 4, 6, or 8 h	Y/N	No effect from RF radiation only on growth and mortality but synergy with nicotine sulphate	
	Iran	frogs	Mortazavi et al. (2015) J. Biomed. Phys. Eng., 5, 121–132	No impact factor	900 MHz radiation emitted from a common cellular phone at a distance of 15 cm from the receiver for 30 min	Y	An effect of RF EMF was found on the pulse height of contractions in the muscle	Questionable experimental set up and relevance to man
Mouse								
Reproduction/ teratology	India	Swiss strain	Shahin et al. (2014) Free Radical Res., 48, 511–525	IF = 2.976	2450 MHz; exposure for 2 h/day over 30 days. SAR estimated to be 0.18 W/kg	Y	Effect on sperm count and testicular function and morphology	

(continued)

Table 5.2 (continued)

	Country	Organism/cells	Author & Journal	IF[a]	Exposure	Effect	Conclusion from author	Remarks
	India	neonatal albino mice	Mugunthan et al. (2015) Int. J. Curr. Res. Rev., 7, 36–47	No impact factor	900 and 1800 MHz; exposure from a cell phone, 48 min/day for 30–180 days. The roof of the mice cage was designed to hang the 2G and 3G cell phone from the distance of five centimetres from the floor which allow the mice to move freely	Y	Histological changes in testes (microscopic changes in the seminiferous epithelium, reduction of serum testosterone level, reduction in the number of Sertoli cells and Leydig cells)	Many questions with experimental set up
	Irak	White mice	Al-Dulamey et al. (2015) Int. J. Adv. Res., 6, 1149–1155	No impact factor	5 GHz; 100, 150 and 200 μW/cm^2, 6 h daily for 45 days	Y	Decreased total number of offspring and greater proportion of female compared to male offspring following exposure	Unacceptable. Too many shortcomings
	China	Adult male ICR mice	Zhu et al. (2015) Mut. Res., 792, 53–57	IF = 2.415	900 MHz; 1.6 mW/cm^2 power intensity (WB SAR of 0.731 W/kg) for 4 h/day for 15 days	N	Absence of mutagenic effect of RF exposure (dominant lethal test conducted over 4 weeks of mating)	
Neurological effects	Korea	Six-week-old male ICR mice	Maskey & Kim (2014) Neurosci. Lett., 564, 78–82	IF = 2.030	835 MHz; SAR of 1.6 W/kg, 8 h/day for 3 months	Y	The results suggest a detrimental effect in the auditory nuclei	Questions on experimental set-up

	India	male Swiss albino mice	Maurya & Upadhyay (2014) G-Journal Environ. Sci. Technol., 1, 103–107	No impact factor	Base station antenna; Mice were kept around mobile tower within 50 m of radius	Y	Effects on memory function and muscle strength	Not well described experiments and therefore not possible to take the results into account
	Australia	female inbred wild type C57BL/6NTac mice	Court-Kowalski et al. (2015) Bioelectromagnetics, 36, 245–250	IF = 1.705	900 MHz; single, far-field whole body exposure at a specific absorption rate of 4 W/kg on 5 successive days per week for 104 weeks	N	No effect on immunoreactivity of astrocytes	
	Korea	C57BL/6 mice	Son et al. (2015) J. Eectromagn. Engin. Sci., 15, 151–157	No impact factor	1950 MHz; SAR = 5.0 W/kg, exposure for 2 h/day for 60 days	N	No effect on memory and morphology of hippocampus	
	Saudi Arabia	Swiss albino mice of both sexes	Bolla (2015) Int. J. Anat. Res., 3, 1216–1224	No impact factor	800-MHz CDMA mobile phone for 15–60 min/day for 30 days	Y	Increased neuronal damage and decreased dendritic arborisation of hippocampal CA3 neurons was found with increasing RF EMF exposure duration	Exposure not well described and probably problematic
Auditory effects	Korea	ICR male mice	Maskey et al. (2014) Int. J. Mol. Med., 34, 409–419	IF = 2.088	835 MHz; SAR of 1.6 W/kg, 8 h/day for 3 months	Y	Immunohistochemical effects in the auditory brainstem region	

(continued)

Table 5.2 (continued)

	Country	Organism/cells	Author & Journal	IF[a]	Exposure	Effect	Conclusion from author	Remarks
	Korea	(Tg)-5xFAD mice	Jeong et al. (2015) Curr. Alzheimer Res., 12, 481–492	IF = 3.889	1950 MHz; SAR = 5.0 W/kg, exposure for 2 h/day for 60 days	Y	Beneficial effect; improvement of Alzheimer's disease-like pathology in the mice via Aβ modulation	
Hematological effects	Italy	female C57BL mice	Rosado et al. (2014) Bioelectromagnetics, 35, 559–567	IF = 1.705	900 MHz; SAR of 2 W/kg, 2 h/day, 5 days/week for 4 consecutive weeks	N	No effect on the ability of bone marrow (BM) precursors to long-term reconstitute peripheral T and B cell compartments	
Kidney	India	Male and female neonatal albino mice	Mugunthan et al. (2014) Int. J. Cur. Res. Rev., 6, 44–49	No impact factor	900 and 1800 MHz; Exposure from a cell phone, 48 min/day for 30–180 days. The roof of the mice cage was designed to hang the 2G and 3G cell phone from the distance of five centimetres from the floor which allow the mice to move freely	Y	Histological changes found in the kidney of exposed mice (microscopic changes in glomerulus, proximal and distal convoluted tubules of the kidney)	This paper has many shortcomings

	India	Male and female neonatal albino mice	Mugunthan et al. (2014) Int. J. Sci. Res., 3, 523–529	No impact factor	900, 1900 MHz; 48 min per day for a period of 30–180 days. The roof of the mice cage was designed to hang the 2G and 3G cell phone from the distance of five centimetres from the floor which allow the mice to move freely	Y	Microscopic changes in the seminiferous tubules, reduction in the number of Sertoli and Leydig cells and decreased serum testosterone level	This paper has many shortcomings
Genetic effects and oxidative damage	China	Adult mice	Zong et al. (2015) Int. J. Radiat. Biol., 91, 270–276	IF = 1.687	900 MHz; 120 μW/cm^2 power density for 4 h/day for 7 days	Y/N	No effects of RF alone but adaptive effect against bleomycine	
Gene expression	China	Male Kunming mice	Zhao et al. (2015) Biomed. Environ. Sci.,28, 558–570	IF = 1.653	2100 MHz; SAR = 0.45 W/kg, 1.8 W/kg, and 3.6 W/kg for 1 h/day for 8 weeks	Y	Long-term, low-level microwave exposure may inhibit learning and memory by affecting protein and energy metabolic processes and signalling pathways relating to neurological functions or diseases	

(continued)

Table 5.2 (continued)

	Country	Organism/cells	Author & Journal	IF[a]	Exposure	Effect	Conclusion from author	Remarks
Cancer	Germany	Male C3H/HeNCrl and female C57Bl/6 N mice	Lerchl et al. (2015) Biochem. Biophys. Res. Comm., 459, 585–590	IF = 2.297	966 MHz; 0.04 & 0.4 W/kg. 24/7 for the entire period (72 weeks) with the exception of half an hour each night at 3 a.m. in order to shut down the generators and to reboot the system	Y	Increased adenoma and carcinoma in the lungs following RF+ENU treatments; i.e. tumour promoting effect of RF	
Guinea Pig								
	Turkey	Guinea pig	Ozgur et al. (2015) Int. J. Radiat. Biol., 91, 187–193	IF = 1.687	900 MHz; SAR of 0,4 W/kg, 10 or 20 min daily for consecutive 7 days	Y	Oxidative effects on liver cells	
Hamster								
Metabolism	Germany	Djungarian hamsters	Taberski et al. (2014) Radiat. Res., 181, 617–622	IF = 2.911	900 MHz; SAR = 0.08, 0.4 or 4 W/kg. Exposures for one week per condition, followed by one week without exposure	Y/N	Effects at highest exposure level: fur temperatures were elevated by approximately 0.5 °C while the temperatures of the eyes' surface were not affected; food consumption was lowered, while water consumption and body weight were not affected; the	

							production of carbon dioxide was lowered during the day and unaffected during the night, while oxygen consumption levels remained unaffected. Finally the respiratory quotient was lower during the day and also somewhat lower during the night (not significant)	
Rabbit								
Apoptosis	Iran	New Zealand white male rabbits	Azadi Oskouyi et al. (2015) Andrologia, 47, 700–705	IF = 1.630	950 MHz; output power of 3 and 6 watts (in order to generate the magnetic field power while talking); exposure for 2 weeks and 2 h a day	Y	Increased apoptosis in epididymis and lower testosteron levels	Inadequate dosimetry
Physiology	Algeria	Male Albino rabbits	Saili et al. (2015) Environ. Toxicol. Pharmacol., 40, 600–605	IF = 2.084	2450 MHz; 1 h exposure to an access point from WIFI device	Y	Effects were found on heart rhythm, blood pressure, and catecholamines indicating that radiofrequency can act directly and/or indirectly on the cardiovascular system	

(continued)

Table 5.2 (continued)

	Country	Organism/cells	Author & Journal	IF[a]	Exposure	Effect	Conclusion from author	Remarks
	China	Rabbits	Zhu et al. (2015) Wei Sheng Yan Jiu., 44, 818–821	No impact factor	2450 MHz; 50, 100, 150 and 200 mW/cm^2 for 20 min	Y	Changes of myocardial tissue and expression of Caspase-3 and HSP 70	Paper in Chinese. No details known (only abstract available); no full evaluation possible
Oxidative damage								
	Switzerland	Cows	Hässig et al. (2014) BMC Vet. Res., 10:136	IF = 1.777	900 MHz-antennas installed at 2 m height behind the animals. The incident field was on average 12 V/m	Y/N	Effects on redox activities in some but not all animals	Questionable for different reasons
Human cytogenet	India	Man	Mahajan et al. (2014) Molec. Cytogenet., 7 (Suppl 1): P49	IF = 2.140	Individuals residing in an area with three base stations close by (150 m) with power density ranging from 172.6 to 550.3μw/m^2	Y	Significant increase in genetic damage (cytome test) assessed in the buccal epithelial cells in individuals residing near mobile phone base stations	Only abstract available; very questionable
	India	Blood	Gandhi et al. (2015) Electromagn. Biol. Med., 34, 344–354	IF = 1.194	People living near telecommunication base stations	Y	Effects with comet assay in the subject's peripheral blood lymphocytes	Questionable procedures and some of the provided info is unlikely

Human other	Iran and Germany	Glioblastoma	Akhavan-Sigari et al. (2014) Rare Tumors, 6, 116–120	No impact factor	Mobile phone using cancer patients	Y	Mutant p53 expression in glioblastoma multiforme tumors of cancer patients is more frequent in patients with higher mobile phone use	Study needs confirmation and shortcomings eliminated

[a]Impact factor from 2014 Thomson Reuters, Journal Citation Reports

sometimes be answered only by a repetition of the circumstances and the observations". Consistency is one of the important criteria that he defined as needed to conclude about 'causation' (here of RF-radiation causing illness). But, "*If many studies show consistent results this is an indication of causality; however, when no consistency is obtained this is no basis to ascertain that causality is not there*".

In conclusion, societal concern about the potential adverse health effects of mobile communication and other EMF-emitting technologies is still important. However, this is largely based on research data reported in often unreliable scientific papers, or at least papers with more or less severe shortcomings. A careful analysis of the research shows that evidence for adverse health effects is not that impressive and hence that exaggerated anxiety should be avoided. As a matter of fact it is virtually impossible to prove the absence of an effect, and hence some doubt will always subside. On the other hand, this does not mean that no protective measures should be taken as lack of consistency does not mean lack of effect and no evidence of harm does not equal evidence of no harm. Measures should however be reasonable.

Acknowledgements This study was partly funded by and carried out on behalf of the 'Environment, Nature and Energy Department' of the Flemish Government (LNE/13113/M&G; subcontract from iMinds VZW, Ghent). The author declares no conflict of interest.

References

Adang D, Vander Vorst A (2006) Results of a long-term epidemiological study on low-level microwave exposed rats. Microwave Symposium Digest IEEE MTT-S International 11:1742–1745

Adang D, Remacle C, Vander Vorst A (2009) Results of a long-term low-level microwave exposure of rats. IEEE Trans Microwave Theor Techn 57:2488–2497

Arendash GW, Mori T, Dorsey M, Gonzalez R, Tajiri N, Borlongan C (2012) Electromagnetic treatment to old Alzheimer's mice reverses β-amyloid deposition, modifies cerebral blood flow, and provides selected cognitive benefit. PLoS One 7:e35751

Aydin D, Feychting M, Schüz J, Andersen TV, Poulsen AH, Prochazka M, Klæboe L, Kuehni CE, Tynes T, Röösli M (2011) Predictors and overestimation of recalled mobile phone use among children and adolescents. Prog Biophys Mol Biol 107:356–361

Baan R, Grosse Y, Lauby-Secretan B, El Ghissassi F, Bouvard V, Benbrahim-Tallaa L, Guha N, Islami F, Galichet L, Straif K et al (2011) WHO International Agency for Research on Cancer Monograph Working Group. Carcinogenicity of radiofrequency electromagnetic fields. Lancet Oncol 12:624–626

Belpomme D, Campagnac C, Irigaray P (2015) Reliable disease biomarkers characterizing and identifying electrohypersensitivity and multiple chemical sensitivity as two etiopathogenic aspects of a unique pathaological disorder. Rev Environ Health 30:251–271

Benson VS, Pirie K, Schuz J, Reeves GK, Beral V, Green J (2013) For the million women study collaborators. Mobile phone use and risk of brain neoplasms and other cancers: prospective study. Int J Epidemiol 42:792–802

Bradford Hill A (1965) The environment and disease: association or causation? Proc Royal Soc Med 58:295–300

Chou C-K (2013) A need to provide explanations for observed biological effects of radiofrequency exposure. Electromagn Biol Med 34:175–179

Clark J, Smith R (2015) Firm action needed on predatory journals. Brit Med J. 350:h210

Coureau G, Bouvier G, Lebailly P, Fabbro-Peray P, Gruber A, Leffondre K, Guillamo JS, Loiseau H, Mathoulin-Pélissier S, Salamon R et al (2014) Mobile phone use and brain tumours in the CERENAT case-control study. Occup Environ Med 71:514–522

Cuppen JJM, Wiegertjes GF, Lobee HWJ, Savelkoul HFJ, Elmusharaf MA, Beynen AC, Grooten HNA, Smink W (2007) Immune stimulation in fish and chicken through weak low frequency electromagnetic fields. Environmentalist. doi:10.1007/s10669-007-9055-2

de Jongh T, Nooijen A, van Barneveld, J.; Zuijdam F (2015) ZonMW Programme evaluation on Electromagnetic Fields & Health (EMF&H). http://www.zonmw.nl/en/programmes/electromagnetic-fields-and-health-research/general-information/

Deltour I, Auvinen A, Feychting M, Johansen C, Klaeboe L, Sankila R, Schüz J (2012) Mobile phone use and incidence of Glioma in the Nordic countries 1979-2008: Consistency check. Epidemiology (23):301–307

de Pomerai D, Daniells C, David H, Allan J, Duce I, Mutwakil M, Thomas D, Sewell P, Tattersall J, Jones D et al (2000) Non-thermal heat-shock response to microwaves. Nature 405:417–418

De Ridder M, Braeckman L (2015) Elektromagnetische overgevoeligheid. Tdschr Geneeskunde, 71:1623–1628 (in Dutch).

Devnani M, Gupta AK (2015) Predatory journals are only part of the problem. Brit Med J 350: h707

De Vocht F, Burstyn I, Cherrie JW (2011) Time trends (1998-2007) in brain cancer incidence rates in relation to mobile phone use in England. Bioelectromagnetics 32:334–339

Diem E, Schwarz C, Adlkofer F, Jahn O, Rudiger H (2005) Non-thermal DNA breakage by mobile-phone radiation (1800 MHz) in human fibroblasts and in transformed GFSH-R17 rat granulosa cells in vitro. Mutat Res (583):178–183

Dye LR (2007) A review of the review process. J Med Toxicol 3:143–145

Elder JA (2003) Survival and cancer in laboratory mammals exposed to radiofrequency energy. Bioelectromagnetics 24(Suppl 6):S101–S106

Elder J (2010) Radiofrequency studies on tumorigenesis and the blood-brain barrier in lab animals support the conclusion of no adverse effects without significant tissue temperature increase. Asia-Pacific International Symposium on Electromagnetic Compatibility. 2010, Beijing, China, pp 13–15

Galeev AL (2000) The effect of microwave radiation from mobile telephones on humans and animals. Neurosci Behav Physiol 30:187–194

Gasparyan AY, Yessirkepoc M, Diyanova SV, Kitas GD (2015) Publishing ethics and predatory practices: a dilemma for all stakeholders of science communication. J Korean Med Sci 30:1010–1016

Hahn RA (1997) The nocebo phenomenon: concept, evidence, and implications for public health. Prev Med 26:607–611

Hansson Mild K, Repacholi M, van Deventer E, Ravazzani P (2006) Electromagnetic Hypersensitivity: proceedings, International Workshop on Electromagnetic Field Hypersensitivity, Prague, Czech Republic, October 25–27, 2004. WHO. 2006, ISBN 92 4 159412 8

Hauri DD, Spycher B, Huss A, Zimmermann F, Grotzer M, von der Weid N, Spoerri A, Kuehni CE, Röösli M, Swiss National Cohort; Swiss Paediatric Oncology Group (2014) Exposure to radio-frequency electromagnetic fields from broadcast transmitters and risk of childhood cancer: a census-based cohort study. Am J Epidemiol 2014(179):843–851

Health Council of the Netherlands (2010) Long-term effects of radiofrequency electromagnetic fields. Assessment of the study of D. Adang. The Hague: Health Council of the Netherlands. 2010/09. http://www.gezondheidsraad.nl/sites/default/files/201009E.pdf

IARC (2013) Non-ionizing radiation, Part II: radiofrequency electromagnetic fields/IARC Working Group on the evaluation of carcinogenic risks to humans (2011: Lyon, France). ISBN 978 92 832 1325 3

Jauchem JR (2008) Effects of low-level radio-frequency (3 kHz to 300 GHz) energy on human cardiovascular, reproductive, immune, and other systems: a review of the recent literature. Int J Hyg Environ Health 211:1–29

Jeong YJ, Kang GY, Kwon JH, Choi HD, Pack JK, Kim N, Lee YS, Lee HJ (2015) 1950 MHz electromagnetic fields ameliorate Aβ pathology in Alzheimer's disease mice. Curr Alzheimer Res 12:481–492

Juutilainen J, Lagroye I, Miyakoshi J, van Rongen E, Saunders R, de Seze R, Tenforde T, Verschaeve L, Veyret B, Xu Z (2009) Exposure to high frequency electromagnetic fields, biological effects and health consequences (100 kHz – 300 GHz). In: Vecchia P, Matthes R, Ziegelberger G, Lin J, Saunders R, Swerdlow A (eds) Review of experimental studies of rf biological effects (100 kHz – 300 GHz), ICNIRP 16/2009, ISBN 978-3-934994-10-2 pp 90–303

Juutilainen J, Lagroye I, Miyakoshi J, van Rongen E, Saunders R, de Seze R, Tenforde T, Verschaeve L, Veyret B, Xu Z (2011) Experimental studies on carcinogenicity of radiofrequency radiation. Crit Rev Environ Sci Technol 41:1664–1695

Leitgeb N (2016) Reply to comment of Sage et al. on SCENIHR's opinion on potential health effects of exposure to electromagnetic fields. Bioelectromagnetics 37:193–194

Lerchl A (2009) Comments on "Radiofrequency electromagnetic fields (UMTS, 1,950 MHz) induce genotoxic effects in vitro in human fibroblasts but not in lymphocytes" by Schwarz et al. (Int. Arch. Occup. Environ. Health 2008: doi:10.1007/s00420-008-0305-5), Int. Arch. Occup. Environ. Health 82:275–278

Little MP, Rajaraman P, Curtis RE, Devesa SS, Inskip PD, Check DP, Linet MS (2012) Mobile phone use and glioma risk: comparison of epidemiological study results with incidence trends in the United States. Brit Med J 344:e1147

Maes A, Van Gorp U, Verschaeve L (2006) Cytogenetic investigation of subjects professionally exposed to radiofrequency radiation. Mutagenesis 21:139–142

Maes A, Anthonissen R, Wambacq S, Verschaeve L (2016) The cytome assay as a tool to investigate the possible association between exposure to extremely low frequency magnetic fields and an increased risk for Alzheimer's disease. J Alzheimers Dis 50:741–749

Mortazavi S, Shojaei-Fard M, Haghani M, Shokrpour N, Mortazavi S (2013) Exposure to mobile phone radiation opens new horizons in Alzheimer's disease treatment. Biomed Phys Eng 3:109–112

Perrin A, Cretallaz C, Collin A, Amourette C, Yardin C (2010) Effects of radiofrequency field on the blood-brain barrier: a systematic review from 2005 to 2009. Comptes Rendus Phys 11:602–612

Porsius JT, Claassen L, Smid T, Woudenberg F, Petrie KJ, Timmermans DRM (2015) Symptom reporting after the introduction of a new high-voltage power line: a prospective field study. Environ Res 138:112–117

Porsius JT, Claassen L, Woudenberg F, Smida T, Timmermans DRM (2016) Nocebo responses to high-voltage power lines: evidence from a prospective field study. Sci Total Environ 543:432–438

Portier JC, Armstrong BK, Baguley BC, Baur X, Belyaev I, Bellé R, Belpoggi F, Biggeri A, Bosland MC, Bruzzi PJ (2016) in press Differences in the carcinogenic evaluation of glyphosate between the International Agency for Research on Cancer (IARC) and the European Food Safety Authority (EFSA). Epidemiol Community Health. doi:10.1136/jech-2015-207005

Repacholi MH, Lerchl A, Röösli M, Sienkiewic Z, Auvinen A, Breckenkamp J, d'Inzeo G, Elliot P, Frei P, Heinrich S et al (2012) Systematic review of wireless phone use and brain cancer and other head tumors. Bioelectromagnetics 33:187–206

Röösli M, Michela G, Kuehnia CE, Spoerri A (2007) Cellular telephone use and time trends in brain tumour mortality in Switzerland from 1969 to 2002. Eur J Cancer Prev 16:77–82

Sage C, Carpenter D, Hardell L (2016) Comment on SCENIHR: opinion on potential health effects of exposure to electromagnetic fields, Bioelectromagnetics 36: 480-484 (2015). Bioelectromagnetics 37:190–192

Salford LG, Brun AE, Eberhardt JL, Malgren L, Persson BR (2003) Nerve cell damage in mammalian brain after exposure to microwaves from GSM mobile telephones. Environ Health Perspect 111:881–883

SCENIHR (2015) Opinion on Potential health effects of exposure to electromagnetic fields (EMF) European Commission, DG Health and Food Safety. http://ec.europa.eu/health/scientific_committees/emerging/docs/scenihr_o_041.pdf

Schwarz C, Kratochvil E, Pilger A, Kuster N, Adlkofer F, Rudiger HW (2008) Radiofrequency electromagnetic fields (UMTS 1,950 MHz) induce genotoxic effects in vitro in human fibroblasts but not in lymphocytes. Int Arch Occup Environ Health 81:755–767

Speit G, Schutz P, Hoffmann H (2007) Genotoxic effects of exposure to radiofrequency electromagnetic fields (RF-EMF) in cultured mammalian cells are not independently reproducible. Mutat Res 626:42–47

Spiegel H (1997) Nocebo: the power of suggestibility. Prev Med 26:616–621

SSM (2014) Recent research on EMF and health risk. Ninth report from SSM's Scientific Council on Electromagnetic Fields. https://www.stralsakerhetsmyndigheten.se/Publikationer/Rapport/Stralskydd/2014/201416

Stam R (2010) Electromagnetic fields and the blood-brain barrier. Brain Res Rev 65:80–87

Szmigielski S (2013) Reaction of the immune system to low-level RF/MW exposures. Sci Total Environ 454-455:393–400

Tafforeau M, Verdu MC, Norris V, White GJ, Cole M, Demarty M, Thellier M, Ripoll C (2004) Plant sensitivity to low intensity 105 GHz electromagnetic radiation. Bioelectromagnetics 25:403–407

van Rongen E, Saunders R, Croft R, Juutilainen J, Lagroye I, Miyakoshi J, de Seze R, Tenforde T, Verschaeve L, Veyret B et al (2009) Effects of radiofrequency electromagnetic fields on the human nervous system. J Toxicol Environ Health B Crit Rev 12:572–597

Verschaeve L (2009) Genetic damage in subjects exposed to radiofrequency radiation. Mutat Res 681:259–270

Verschaeve L (2012) Evaluations of international expert group reports on the biological effects of radiofrequency fields. In: Ali E (ed) Wireless Communications. Chapter 20, INTECH Publ, pp 523–546

Verschaeve L (2013) Environmental impact of radiofrequency fields from mobile phone base stations. Crit Rev Environ Sci Technol 44:1313–1369

Verschaeve L, Juutilainen J, Lagroye I, Miyakoshi J, van Rongen E, Saunders R, de Seze R, Tenforde T, Veyret B, Xu Z (2010) In vitro and in vivo genotoxicity of radiofrequency fields. Mutat Res 705:252–268

Vian A, Faure C, Girard S, Davies E, Hallé F, Bonnet P, Ledoigt G, Paladian F (2007) Yes, plants do respond to GSM-like radiations!!! Plant Signal Behav 2:e1–e3

Vijayalaxmi, Prihoda TJ (2012) Genetic damage in human cells exposed to non-ionizing radiofrequency fields: a meta-analysis of the data from 88 publications (1990-2011). Mutat Res 749:1–16

Vijayalaxmi, Prihoda TJ (2014) Mobile phones, non-ionizing radiofrequzncy fields and brain cancer: is there an adaptive response? Dose-Response 12:509–514

Vijayalaxmi, Scarfi MR (2014) International and national expert group evaluations: biological/health effects of radiofrequency fields. Int J Environ Res Public Health. 11, 9376–9408.

Vijayalaxmi, McNamee JP, Scarfi MR (2006) Comments on: "DNA strand break" by Diem et al. [Mutat. Res. 583 (2005) 178–183] and Ivancsits et al. [Mutat. Res. 583 (2005) 184–188], Mutat Res. 603, 104–106

WHO (2005) Electromagnetic fields and public health: Electromagnetic hypersensitivity. Fact sheet N°296. http://www.who.int/peh-emf/publications/facts/fs296/en/

Chapter 6
Application of Microwave Irradiation to Bone Decalcification and Its Effect on DNA Quality

Kazuhiko Imaizumi

Abstract Microwave (MW) irradiation has been widely used in the field of histopathology to reduce the total time to prepare tissue sections for microscopy. The processes before tissue sectioning include fixation, dehydration, and embedding in paraffin, and MW irradiation accelerates these reactions. In histological studies of bone, a process to soften compact bone, "decalcification", is necessary. Many studies have applied MW irradiation to this time-consuming process, and positive results of MW irradiation for bone decalcification are accumulating. In recent years, genetic analysis has become popular in this field through the techniques of in situ hybridization and reverse transcription-polymerase chain reaction as well as genotyping of DNA extracted from decalcified specimens. The effect of MW irradiation on the quality of nucleic acids is now a major consideration. This chapter presents the MW effects on bone decalcification in terms of bone histology, the progress of decalcification techniques based on MW, and trends concerning the applicability of specimens decalcified with MWs for genetic analysis.

Keywords Decalcification • Compact bone • Microwave acceleration • DNA deterioration • Forensic anthropology

6.1 Bone as the Source for DNA Identification

Particularly in forensics, we often encounter cases where severely decomposed skeletal remains are submitted for identification. Unlike specimens from common biopsy cases and experimental samples, bone in forensic cases involves problems that make DNA testing difficult because of exogenous DNA decomposition following initial degradation by postmortem autolysis. Figure 6.1 shows histological images of decomposed bone that had been buried for many years. The surface area

K. Imaizumi (✉)
National Research Institute of Police Science, 6-3-1, Kashiwanoha, Kashiwa, Chiba 277-0882, Japan
e-mail: imaizumi@nrips.go.jp

C.D. Geddes (ed.), *Microwave Effects on DNA and Proteins*,
DOI 10.1007/978-3-319-50289-2_6

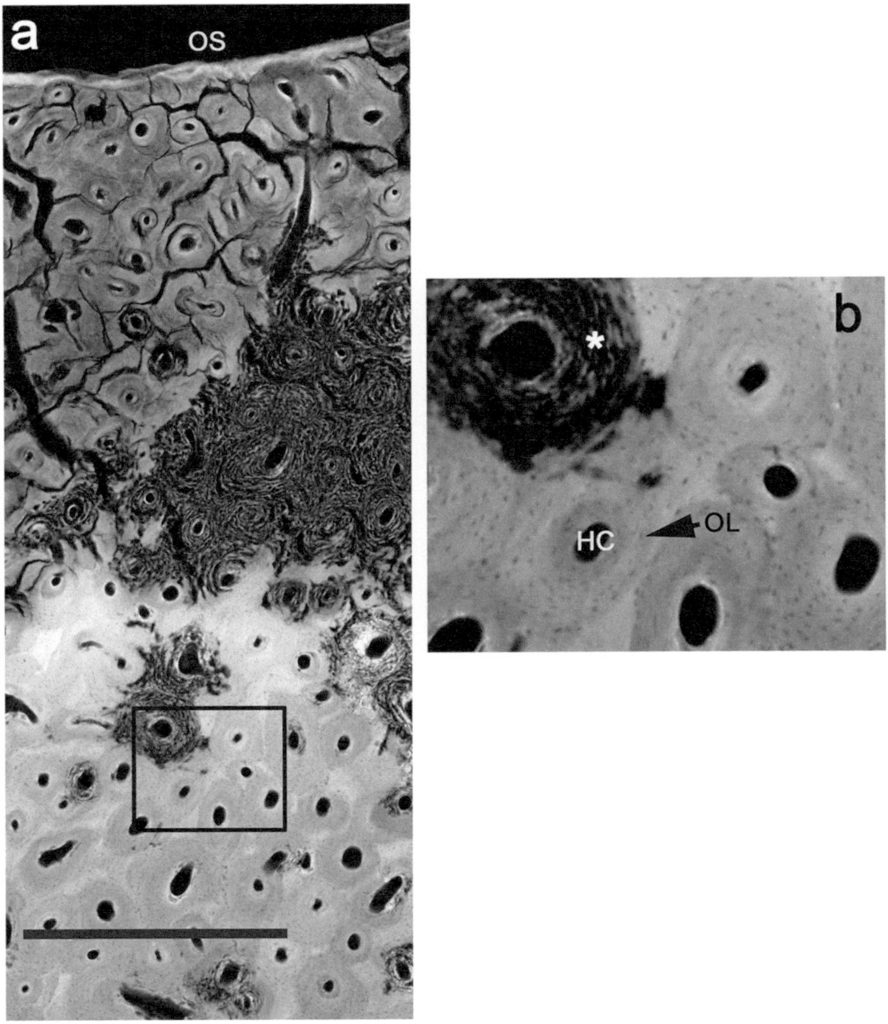

Fig. 6.1 Microscopic images of compact bone buried underground. Observations were made by X-ray microradiography. (**a**) The area near the *outer* surface (OS) of the bone was degraded severely. Bar: 1 mm. (**b**) Enlarged image of the *boxed* area in (**a**). *Asterisk* indicates the area degraded by microbial invasion. *HC* haversian canal, *OL* osteocyte lacuna

of the compact bone was severely invaded by microorganisms. These degraded areas were not appropriate for bone DNA extraction. Such microbial invasion was also observed inside the compact bone, and the area around the Haversian canal was more severely invaded compared with the remaining area that was enriched with osteocyte lacunae. The lacuna is a hollow compartment where a single osteocyte had been located. It retains a certain degree of remnant DNA from the osteocyte even after complete skeletonization.

According to Ortner and Turner-Walker (2003), compact bone consists of 62% bone minerals by weight and 24% organic matrix with the remaining 14% as water. Over 90% of the organic matrix is collagen that forms the fibrous component, collagen fibers. The major bone mineral is calcium phosphate existing as the crystalized form similar to hydroxyapatite. Each collagen fiber is coated with these crystals, resulting in mineralization of the compact bone (Weiner and Wagner 1998; Cui et al. 2007; Olszta et al. 2007). Scanning electron microscopy of compact bone by Boyde and Jones (1996) showed an osteocyte lacuna surrounded by tightly tangled, mineralized collagen fibers. These findings suggest that the osteocyte lacuna is well protected not only from microbes but also physiochemical alterations in moisture, ultraviolet light, and pH that are thought to deteriorate DNA. To obtain DNA from the osteocyte lacuna, the collagen fibers, which comprise a large part of compact bone, must be digested with a proteolytic enzyme (usually proteinase K). The enzyme does not lyse the collagen fibers coated with crystals. Therefore, the bone minerals should be removed by decalcification (a term synonymous with demineralization) prior to digestion.

6.2 Decalcification of Compact Bones

Decalcification of bone has been discussed for many years in the field of histopathology. As a hard substance, bone has to be softened by decalcification to be cut into thin sections for microscopy. The methods for decalcification vary considerably in terms of the reagents and conditions for processing (e.g. time and temperature). Each decalcification method has been validated according to aspects such as influences on histological structures and tissue stainability and applicability to in situ hybridization.

Decalcification reagents are roughly classified into three groups (Jimson et al. 2014), strong (inorganic) acids [nitric acid and hydrochloric acid (HCl)], weak (organic) acids (formic, acetic and picric acids), and chelating agents [ethylenediaminetetraacetic acid, (EDTA)]. In addition to singular use of these chemicals, many kinds of premixed reagents specialized for decalcification are commercially available. Strong acids decalcify compact bone faster than weak acids or chelating agents. Based on a study by Callis and Sterchi (1998) using 6 mm-thick sections of canine humeri, complete decalcification was achieved in 2, 7, and 13 days at room temperature with HCl, nitric acid, and formic acid, respectively, whereas EDTA required 40 days. Such lengthy decalcification with EDTA has been reported in other studies (Jimson et al. 2014; González-Chávez et al. 2013; Sangeetha et al. 2013; Neves et al. 2011; Yamamoto-Fukuda et al. 2000).

Although strong acids decalcify compact bone faster, they cause deteriorative alterations to histological structures, which negate microscopic observations. Tissue swelling (Jimson et al. 2014) and shrinkage (González-Chávez et al. 2013) are severe in specimens decalcified with nitric acid. Weak stainability in the cytoplasm

(González-Chávez et al. 2013) and nucleus (Jimson et al. 2014; González-Chávez et al. 2013; Neves et al. 2011) has been observed in specimens decalcified with nitric acid or HCl. The latter finding suggests that acidic reagents result in deterioration of DNA. Thomas and Doty (1956) analyzed the process of DNA degradation under acidic conditions and found immediate degradation of DNA in relatively mild acidic conditions (pH 2.6). Yamamoto-Fukuda et al. (2000) evaluated DNA retention in cell nuclei of mouse mandible tissue decalcified with several kinds of reagents. Nuclear DNA staining was conducted with methyl green-pyronine Y (MPY) or 4′ 6-diamidino-2-phenylindole (DAPI). No staining was found in specimens decalcified with nitric acid and HCl, whereas those decalcified with EDTA or other acidic reagents (formic and trichloroacetic acids) showed intense staining. Alers et al. (1999) performed in situ hybridization with an oligonucleotide probe in tissue sections of formalin-fixed human vertebra decalcified with EDTA- or HCl-based decalcification reagents. No signal was found in tissues applied to the HCl-based decalcification. They extracted DNA from decalcified tissue blocks and performed electrophoresis in agarose gels. No visible DNA was found in the specimen subjected to HCl-based decalcification, whereas EDTA decalcification retained DNA observed as a smeared (degraded) but intense image. Sarsfield et al. (2000) compared the quality of DNA obtained from decalcified bone marrow trephines. Higher molecular weight DNA was observed in the specimen decalcified with EDTA compared with the specimen decalcified with formic acid. Singh et al. (2013) examined the applicability of DNA extracted from decalcified bones to polymerase chain reaction (PCR) analysis. The comparison was made using several kinds of commercially available decalcification reagents in terms of DNA integrity using real-time PCR analysis. The results clearly showed that reagents containing HCl or nitric acid prominently reduced PCR applicability. Considering such acidity-driven DNA deterioration, EDTA is the most appropriate decalcification reagent to be applied to DNA extraction from compact bones.

Decalcification with EDTA is a generally accepted process for bone DNA extraction (Hochmeister et al. 1991; Imaizumi et al. 2002; Rohland and Hofreiter 2007a), and its effects have been investigated in DNA retrieval and PCR applicability (Imaizumi et al. 2005; Evison et al. 1997; Rohland and Hofreiter 2007b; Kuś et al. 2016; Balayan et al. 2015). Protocols for decalcification slightly vary by laboratory in terms of the EDTA concentration and incubation conditions. However, generally, compact bone is first pulverized into fine powder and then treated with 0.5 M neutral EDTA at 4 °C for 3–5 days (Hochmeister et al. 1991) or 56 °C overnight (Imaizumi et al. 2005). In either case, EDTA decalcification requires a certain amount of time. Therefore, it would be beneficial to increase the rate of the decalcification process.

6.3 Accelerating Decalcification by Microwave (MW) Irradiation

With the expectation of energetic enhancement of chemical reactions, applications of MW irradiation to tissue preparation for microscopy (histoprocessing) have been studied for more than 20 years. In addition to reducing the decalcification time, MW irradiation can be applied to shorten fixation (Login et al. 1991; Login et al. 1998; Giberson and Demaree 1995, 1999; Roncaroli et al. 1991), dehydration, and paraffin embedding (Kovács et al. 1996; Bödör et al. 2007) times and produce clearer signals in immunohistochemical staining by unmasking antigens (Yamashita et al. 1990; McCluggage et al. 1995; Erber et al. 1996; Man and Burgar. 2003).

Studies on the application of MW irradiation to bone decalcification have been conducted by many researchers. Table 6.1 summarizes their results regarding the effects of MW irradiation on reducing decalcification time. As shown in the table, specimen types, the composition of decalcification solutions, and reaction temperatures vary considerably by study. In addition to these factors, the irradiation program (intermittent or not and if so, irradiation frequency), methods to evaluate the progress of decalcification (analysis of calcium decrease radiographically or chemically), and stringency of temperature control also vary. Therefore, we should be careful when interpreting the results shown in Table 6.1. Among the factors varied by study, temperature control is the most crucial to obtain comparable results for validation. A solution in a MW oven is immediately heated by the irradiation. Therefore, feedback control of the reaction temperature by adjusting the frequency of irradiation cycles together with water bath cooling is necessary. Specifically designed MW ovens for these functions are commercially available (Kovács et al. 1996; Bödör et al. 2007; Imaizumi et al. 2013; Ekuni et al. 2006). The results shown in Table 6.1 strongly suggest that MW irradiation has a positive effect on reducing the decalcification time regardless of the type of reagent. However, the effect varies in degree. In EDTA decalcification, the maximum improvement was found in a report by Madden and Henson (1997), showing that MW irradiation reduced the time from 2 months to 15 h, which is equivalent to shortening to one-hundredth of the original time. In contrast to such a drastic improvement, some studies found slight effects (see Table 6.1, Vongsavan et al. 1990; Izumi and Ozawa 1995). Petrović et al. (2015) compared the weight decrease of teeth due to calcium release during EDTA decalcification between MW-irradiated and non-irradiated specimens. Although they found slightly accelerated decalcification in the irradiated group until reaching half the weight, the speed was equivalent to that in the non-irradiated group thereafter. They concluded that EDTA decalcifies teeth independently of MW irradiation. Imaizumi et al. (2013) examined the effect of MW irradiation on EDTA decalcification using bovine compact bone segments standardized in size (cube 2.5 mm on a side). Intermittent MW irradiation was conducted under strict temperature control (56 °C) using an oven specialized for MW histoprocessing. The degrees of decalcification were quantitatively analyzed by precise volume measurement of the undecalcified area using micro-X-ray

Table 6.1 Summary of results on the effects of microwave (MW) irradiation on reducing decalcification time

References	Specimen	Main chemical reagent in the decalcification solution	Maximum temperature at decalcification (°C)	Time needed for complete decalcification		Decrease of decalcification time (%)
				MW (−)[a]	MW (+)[b]	
Vongsavan et al. (1990)	Rat teeth	0.1 M EDTA	39		No improvement	–
Ng and Ng. (1992)	Rabbit femur	22.5% Formic acid		75 h	12 h	16.0
Faria et al. (1992)	Rat mandible	7% EDTA	40	40 days	19 h	2.0
Hellström and Nilsson (1992)	Rat temporal bone	EDTA	35	10 days	6 h	2.5
Fujiyoshi et al. (1992)	Monkey temporal bone	10% EDTA	37	28 days	30 h	4.4
Izumi and Ozawa (1995)	Rat tibia	4.13% EDTA	40	9 days	7 days	77.8
Madden and Henson (1997)	Monkey temporal bone	0.1 M EDTA	45	2 months	15 h	1.0
Kaneko et al. (1999)	Mouse tibia	20% EDTA	50	8.25 days	3.25 days	39.4
Keithley et al. (2000)	Guinea pig tibia	10% EDTA	45	23 days	17 h	3.1
Weisberger et al. (2001)	Human mandible	HCl	40	7 days	3 h	1.8
Cunningham et al. (2001)	Human temporal bone	0.35 M EDTA	42	35 days	80 h	9.5
Tinling et al. (2004)	Mongolian gerbil femur	EDTA	20	45 h	21 h	46.7
Ekuni et al. (2006)	Rat mandible	10% EDTA	45	12 days	6 days	50.0
Pitol et al. (2007a)	Rat maxilla	8.5% EDTA	37	45 days	48 h	4.4

Pitol et al. (2007b)	Dog teeth	8.5% EDTA	38	150 days	35 days	23.3
		5% Nitric acid	38	65 days	15 days	23.1
Marr and Wong (2009)	Rat sternum	Formic acid	37	8 h	4 h	50.0
Sangeetha et al. (2013)	Human mandible	14% EDTA	43	57 days	9 days	15.8
		5% Nitric acid	43	22 days	2 days	9.1
		5% Citric acid	43	33 days	6 days	18.2
	Human premolor	14% EDTA	43	85 days	20 days	23.5
		5% Nitric acid	43	35 days	4 days	11.4
		5% Citric acid	43	42 days	9 days	21.4

[a]No microwave (MW) irradiation
[b]MW irradiated

computed tomography. Although MW irradiation showed a statistically significant effect on the decalcification, the ratio of the undecalcified area at 11 h differed slightly (11.9% with irradiation and 18.6% without irradiation).

Based on the results in these reports, it is reasonable to conclude that MW irradiation is effective to accelerate decalcification. However, the degrees of improvement differ considerably by report because of the lack of a standardized method that produces reliable results comparable with each other, particularly in terms of the MW irradiation conditions. Further studies to overcome these problems should be conducted in the future.

6.4 Deterioration of Nucleic Acids by MW Irradiation

When applying MW irradiation to shortening the bone DNA extraction process, we need to determine whether MW irradiation has an effect on the quality of DNA. Unfortunately, research concerning this issue is currently very scarce. There are two reports that indirectly suggest the effects of MW irradiation on DNA through the results of in situ hybridization for RNA detection in tissue. Note that we regard the two kinds of nucleic acids, DNA and RNA, as equivalent in regard to deterioration in this chapter. Kaneko et al. (1999) decalcified a mouse tibia in a 20% EDTA solution using MW irradiation. Three days were needed for complete decalcification with 2 h of irradiation at 50 °C each day. The bone was sectioned and in situ hybridization was performed to detect mouse α1 chain of procollagen type I mRNA. The intensity of the signal was equivalent to that in the specimen without irradiation. They also showed that excessive MW irradiation (2 h of irradiation each day for 30 days) resulted in signal disappearance. These results suggest that MW irradiation mildly deteriorates RNA. Another study reported in situ reverse transcription-polymerase chain reaction (in situ RT-PCR) analysis. Ekuni et al. (2006) decalcified a rat mandible in a 10% EDTA solution with MW irradiation. The sample was completely decalcified in 6 days with 6 h of irradiation at 45 °C each day. The results of in situ RT-PCR targeting three kinds of tissue-specific genes showed no significant difference between irradiated and non-irradiated specimens in terms of signal intensities. This result suggests that RNA retains a certain quality after MW irradiation. Simultaneously, they directly estimated the quality of RNA by electrophoresing RNA extracted from the processed tissues. Although the RNA obtained from both irradiated and non-irradiated tissues was severely degraded, no significant difference was found in its degree. The authors regarded that this degradation was largely caused by fixation with paraformaldehyde. It is well known that tissue fixatives such as formalin (37% formaldehyde) and paraformaldehyde cause DNA deterioration because of their major action as fixatives, cross-linking DNA molecules with nucleoproteins (Srinivasan et al. 2002; Vincek et al. 2003; Paireder et al. 2013; Bonin et al. 2003; Santos et al. 2009; Gilbert et al. 2007; Greer et al. 1991), and lowering the pH. Even in experiments to validate decalcification, many researchers fix a specimen with formaldehyde prior to decalcification. We should bear this fact in mind to interpret the results regarding the integrity of DNA or RNA obtained from tissues that had been formalin-fixed to any degree.

Several reports have directly examined the quality of DNA extracted from MW-irradiated specimens. Hsu et al. (1991) applied MW irradiation to fix human liver tissue as an alternative to the conventional method with a fixative. They immersed the tissue in phosphate-buffered saline (PBS) and immediately irradiated by MWs for 20 s, resulting in elevation of the reaction temperature to 50–70 °C. Koshiba et al. (1993) also fixed mouse liver in PBS by irradiation with MW until the reaction temperature reached 60 °C. In both studies, intact high molecular weight DNA was observed by electrophoresis. However, the times of the MW irradiation were both too short to conclude that MW irradiation has no deteriorative effect on DNA. Bödör et al. (2007) processed soft tissues from fixation to paraffin embedding using two different protocols. One was semi-automatic conventional histoprocessing without MW irradiation and the other employed an automatic vacuum-assisted tissue processor with MW irradiation at every step in the histoprocessing (fixation, dehydration, and embedding). Formalin was used for the fixation in both protocols. The protocol with MW irradiation reduced the total time for histoprocessing and the DNA from the specimens had high amounts of PCR-amplified products of the human β-globin gene, whereas slight or intermediate amplification was observed in specimens processed with the conventional protocol. The decrease of amplifiable DNA in the latter protocol might be due to longer fixation with formalin. The results obtained in this study at least indicate that a considerable amount of amplifiable DNA remains in tissue after MW irradiation. PCR success of DNA extracted from tissue sections exposed to MW irradiation was examined by Man and Burgar (2003). Sections prepared from formalin-fixed, paraffin-embedded lung tissue blocks were immersed in 10 mM citrate buffer and exposed to five MW irradiations for 2 min each. The portion enriched with cells was microdissected and subjected to DNA extraction. A total of 15 PCR targets from 95 to 290 base pairs were amplified. Targets with a long length failed to be amplified, whereas control sections (no irradiation or immersion in citrate buffer) resulted in large amounts of PCR products for all targets. Although this result suggests a deteriorative effect of MWs on DNA, it is possible that citrate buffer might contribute to DNA deterioration. The previously discussed study by Imaizumi et al. (2013) performed PCR targeting a bovine mitochondrial DNA D-loop region in DNA extracted from standardized cubic compact bone segments. By comparing PCR efficiencies between MW-irradiated and non-irradiated specimens both incubated at 56 °C, the irradiated specimen showed poorer DNA amplification at every incubation period, 3, 7, 11 and 24 h. This result affirmatively suggests DNA deterioration by MW irradiation. Finally, in vitro studies were performed to examine the effect of MWs on DNA quality. Debode et al. (2007) extracted DNA from soybean powder. An aliquot of the DNA solution was subjected to four different physical treatments (MW irradiation, heating at 99 °C, sonication, and autoclaving) that deteriorate DNA. In the MW experiment, the DNA solution in Eppendorf tubes was irradiated using a conventional MW oven for

0–15 min and then electrophoresed. It was shown that MW irradiation severely degraded DNA to a considerable degree. The authors mentioned that the irradiation time was limited to 15 min at maximum because of the durability of the tube against heat produced during the MW irradiation. This finding suggests that a DNA solution might reach a high temperature even when irradiated for less than 15 min. The results from the heating experiment conducted simultaneously clearly showed time-dependent DNA degradation at 99 °C. Therefore, some DNA degradation by MW irradiation might be caused by the temperature increase.

As discussed above, there have been several studies that examined the effects of MW irradiation on the quality of nucleic acids. Each examination was well designed to reveal the issues of interest in each field. However, findings have not accumulated enough to conclude the effect of MWs as a general property, regardless of positive or negative effects. The most crucial problem is the temperature increase along with MW irradiation. This makes it difficult to estimate MW effects independently of heat-induced effects. A study designed to obtain generally acceptable results should be performed to resolve this issue.

6.5 Conclusion

MW irradiation is a powerful treatment to improve the rate of histoprocessing as introduced by many researchers. The range of its application is wide because of the versatile enhancement of water-mediated reactions. Currently, many laboratories routinely use MW irradiation to fix tissue blocks and unmask antigens in immunohistochemistry, and bone decalcification to section compact bones with certain reliability. In addition to these uses, MW irradiation is considered to have huge potential to enhance reactions in genetic analyses. Although the reports introduced in this chapter were limited to studies that conducted genetic analyses of specimens decalcified with MW irradiation, direct applications to other genetic analyses, such as PCR itself, DNA-specific staining, and other reactions in genetic engineering may be possible. For such studies, establishing a system to enable strict control of irradiation conditions is a matter of high priority.

References

Alers JC, Krijtenburg PJ, Vissers KJ, van Dekken H (1999) Effect of bone decalcification procedures on DNA in situ hybridization and comparative genomic hybridization. EDTA is highly preferable to a routinely used acid decalcifier. J Histochem Cytochem 47(5):703–710

Balayan A, Kapoor A, Chaudhary G, Raina A (2015) Evaluation of techniques for human bone decalcification and amplification using sixteen STR markers. Egypt J Forensic Sci 5(1):30–35

Bödör C, Schmidt O, Csernus B, Rajnai H, Szende B (2007) DNA and RNA isolated from tissues processed by microwave-accelerated apparatus MFX-800-3 are suitable for subsequent PCR and Q-RT-PCR amplification. Pathol Oncol Res 13(2):149–152

Bonin S, Petrera F, Niccolini B, Stanta G (2003) PCR analysis in archival postmortem tissues. Mol Pathol 56(3):184–186

Boyde A, Jones SJ (1996) Scanning electron microscopy of bone: instrument, specimen, and issues. Microsc Res Tech 33(2):92–120

Callis G, Sterchi D (1998) Decalcification of bone: literature review and practical study of various decalcifying agents. Methods, and their effects on bone histology. J Histotechnol 21:49–58

Cui F, Li Y, Ge J (2007) Self-assembly of mineralized collagen composites. Mater Sci Eng R 57:1–27

Cunningham CD 3rd, Schulte BA, Bianchi LM, Weber PC, Schmiedt BN (2001) Microwave decalcification of human temporal bones. Laryngoscope 111(2):278–282

Debode F, Janssen E, Berben G (2007) Physical degradation of genomic DNA of soybean flours does not impair relative quantification of its transgenic content. Eur Food Res Technol 226(1):273–280

Ekuni D, Firth JD, Putnins EE (2006) RNA integrity and in situ RT-PCR in dento-alveolar tissues after microwave accelerated demineralisation. Arch Oral Biol 51(2):164–169

Erber WN, Gibbs TA, Ivey JG (1996) Antigen retrieval by microwave oven heating for immunohistochemical analysis of bone marrow trephine biopsies. Pathology 28(1):45–50

Evison MP, Smillie DM, Chamberlain AT (1997) Extraction of single-copy nuclear DNA from forensic specimens with a variety of postmortem histories. J Forensic Sci 42(6):1032–1038

Faria MR, Dechichi P, Rode SM (1992) Decalcification of mineralized rat mandible tissues with the aid of microwaves. Rev Paul Med 110(6):283–284

Fujiyoshi T, Mogi G, Watanabe T, Matsushita F (1992) Undecalcified temporal bone morphology: a methodology useful for gross to fine observation and three-dimensional reconstruction. Acta Otolaryngol Suppl 493:7–13

Giberson RT, Demaree RS Jr (1995) Microwave fixation: understanding the variables to achieve rapid reproducible results. Microsc Res Tech 32(3):246–254

Giberson RT, Demaree RS Jr (1999) Microwave processing techniques for electron microscopy: a four-hour protocol. Methods Mol Biol 117:145–158

Gilbert MT, Haselkorn T, Bunce M, Sanchez JJ, Lucas SB, Jewell LD, Van Marck E, Worobey M (2007) The isolation of nucleic acids from fixed, paraffin-embedded tissues-which methods are useful when? PLoS One 2(6):e537

González-Chávez SA, Pacheco-Tena C, Macías-Vázquez CE, Luévano-Flores E (2013) Assessment of different decalcifying protocols on Osteopontin and Osteocalcin immunostaining in whole bone specimens of arthritis rat model by confocal immunofluorescence. Int J Clin Exp Pathol 6(10):1972–1983

Greer CE, Lund JK, Manos MM (1991) PCR amplification from paraffin-embedded tissues: recommendations on fixatives for long-term storage and prospective studies. PCR Methods Appl 1(1):46–50

Hellström S, Nilsson M (1992) The microwave oven in temporal bone research. Acta Otolaryngol Suppl 493:15–18

Hochmeister MN, Budowle B, Borer UV, Eggmann U, Comey CT, Dirnhofer R (1991) Typing of deoxyribonucleic acid (DNA) extracted from compact bone from human remains. J Forensic Sci 36(6):1649–1661

Hsu HC, Peng SY, Shun CT (1991) High quality of DNA retrieved for Southern blot hybridization from microwave-fixed, paraffin-embedded liver tissues. J Virol Methods 31:251–261

Imaizumi K, Saitoh K, Sekiguchi K, Yoshino M (2002) Identification of fragmented bones based on anthropological and DNA analyses: case report. Leg Med (Tokyo) 4(4):251–256

Imaizumi K, Noguchi K, Shiraishi T, Sekiguchi K, Senju H, Fujii K, Yoshida K, Kasai K, Yoshino M (2005) DNA typing of bone specimens–the potential use of the profiler test as a tool for bone identification. Leg Med (Tokyo). 7(1):31–41

Imaizumi K, Taniguchi K, Ogawa Y (2013) An evaluation of the effect of microwave irradiation on bone decalcification aimed to DNA extraction. Leg Med (Tokyo). 15(5):272–277. doi:10.1016/j.legalmed.2013.06.001

Izumi N, Ozawa H (1995) Rapid microwave decalcification for bone tissues. Saibo 27:141–145 (in Japanese)

Jimson S, Balachander N, Masthan KMK, Elumalai R (2014) A comparative study in bone decalcification using different decalcifying agents. Int J Sci Res 3:1226–1229

Kaneko M, Tomita T, Nakase T, Takeuchi E, Iwasaki M, Sugamoto K, Yonenobu K, Ochi T (1999) Rapid decalcification using microwaves for in situ hybridization in skeletal tissues. Biotech Histochem 74(1):49–54

Keithley EM, Truong T, Chandronait B, Billings PB (2000) Immunohistochemistry and microwave decalcification of human temporal bones. Hear Res 148(1–2):192–196

Koshiba M, Ogawa K, Hamazaki S, Sugiyama T, Ogawa O, Kitajima T (1993) The effect of formalin fixation on DNA and the extraction of high-molecular-weight DNA from fixed and embedded tissues. Pathol Res Pract 189(1):66–72

Kovács L, Szende B, Elek G, Lapis K, Horváth O, Hiszek I, Tamási A, Schmidt O (1996) Working experience with a new vacuum-accelerated microwave histoprocessor. J Pathol 180 (1):106–110

Kuś M, Ossowski A, Zielińska G (2016) Comparison of three different DNA extraction methods from a highly degraded biological material. J Forensic Legal Med 40:47–53. doi:10.1016/j.jflm.2016.03.002

Login GR, Schnitt SJ, Dvorak AM (1991) Microwave fixation provides rapid, primary fixation for light and electron microscopy and for immunohistochemistry and immunocytochemistry. Eur J Morphol 29(3):206–210

Login GR, Leonard JB, Dvorak AM (1998) Calibration and standardization of microwave ovens for fixation of brain and peripheral nerve tissue. Methods 15(2):107–117

Madden VJ, Henson MM (1997) Rapid decalcification of temporal bones with preservation of ultrastructure. Hear Res 111(1–2):76–84

Man YG, Burgar A (2003) An antigen unmasking protocol that satisfies both immunohistochemistry and subsequent PCR amplification. Pathol Res Pract 199(12):815–825

Marr AL, Wong A (2009) Effects of microwave fixation and decalcification on rodent tissue. J Histotechnol 32(4):190–192

McCluggage WG, Roddy S, Whiteside C, Burton J, McBride H, Maxwell P, Bharucha H (1995) Immunohistochemical staining of plastic embedded bone marrow trephine biopsy specimens after microwave heating. J Clin Pathol 48(9):840–844

Neves Jdos S, Omar NF, Narvaes EA, Gomes JR, Novaes PD (2011) Influence of different decalcifying agents on EGF and EGFR immunostaining. Acta Histochem 113(4):484–488. doi:10.1016/j.acthis.2010.04.006

Ng KH, Ng LL (1992) Microwave-stimulated decalcification of compact bones. Eur J Morphol 30(2):150–155

Olszta MJ, Cheng X, Jee SS, Kumar R, Kim Y, Kaufman MJ, Douglas EP (2007) Gower LB bone structure and formation: a new perspective. Mater Sci Eng R 58:77–116

Ortner DJ, Turner-Walker G (2003) The biology of skeletal tissues. In: Ortner DJ (ed) Identification of pathological conditions in human skeletal remains, 2nd edn. Academic Press, London, pp 11–35

Paireder S, Werner B, Bailer J, Werther W, Schmid E, Patzak B, Cichna-Markl M (2013) Comparison of protocols for DNA extraction from long-term preserved formalin fixed tissues. Anal Biochem 439(2):152–160. doi:10.1016/j.ab.2013.04.006

Petrović V, Nikolić I, Tijanić M, Obradović R, Jović M, Janjić M, Jordačević J, Petrović A (2015) The influence of different agitation techniques on the results of teeth demineralization by alkaline solution of EDTA. Acta Medica Median 54(4):24–31

Pitol DL, Caetano FH, Lunardi LO (2007a) Microwave-induced fast decalcification of rat bone for electron microscopic analysis: an ultrastructural and cytochemical study. Braz Dent J 18(2):153–157

Pitol DL, Issa JPM, Caetano FH, Lunardi LO (2007b) Decalcification dynamic of dog mineralized tissue by microwaves. Int J Morphol 25(2):309–313

Rohland N, Hofreiter M (2007a) Ancient DNA extraction from bones and teeth. Nat Protoc 2 (7):1756–1762

Rohland N, Hofreiter M (2007b) Comparison and optimization of ancient DNA extraction. Biotechniques 42(3):343–352

Roncaroli F, Mussa B, Bussolati G (1991) Microwave oven for improved tissue fixation and decalcification. Pathologica 83(1085):307–310

Sangeetha R, Uma K, Chandavarkar V (2013) Comparison of routine decalcification methods with microwave decalcification of bone and teeth. J Oral Maxillofac Pathol 17(3):386–391. doi:10.4103/0973-029X.125204

Santos S, Sá D, Bastos E, Guedes-Pinto H, Gut I, Gärtner F, Chaves R (2009) An efficient protocol for genomic DNA extraction from formalin-fixed paraffin-embedded tissues. Res Vet Sci 86 (3):421–426. doi:10.1016/j.rvsc.2008.08.007

Sarsfield P, Wickham CL, Joyner MV, Ellard S, Jones DB, Wilkins BS (2000) Formic acid decalcification of bone marrow trephines degrades DNA: alternative use of EDTA allows the amplification and sequencing of relatively long PCR products. Mol Pathol 3(6):336

Singh VM, Salunga RC, Huang VJ, Tran Y, Erlander M, Plumlee P, Peterson MR (2013) Analysis of the effect of various decalcification agents on the quantity and quality of nucleic acid (DNA and RNA) recovered from bone biopsies. Ann Diagn Pathol 17(4):322–326. doi:10.1016/j.anndiagpath.2013.02.001

Srinivasan M, Sedmak D, Jewell S (2002) Effect of fixatives and tissue processing on the content and integrity of nucleic acids. Am J Pathol 161(6):1961–1971

Thomas CA, Doty P (1956) The mild acidic degradation of desoxyribose nucleic acid. J Am Chem Soc 78(9):1854–1860

Tinling SP, Giberson RT, Kullar RS (2004) Microwave exposure increases bone demineralization rate independent of temperature. J Microsc 215(Pt 3):230–235

Vincek V, Nassiri M, Nadji M, Morales AR (2003) A tissue fixative that protects macromolecules (DNA, RNA, and protein) and histomorphology in clinical samples. Lab Investig 83(10):1427–1435

Vongsavan N, Matthews B, Harrison GK (1990) Decalcification of teeth in a microwave oven. Histochem J 22:377–380

Weiner S, Wagner HD (1998) The material bone: structure-mechanical function relations. Annu Rev Mater Sci 28:271–298

Weisberger EC, Hilburn M, Johnson B, Nguyen C (2001) Intraoperative microwave processing of bone margins during resection of head and neck cancer. Arch Otolaryngol Head Neck Surg 127(7):790–793

Yamamoto-Fukuda T, Shibata Y, Hishikawa Y, Shin M, Yamaguchi A, Kobayashi T, Koji T (2000) Effects of various decalcification protocols on detection of DNA strand breaks by terminal dUTP nick end labelling. Histochem J 32(11):697–702

Yamashita S, Katsumata O, Sekine R (1990) Evaluation of microwave irradiation in immunohistochemical reactions. Acta Histochem Cytochem 23:553–562

Chapter 7
Ultra-High-Frequency Electromagnetic Radiation and Reactive Species in Mammals

Orlando Vieira Furtado-Filho

Abstract The increasing emission of electromagnetic waves by various devices has increased concerns of possible health effects. Exposure to ultra-high-frequency (UHF) radiation is increasing due to the development of wireless communication technologies. This type of radiation has been linked to changes in organisms through the generation of RS (Oxygen and Nitrogen). The aim of this review was to discuss the effects of ultra-high-frequency electromagnetic radiation (UHF-EMR) on the metabolism of RS and its consequences for mammals. It was discussed the relationship of UHF electromagnetic waves with oxidants and antioxidants in mammals. After reviewing several studies, it is concluded that UHF-EMR can cause changes in biomarkers of oxidative/nitrosative damage and can cause decreases in antioxidant defenses, which may result from these biomarkers' fight against the oxidative damage to macromolecules induced by ROS/RNS.

Keywords Electromagnetic Radiation • ROS • RNS • Oxidative Damage • Nitrosative Damage • Antioxidant • Oxidative Stress • Nitrosative Stress

7.1 Introduction

Technological advancement has increased the exposure of organisms to electromagnetic waves. Electrical networks, broadcast radio and television stations, aircraft radar, microwave ovens, computer monitors and lights are some examples of objects that produce electromagnetic waves. Recently, the world has seen an increase in the number of mobile phones and base stations. These devices produce ultra-high-frequency electromagnetic radiation (UHF-EMR). Currently, this type of EMR has been linked to changes in biological systems. Physiological, molecular and behavioral changes have been observed in various organisms (D'Andrea et al.

O.V. Furtado-Filho (✉)
Department of Biophysics and Biotechnology Center, Federal University of Rio Grande do Sul (UFRGS), Bento Gonçalves, 9500, 15005 Porto Alegre, RS, Brazil

Military School of Porto Alegre (CMPA), Brazilian Army, Porto Alegre, Brazil
e-mail: orlandovff@gmail.com

C.D. Geddes (ed.), *Microwave Effects on DNA and Proteins*,
DOI 10.1007/978-3-319-50289-2_7

2003; Paulraj and Behari 2006; Silva et al. 2015). In this chapter, it was analyzed the data reported in the literature on the effect of UHF-EMR on the metabolism of reactive species of oxygen and nitrogen (ROS and RNS) and its consequences for mammals.

7.2 Electromagnetic Radiation

In the nineteenth century, James Maxwell expressed the existence of electromagnetic waves by expressing mathematically the laws of electricity and magnetism in a formula (apud Hall 2008). Later that century, Heinrich Hertz produced electromagnetic waves for the first time in his laboratory (apud Sengupta and Sarkar 2003). Since these two researchers' discoveries, the use of electromagnetic radiation has increased due to the development of new technologies, including wireless and mobile telephony.

In addition to the technologies mentioned above, devices such as radio, TV, the electric transmission network, visible lights and microwaves emit non-ionizing radiation. These emissions have energies lower than 10 eV and wavelengths greater than 200 nm. They have insufficient energy to produce electron emission (Eblem 2011; Júnior and Martin 2006). The frequency range from 800 to 2200 MHz is used for mobile telephony. These frequencies are in the range classified as UHF (from 300 MHz to 3 GHz) (Fig. 7.1) (Verschaeve and Maes 1998; Figura and Teixeira 2007).

The scientific literature has shown that, given certain conditions, UHF-EMR induces ROS and RNS production in several biological systems (Moustafa et al. 2001; Zmyslony et al. 2004; Simko et al. 2006; Yurekli et al. 2006; Arthur 2007; Valko et al. 2007; Phillips et al. 2009; Olgar et al. 2015).

7.3 Reactive Oxygen Species (ROS) and Reactive Nitrogen Species (RNS)

A chemical species capable of independent existence that contains one or more unpaired electrons is defined as free radical. An unpaired electron is one that occupies an atomic or molecular orbital alone (Cadenas 1995; Halliwell and Gutteridge 2007). The superoxide and hydroxyl radicals are examples of oxygen free radicals. Nitric oxide and nitrogen dioxide are nitrogen free radicals.

ROS and RNS are composed of free radicals and molecules that can generate radicals. Hydrogen peroxide and hydroxyl are examples of ROS. Peroxynitrite and nitric oxide are examples of RNS.

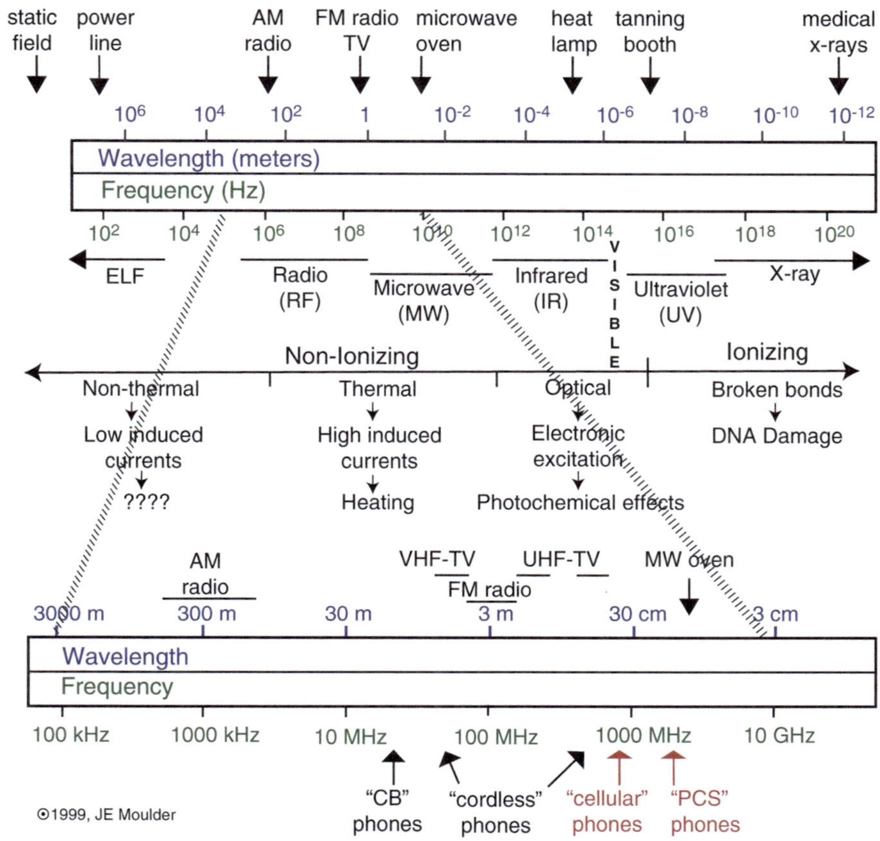

Fig. 7.1 Electromagnetic spectrum (Moulder 2000)

7.3.1 Biochemistry of RS

7.3.1.1 ROS

During oxidative energy production in mitochondria, dioxygen is reduced by four electrons from cytochrome oxidase, forming two molecules of water. However, in mammals, a leak of electrons from the respiratory chain can result in oxygen being reduced by one electron, producing superoxide ($O_2^{-\bullet}$) (Lenaz 2012). The generation of ROS by mitochondria depends on the tension of O_2 and its concentration in the tissue.

The rotation of electrical charges generates a magnetic field. The pairing of electrons of opposite spins counteracts this effect, but an unpaired electron causes a molecule to be paramagnetic (slightly attracted by a magnetic field) and possibly highly reactive. Oxygen gas is paramagnetic, containing two unpaired electrons with the same spin (Cadenas 1995; Fridovich 1998; Lenaz 2012). Thus, the O_2 electrons tend to occupy these spins through monoelectronic reduction reactions leading to the formation of species $O_2^{-\bullet}$ and O_2^{2-} (superoxide and peroxide, respectively, i.e., reactions 7.1 and 7.2).

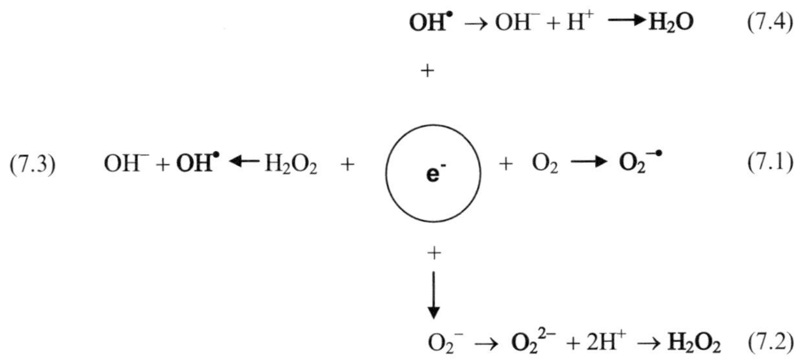

The monoelectronic reduction of H_2O_2 occurs by a heterolytic fission of the O-O bond leading to the formation of a hydroxyl radical ($OH^\bullet$) and a hydroxyl anion (OH^-) (reaction 7.3). When the hydroxyl radical is reduced, it forms the hydroxide anion, which can react with a proton to form water (reaction 7.4). In vivo, the electron donors for H_2O_2 are usually transition metals such as iron and copper through the Fenton reaction (reaction 7.5). Fe (III) ions produced by the reaction can be recycled by reaction with reducing agents such as $O_2^{-\bullet}$ to generate Fe (II) (reaction 7.6). The sum of reactions 7.5 and 7.6 results in the Haber-Weiss reaction (reaction 7.7) (Cadenas 1995; Mccord and Day 1978).

$$\mathrm{Fe(II)} + \mathbf{H_2O_2} \rightarrow \mathrm{Fe(III)} + \mathrm{OH^-} + \mathbf{OH^\bullet} \quad (7.5) \text{ (reaction Fenton)}$$
$$\mathrm{Fe(III)} + \mathbf{O_2^{-\bullet}} \rightarrow \mathrm{Fe(II)} + \mathrm{O_2} \quad (7.6)$$
$$\mathbf{O_2^{-\bullet}} + \mathrm{H_2O_2} \rightarrow \mathrm{O_2} + \mathrm{OH^-} + \mathbf{OH^\bullet} \quad (7.7) \text{ (reaction Haber-Weiss)}$$

The superoxide radical is generated in all cellular compartments. It can be produced by the autoxidation of small molecules such as catecholamines and hydroquinones, activation of macrophages and neutrophils (the action of NADPH oxidase), the activity of various oxidases such as xanthine oxidase, cytochrome P450 and the electron transport chain of mitochondria and chloroplasts through Fe-S protein and coenzyme Q (Cadenas 1995; Halliwell and Gutteridge 2007).

The superoxide radical, at neutral pH, has a half-life of seconds. It has low reactivity and is unable to cross membranes. However, it is potentially harmful, as it is able to diffuse considerable distances to find a target. Diffusion occurs because superoxide can pass through ion channels. This radical can cause significant oxidative damage in subcellular structures through the formation of hydroxyl radicals via the Haber-Weiss reaction. Some enzymes such as aconitase can be inactivated directly by $O_2^{-\bullet}$, and the low activity of this enzyme is considered an indicator of oxidative stress in vivo (Cadenas 1995; Lenaz 2012; Anderson et al. 1998).

Hydrogen peroxide is produced mainly by the dismutation of superoxide, but there are enzymes that produce it without the intermediary of $O_2^{-\bullet}$ (e.g., glycolate oxidase, D-amino acid oxidase, urate oxidase). Although H_2O_2 is a weak oxidizing agent, it is capable of inactivating some enzymes directly, usually by oxidation of the thiol groups (such as glyceraldehyde-3-phosphate dehydrogenase) (Halliwell and Gutteridge 2007). Sohal and Weindruch found that the higher the concentration of superoxide and hydrogen peroxide in the mitochondria, the lower the lifespan of some animals (Sohal and Weindruch 1996).

The toxicity of superoxide and hydrogen peroxide is dependent on the presence of iron or copper, which leads to the formation of hydroxyl radicals via the Haber-Weiss reaction (reaction 7.7). Thus, the nature and location of the hydroxyl radical damage depends on the presence of transition metals. The hydroxyl radical is one of the most reactive chemical species known; it can rapidly react with all types of cellular molecules (sugars, amino acids, proteins, nucleic acids and lipids, especially those containing polyunsaturated fatty acids) (Halliwell and Gutteridge 2007). The rate constants for the reaction of hydroxyl radical with biomolecules are in the range of $10^9\ M^{-1}\ s^{-1}$ (Ischiropoulos 1992). Due to its high reactivity, the hydroxyl radical has a small diffusion radius inside the cell, causing damage near the sites where it is generated.

7.3.1.2 RNS

During some biological processes such as muscle relaxation, neurotransmission and immune regulation, there is the action of $NO^\bullet$ which is in large quantity. The RNS is synthesized by the enzyme nitric oxide synthase (NOS) which convert arginine into nitric oxide and citrulline (reaction 7.8).

$$\text{L-arginine} \xrightarrow{\text{NOS}} NO^\bullet + \text{L-citrulline} \tag{7.8}$$

Once synthesized, this free radical nitrogen ($NO^\bullet$) diffuses rapidly. If exposed to air, reacts with oxygen and produces nitrogen dioxide $NO^\bullet_2$ (reaction 7.9). This is a potent initiator of lipid peroxidation in organisms (Vasconcelos et al. 2007).

$$2NO^\bullet + O_2 \rightarrow 2NO^\bullet{}_2 \tag{7.9}$$

When $NO^\bullet$ is produced in a molecular environment that exists $O_2^{\bullet-}$, there is the formation of peroxynitrite ($ONOO^-$) (reaction 7.10). This RS causes damage to proteins, especially the sulphydryl group present in some residues amino acids. The rate constant for the reaction of $NO^\bullet$ with $O_2^{\bullet-}$ is in the range of $6 \times 10^9\ M^{-1}\ s^{-1}$.

$$NO^{\bullet} + O_2^{\bullet -} \rightarrow ONOO^{-} \quad (7.10)$$

In the presence of hydrogen protons, peroxynitrite can generate the hydroxyl radical, even without the presence of transition metals (reação11) (Beckman and Koppenol 1996).

$$ONOO^{-} + H^{+} \rightarrow OH^{\bullet} + NO^{\bullet}{}_2 \quad (7.11)$$

7.4 Production of ROS and RNS by EMR

7.4.1 ROS

ROS are generated naturally by cellular metabolism or induced by exogenous agents, such as parasites, pollutants and radiation. Electromagnetic radiation can increase ROS by the action of NADH oxidase, via the Fenton reaction or by altering the expression or activity of antioxidant enzymes.

7.4.1.1 Via NADH Oxidase

NADH oxidase is a plasma membrane protein responsive to exogenous agents. This oxidase is present in both animals and plants (Castillo-Olivares et al. 2000; Ushio-Fukai 2006). Its enzymatic activity is stimulated by hormones and growth factors in rat liver and by LPS (lipopolysaccharide) in mouse cardiomyocytes (Bruno et al. 1992; Vasconcelos et al. 2007; Beckman and Koppenol 1996). In the latter work, the increased activity of NADH oxidase induced higher expression of COX-2 (cyclooxygenase 2). Once stimulated, the enzyme produces the free radical $O_2^{\bullet -}$ (Peng et al. 2005). Some studies have shown that non-ionizing electromagnetic waves at UH frequencies induce the formation of ROS through the activation of plasma membrane NADH oxidase (Friedman et al. 2007; Desai et al. 2009). The ROS then activate extracellular matrix metalloproteinases that form Hb-EGF (heparin-binding epidermal growth factor) after cleavage of pro-Hb-EGF from the plasma membrane. This epidermal growth factor (EGF) stimulates the EGF receptor that triggers the activation of MAPK (mitogen-activated protein kinase), especially ERK (extracellular-signal-regulated kinase) (Friedman et al. 2007) (Fig. 7.2).

7.4.1.2 Via the Fenton Reaction

In 1876, Fenton found that the reaction of ferrous iron with hydrogen peroxide oxidizes organic compounds (Koppenol 2001). Today, the Fenton reaction is used

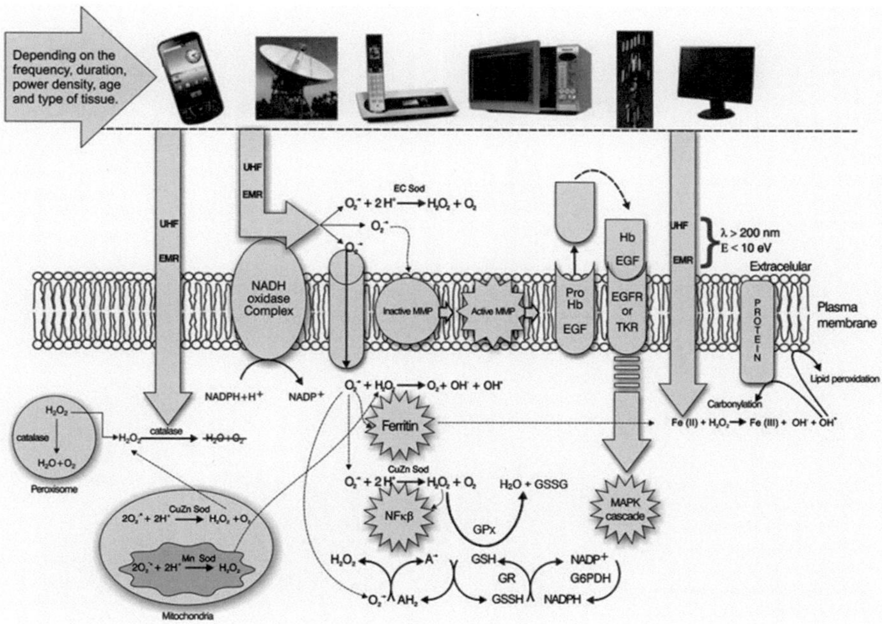

Fig. 7.2 Mechanisms by which UHF-EMR generates ROS in the plasma membrane and cytoplasm. This figure also shows the consequences for cell signaling and antioxidant defense systems in addition to oxidative damage. *UHF-EMR* ultra-high-frequency electromagnetic radiation, *EC SOD* extracellular superoxide dismutase, *CuZn SOD* copper-zinc superoxide dismutase, *Mn SOD* manganese superoxide dismutase, *MMP* extracellular matrix metalloproteinases, *EGF* epidermal growth factor, *Hb-EGF* heparin-binding epidermal growth factor, *MAPK* mitogen-activated protein kinase, *NF-κB* nuclear factor kappa-light-chain-enhancer of activated B cells, *GR* glutathione reductase, *G6PDH* glucose-6-phosphate dehydrogenase, *GSH* reduced glutathione, *GSSG* oxidized glutathione, AH_2 ascorbic acid, $A^{-\bullet}$ ascorbyl radical; $O_2^{-\bullet}$ superoxide radical, H_2O_2 hydrogen peroxide, $OH^{\bullet}$ hydroxyl radical, OH^{-} hydroxyl anion

to destroy pollutants and to improve the biodegradability of organic molecules (Mackul'Ak and Prousek 2011). This reaction is also related to a number of neurodegenerative (Mackul'Ak and Prousek 2011; Halliwell 2006; Altamura & Muckenthaler 2009; Uttara et al. 2009) and cardiovascular (Madamanchi et al. 2005) diseases. As described previously in this review, the products of this reaction are ferric iron, hydroxyl anion and hydroxyl free radical (Halliwell and Gutteridge 2007). This is extremely reactive with a rate constant of nonspecific reaction of 10^9 M^{-1} S^{-1} (Augusto 2006). Some studies show that the interaction of electromagnetic fields with the Fe^{+2} ion in the Fenton reaction (Phillips et al. 2009), leads to the formation of hydroxyl radical. Céspedes and Ueno (2009) discovered the functional alteration of the protein ferritin, in organisms with large amounts of iron, after exposure to magnetic field (Céspedes and Ueno 2009). Depending on the molecular environment, the release of iron by ferritin may lead to the formation of ROS (Fig. 7.2).

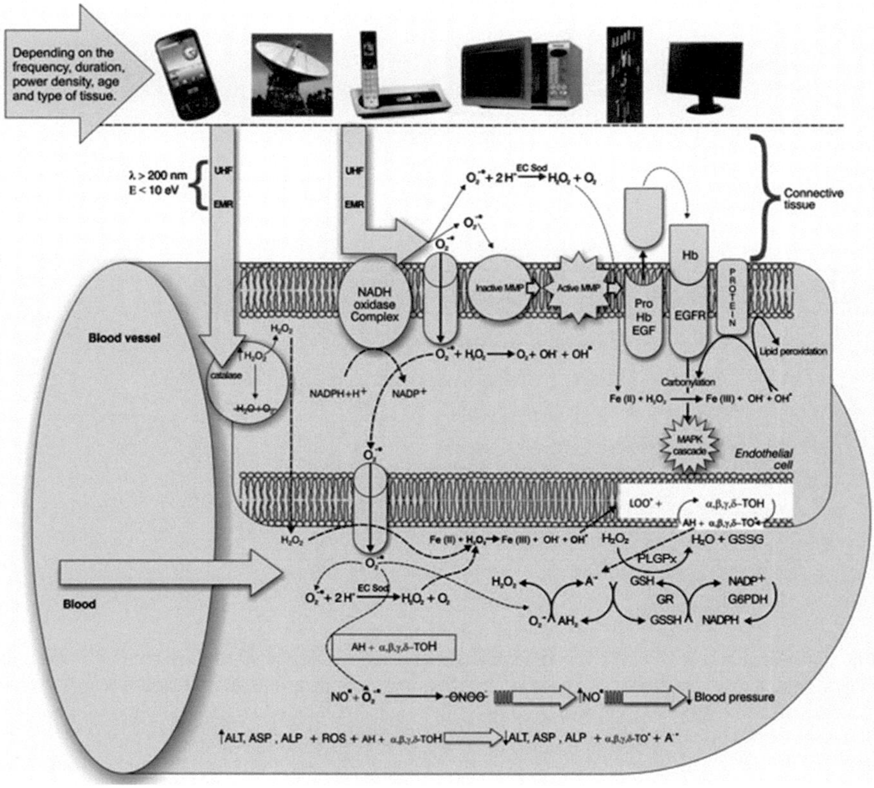

Fig. 7.3 Mechanisms proposed to explain the effects of UHF-EMR on the metabolism of ROS in epithelial cells and inside blood vessels. This figure also shows the consequences for antioxidant defense systems in addition to oxidative damage in epithelial cells and blood vessels. *α,β,γ,δ-TOH* vitamin E, *α,β,γ,δ-TO•* tocopheryl radical, *LOO•* lipoperoxyl radical, *NO•* nitric oxide, *ONOO⁻* peroxynitrite, *ALT* alanine aminotransferase, *ASP* aspartate aminotransferase, *ASP* alkaline phosphatase.

7.4.1.3 Via Alteration of Antioxidant Enzymes

A decrease in antioxidants is one way to cause oxidative stress (Sies 1993). An increase in ROS by UHF-EMR via NADH oxidase, can lead to the inactivation of antioxidant enzymes. The decrease in enzyme activity may be caused by the oxidation of sulfhydryls or by the loss of metals in metalloproteins, leading to structural changes in these proteins (Dean et al. 1997).

The interaction of electromagnetic waves with metals present in enzymes can lead to structural changes of these molecules, altering the activity of some antioxidant enzymes (Kesari and Behari 2008). The presence of iron in catalase and copper in superoxide dismutase could make these proteins more susceptible to structural changes induced by UHF-EMR, leading to changes in enzyme activity (Fig. 7.3).

7.4.2 RNS

There are few studies that relate the UHF-EMR effects on the metabolism of RNS specifically. The generation of RNS may occur after formation of superoxide, via NADH oxidase, or by the action of electromagnetic radiation on NOS.

7.4.2.1 Via NADH Oxidase

The RNS increase their number when there is an increase in ROS (Rosales 2014). Thus, when the UHF-EMR acts on NADH oxidase (Fig. 7.1), this produces superoxide can react with nitric oxide, peroxynitrite generating (reaction 7.11).

7.4.2.2 Via NOS (Nitric Oxide Synthase)

NOS generates nitric oxide when it converts L-arginine into L-citrulline (reaction 7.8). In its dimer form, NOS oxidase domain has a heme group (Fig. 7.4). It is likely iron presence in this area can make the most sensitive enzyme to the effect of UHF-EMR, increasing its activity. Some studies are showing increased levels of $NO^{\bullet}$ in various mammalian organs after continuous exposure UHF-EMR (Olgar et al. 2015; Bilgici et al. 2013).

7.5 Oxidative and Nitrosative Damages to Biomolecules

7.5.1 Lipids

The action of ROS/RNS on biological membranes promotes chain reactions that can lead to peroxidation of membrane phospholipids containing unsaturated fatty acids, generating peroxyl radicals, alkoxyl and lipid hydroperoxides. The peroxidation of membrane lipids alters membrane fluidity and can lead to rupture. In the case of membrane rupture, organelles could release proteases, lipases and ions. Calcium, for example, can activate a series of Ca^{2+}-dependent enzymes and several signal transduction pathways. Lipid hydroperoxides can be broken down into low molecular weight molecules (such as alkanes, ketones and aldehydes), which are also toxic to the cell (Halliwell and Gutteridge 2007; Marban et al. 1994; Chehen et al. 1995; Hermes-Lima et al. 2004). Peroxidation of liver membranes of mitochondria-mediated iron ions is an important event in the pathophysiology of hemochromatosis (Hermes-Lima et al. 1995; Santos 1998). The peroxidation of low-density lipoprotein (LDL) is an important phenomenon in the process of atherogenesis (Darley-Usmar et al. 1995; Nageswara et al. 2005).

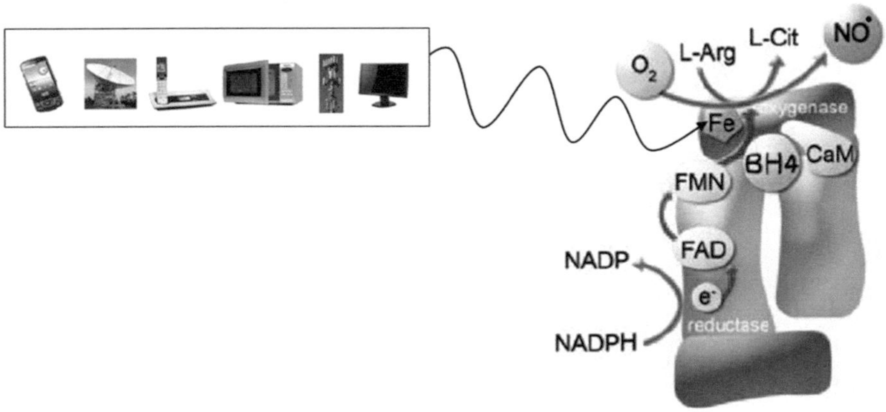

Fig. 7.4 Scheme of the dimer form of the enzyme nitric oxide synthase (Modified Rosales 2014) showing the synthesis of nitric oxide after induction by UHF-EMR

Lipid peroxidation is initiated by the attack on the lipid bilayer by any species sufficiently reactive to remove a hydrogen atom from a polyunsaturated fatty acid (PUFA). The removal of hydrogen generates an unpaired electron on a carbon atom, which reacts with molecular oxygen, forming peroxyl radical ($LOO^•$). This reaction is capable of reacting with another PUFA, initiating a new chain of oxidation. The peroxyl radical combines with the abstracted hydrogen atom, forming a lipid hydroperoxide (LOOH) (Hermes-Lima et al. 2004; Vaca et al. 1988). This hydroperoxide is unstable in the presence of transition metals, for example, iron and copper, which react with it to form alkoxyl and peroxyl radicals (Halliwell and Gutteridge 2007). The alkoxyl radical can remove hydrogen atoms from other PUFA and lipid hydroperoxides to form lipid and peroxyl radicals, thus contributing to the propagation of lipid peroxidation. The decomposition of LOOH generates not only radicals that propagate lipid peroxidation but also radical products such as aldehydes and ketones among others (Fig. 7.5).

RNS can also cause damage to lipids. Nitrogen dioxide, formed when nitric oxide is exposed to air (reaction 7.9) or the peroxynitrite is protonated (reaction 7.11), is a potent initiator of lipid peroxidation in biological fluids. In addition, peroxynitrite can generate the hydroxyl radical even without being in the presence of transition metals (Beckman and Koppenol 1996) (Fig. 7.5)

Aldehydes are a biomarker of oxidative damage to lipids. These end products of lipid peroxidation are used as markers of the action of some exogenous agents. Electromagnetic radiation can induce an increase in aldehydes. Of these, the one most often measured is MDA (malondialdehyde). Most studies published in the literature show an increase in tissue MDA irrespective of the tissue analyzed (Moustafa et al. 2001; Ozguner et al. 2005; Oktem et al. 2005; Meral et al. 2007; Ozgur et al. 2007; Sokolovic et al. 2008; Moussa 2009; Dindic et al. 2010; Hassan et al. 2010; Ragy 2015; Yilmaz et al. 2016) (Figs. 7.1, 7.2 and 7.3).

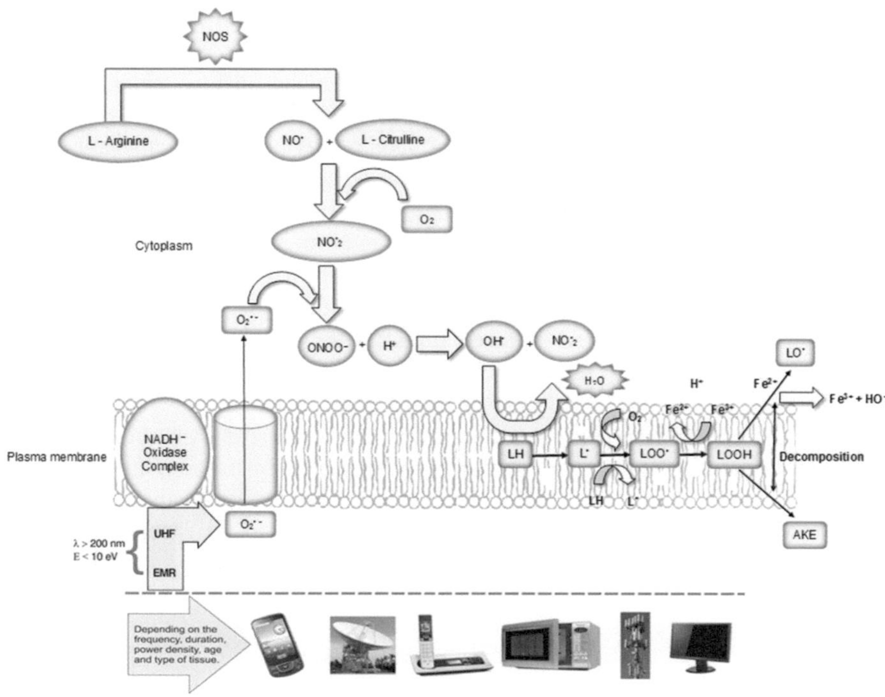

Fig. 7.5 General representation of the RNS and the formation of phases of lipid peroxidation induced by UHF-EMR. *NOS* nitric oxide synthase, *NO*˙ nitric oxide, $NO^{\bullet}_2$ nitrogen dioxide, $ONOO^-$ peroxynitrite, *LH* unsaturated fatty acid, *L*˙ alkyl radical, *LO*˙ alkoxyl radical, *LOO*˙ radical lipoperoxil, *LOOH* lipid hydroperoxide, *AKE* aldehydes, ketones, epoxides

Another biomarker of oxidative damage to lipids is lipoperoxide. There are few studies quantifying lipoperoxide in relation to EMR. In one, Tomruk and colleagues found increased levels of FOX (Ferrous Oxidation in Xylenol Orange) in the liver of rabbits after exposure to UHF-EMR (Ozgur et al. 2007).

The increase in lipoperoxides and aldehydes resulting from exposure to EMR is explained by the fact that radiation generates ROS. These reactive species attack PUFAs, leading to the formation of intermediate products, and finally, lipoperoxidation.

7.5.2 Proteins

The introduction of carbonyl groups into proteins is a hallmark of oxidative modification and has been linked to oxidative damage via hydroxyl radical reaction with portions of the amino-terminal peptide (Stadman 1992). Oxidation – mediated by ROS – of some amino acid residues such as lysine, arginine and proline leads to the formation of carbonyl groups (Stadman 1992; Kristal and Yu 1992). These groups can also be formed as the result of secondary reactions of some amino acid side chains with lipid oxidation products, or even reduced sugars or their oxidation products (Kristal and Yu 1992). Proteins can undergo other oxidative changes, for example, the oxidation of sulfhydryl groups, leading to conformational changes and the alteration of enzyme activity, cleavage of peptide bonds, modification of carbohydrates in glycoproteins and the loss of metals in metalloproteins (Dean et al. 1997; Stadman 1992). Oxidatively damaged proteins are degraded by cytosolic proteases (Stadman 1992). Proteins containing oxidized amino acids are recognized and degraded by ubiquitin-dependent proteolytic systems. The protein complexes that degrade these oxidized proteins are called proteasomes and are present in many copies in the cytosol (Hershko and Ciechanover 1992) and cell nucleus (Peters 1994).

The oxidation of proteins is associated with oxidative stress situations, such as ischemia-reperfusion, hyperoxia, smoking, estrogen administration, artificial ventilation, forced exercise and in cultured cells exposed to H_2O_2 or xanthine oxidase. High levels of carbonyl groups are associated with various diseases such as Alzheimer's, Parkinson's, diabetes, rheumatoid arthritis and muscular dystrophy, and can induce kidney tumors (Hermes-Lima et al. 2004). In addition, protein carbonyl levels are directly associated with the natural aging of various species (Hermes-Lima et al. 2004; Stadman and Levine 2000).

RNS can damage some amino acids. Peroxynitrite and nitrogen dioxide can cause damage to tyrosine, forming nitrotyrosine (Beckman and Koppenol 1996; Augusto 2006). Nitric oxide acts directly on the cysteine forming the S-nitrosylation with nitrosonium ion transfer to the sulfhydryl group of cysteine (Hess et al. 2005). The increase of nitrotyrosine and nitric oxide levels are related to inflammatory processes and various types of skin diseases (cancers of the skin, systemic lupus erythematosus, psoriasis, urticaria, and atopic dermatitis) (Pacher et al. 2007).

There are few studies with measurements of protein carbonyls as a function of exposure to ultra-high-frequency electromagnetic waves. The research groups of Sokolovic et al. (2008) and Djordjevic et al. (2015) showed increased levels of carbonylated proteins after exposure to UHF-EMR on rats (Sokolovic et al. 2008; Djordjevic et al. 2015). Arendash et al. (2010) found that even with a decrease in MnSOD activity, protein carbonyl content was not changed in the hippocampus of mice (Arendash et al. 2010).

The increase in protein carbonylation may be due to the direct attack of ROS/RNS on the amino acids of the polypeptide. Indirectly, the protein may be carbonylated by reaction with aldehydes generated by lipid peroxidation (Figs. 7.2 and 7.3).

7.5.3 Nucleic Acids

DNA is also damaged by ROS, lipid peroxidation products and RNS (Fig. 7.6). These molecules can cause the DNA strand breaks and oxidative modifications to bases, creating a series of products (8-hydroxyguanine, glycol 5-hydroxymethyl-glutaryl and cytosine) that can lead to mutagenic and carcinogenic processes (Halliwell and Gutteridge 2007; Hermes-Lima et al. 1998b; Saffi 2003). Free radical attack may also lead to cross-reactions between DNA and proteins. Damage to the sugar moiety is also frequently observed. These multiple types of DNA damage (i.e., all the injuries in regions close to the DNA) together with the double-strand break can be genotoxic, mutagenic and carcinogenic (Saffi 2003).

The ROS superoxide and hydrogen peroxide do not attack the DNA directly, whereas hydroxyl radicals and singlet oxygen can cause direct injury. The first reacts with the four nitrogenous bases, generating modified bases, breaking DNA chains and forming apurinic and apyrimidinic (AP) sites. Singlet oxygen can cause cyclo-addition to the carbon double bond of the imidazole ring, generating alkali-labile AP sites and breaks in the chain next to guanines. The preferred target of ROS is guanine and most common product is 7.8-hydroxy-8-oxo-2 deoxyguanine (8-OHdG) (Saffi 2003).

Damage to DNA can be measured by the comet assay, by monitoring levels of 8-OHdG and by PFGE (Pulsed-Field Gel Electrophoresis). The results of the experiments using continuous UHF-EMR showed no differences in DNA fragmentation in rat brains, rat glial cell cultures and in rabbit livers (Tomruk et al. 2010; Belyaev 2005; Campisi et al. 2010; Di Mascio et al. 1991). However, some researchers observed an increase in oxidative damage to DNA at the beginning of exposure to the UHF-EMR. This damage decreases with time, showing an adaptation and increased DNA repair mechanisms (Sahin et al. 2016).

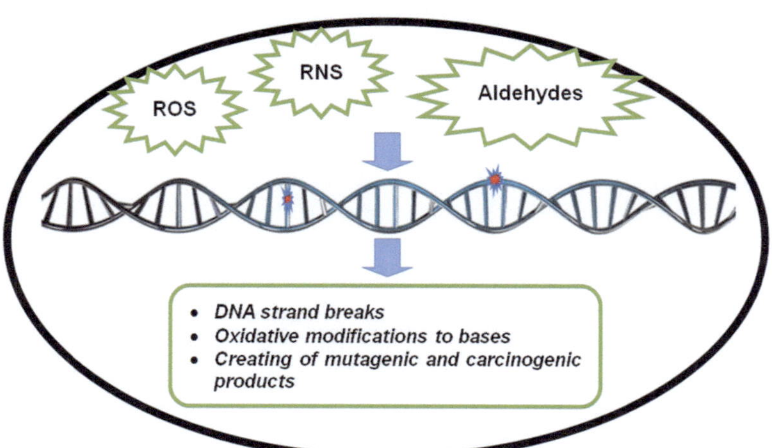

Fig. 7.6 General representation of DNA damage (Modified Cooke and Evans 2005)

7.6 Antioxidant Defenses

Any substance which, when present in low concentrations compared with the oxidant, slows or inhibits the rate of oxidation of a substrate is considered an antioxidant (Halliwell and Gutteridge 2007).

Aerobic organisms evolved enzymatic antioxidant defenses and related non-enzymatic adaptations to environmental conditions, due to survival by those most able to defend themselves from damage caused by ROS.

7.6.1 Non-enzymatic

The non-enzymatic defenses are mainly composed of low molecular weight molecules. They include vitamins C and E, beta-carotene, glutathione, bilirubin, lactate and uric acid. Most non-enzymatic antioxidants are found in water (plasma, cytoplasm, mitochondrial matrix, etc.); vitamin E and carotenoids are found in the membranes. The mechanism of action of these defenses is to interrupt the chain of oxidative reactions, stopping the spread of radical reactions (Cadenas 1995; Di Mascio et al. 1991).

7.6.1.1 Vitamin A

Vitamin A (retinol) is formed from carotenoids. In this process, the reaction of the carotenoid with O_2 is catalyzed by dioxygenase to produce two molecules of retinol (Halliwell and Gutteridge 2007). The antioxidant activity of vitamin A (RetOH) is due to polyene units that act on singlet oxygen and peroxyl free radical ($ROO^{\bullet}$) (reaction 7.12). In addition to these ROS, retinol also acts on tiils radicals. Depending on the reactive species, retinol can act as a "quencher" (singlet oxygen) or "scavenger" (peroxyl radical). In addition to the antioxidant activities mentioned, Barber and colleagues found that mice deficient in vitamin A had a decreased ratio of GSH / GSSG in rat liver mitochondria (Barber et al. 2000). These researchers also observed an increase in the level of MDA and 8-oxodG. These results were reversed after re-feeding with vitamin A.

The discovery that electromagnetic radiation can generate oxidative/nitrosative stress (OS) (Elhag et al. 2007) has led to experimental designs that include non-enzymatic antioxidants, dietary or not, along with radiation. Retinol levels were lower (relative to the control group) due to oxidative damage induced by UHF-EMR in the heart and plasma of rats (Elhag et al. 2007; Türker et al. 2011). However, the same result was not found in the plasma and brain of guinea pigs (Meral et al. 2007).

Vitamin A is consumed to reduce oxidative damage by acting as a quencher or scavenger of ROS generated by electromagnetic radiation (Fig. 7.3).

Reaction

$$\mathbf{RetOH + ROO^{\bullet} \rightarrow RetO^{\bullet} + ROOH} \tag{7.12}$$

7.6.1.2 Vitamin C

Vitamin C (ascorbic acid or AH_2) is formed from glucose in many organisms. Monkeys do not produce this vitamin. This is because the enzyme responsible for the synthesis of ascorbic acid (L-gulono-1, 4-lactone oxidase) is not transcribed (Cerqueira et al. 2007). Therefore, many organisms need to acquire this acid in their diet because it has many biological functions. Some of these are the post-translational modification of collagen, the production of carnitine, the transformation of dopamine into norepinephrine, and the metabolism of tyrosine. In addition to its function as a cofactor, vitamin C participates in the transport of iron, and, depending on its concentration, it is an important antioxidant.

AH_2 is water-soluble and is a strong reducing agent (Cerqueira et al. 2007). These characteristics make it an excellent antioxidant in aqueous intra- and extra-cellular milieus. It acts directly as an antioxidant when it reduces ROS, e.g., superoxide, hydroxyl radical and nitrogen dioxide. Vitamin C acts indirectly as an antioxidant in alpha-tocopherol recycling. This recycling protects membranes against lipid peroxidation. After recycling of alpha-tocopherol (vitamin E) from the tocopheryl radical, vitamin C becomes the ascorbyl radical ($A^{-\bullet}$) (reaction 7.13). This radical has low reactivity and can be transformed back into ascorbic acid by the action of NADH-dependent reductases or by the action of GSH.

Reaction

$$\mathbf{AH_2 + O_2^{-\bullet} \rightarrow A^{-\bullet} + H_2O_2} \tag{7.13}$$

The levels of vitamin C have been monitored in experiments with electromagnetic radiation. Türker et al. (2011) found lower levels of vitamin C in the heart tissue of rats exposed to UHF-EMR of 2.45 GHz (Türker et al. 2011). Elhag et al. (2007) also observed decreased levels of vitamin C in the plasma of rats exposed to 900 MHz UHF-EMR (Elhag et al. 2007). The decrease in content of vitamin C comes from its use as a reducing agent of ROS, especially superoxide generated by EMR (Figs. 7.2 and 7.3).

7.6.1.3 Vitamin E

The tocopherols and tocotrienols together constitute vitamin E ($\alpha,\beta,\gamma,\delta$-TOH). These two groups of molecules are formed by a chromanol nucleus and an aliphatic chain. This formation facilitates the insertion of the vitamin in the phospholipid layer, while the core chromanol ring is responsible for the lipid radical ($LOO^{\bullet}$), interrupting the lipid peroxidation chain and forming the tocopheryl radical (α,β,γ,δ-$TO^{\bullet}$) (reaction 7.14). This layer is recycled to the alpha-tocopherol by ascorbic acid (AH_2) (reaction 7.15).

Reactions

$$\alpha, \beta, \gamma, \delta - \mathbf{TOH} + \mathbf{LOO}^{\bullet} \rightarrow \alpha, \beta, \gamma, \delta - \mathbf{TO}^{\bullet} + \mathbf{LOOH} \quad (7.14)$$

$$\alpha, \beta, \gamma, \delta - \mathbf{TO}^{\bullet} + \mathbf{AH_2} \rightarrow \alpha, \beta, \gamma, \delta - \mathbf{TOH} + \mathbf{A}^{-\bullet} \quad (7.15)$$

Türker et al. (2011) observed a decrease in the levels of this vitamin in rats exposed to 2.45 GHz UHF-EMR (Türker et al. 2011). The activity of the enzymes alanine aminotransferase (ALT), aspartate aminotransferase (ASP) and alkaline phosphatase (ALP) indicate the functional state of the liver. The activity of these enzymes increases when there is illness or injury by toxic agents to the liver (Aragon and Younossi 2010). Exposure to UHF electromagnetic waves increases the activity of ALT, ASP and ALP in the blood and liver of rats (Moussa 2009; Yilmaz et al. 2016; Aziz et al. 2010). When rats were exposed to UHF-EMR and received vitamins C and E, the enzyme activity decreased (Aziz et al. 2010). This can be explained due to the antioxidant action of those compounds that decreased the liver damage caused by ROS (Fig. 7.3). The reduction of damage led to a decrease of measured enzyme activities.

Another action of vitamin E and C is related to the cardiovascular system. The use of these vitamins caused decreases in the blood pressure of rats exposed to EMR (Aweda et al. 2011). The antioxidant capacity of these vitamins decreased the ROS formed by the UHF-EMR. This led to the increased availability of nitric oxide, which lowered the blood pressure of the rats (Fig. 7.3).

7.6.1.4 GSH

Reduced glutathione (GSH) (γ-glutamyl-cysteinyl-glycine) is present in most cells and it is the most abundant thiol in the intracellular environment (Meister and Anderson 1983). This tripeptide is free in animal tissues, its concentration ranges from 0.5 to 10×10^{-3} M. Its reductive capacity is determined by the sulfhydryl (−SH) group, of the amino acid cysteine.

Glutathione is an important non-enzymatic antioxidant defense and participates in the regeneration of ascorbate and alpha-tocopherol (reactions 7.13 and 7.14, respectively) and is a co-substrate of GPx enzyme reaction (7.14 and 7.15) and glutathione S-transferase. GSH can be considered one of the most important agents of the antioxidant defense system of cells, protecting them against injury that results from exposure to agents such as iron ions (Galleano and Puntarulo 1995), hydroxyl radicals (reaction 7.10), and alkoxyl and peroxyl ($ROO^{\bullet}$) radicals (reaction 7.11), as well as superoxide, hyperbaric oxygen, radiation and ultraviolet light (Deneke and Fanburg 1989). In addition, GSH decreases the susceptibility to renal injury caused by ischemia and reperfusion (Shan et al. 1990), and it acts as a carrier and reservoir of cysteine and participates in the detoxification of xenobiotics and lipid peroxidation products (Ferreira and Matsubara 1997). An increase in the generation of ROS in the cell leads to an increased ratio of GSSG/GSH (glutathione disulfide/reduced glutathione), and this ratio used as an important indicator of oxidative stress (Figs. 7.2 and 7.3).

Reactions

$$\alpha, \beta, \gamma, \delta - \mathbf{TO}^{\bullet} + \mathbf{AH_2} \rightarrow \alpha, \beta, \gamma, \delta - \mathbf{TOH} + \mathbf{A}^{-\bullet} \tag{7.16}$$

$$\mathbf{A}^{-\bullet} + 2\mathbf{GSH} \rightarrow \mathbf{AH_2} + \mathbf{GSSG} \tag{7.17}$$

$$\mathbf{GSH} + \mathbf{OH}^{\bullet} \rightarrow \mathbf{GS}^{\bullet} + \mathbf{H_2O} \tag{7.18}$$

$$\mathbf{GSH} + \mathbf{RO_2}^{\bullet} \rightarrow \mathbf{GS}^{\bullet} + \mathbf{ROOH} \tag{7.19}$$

$$\mathbf{GS}^{\bullet} + \mathbf{GS}^{\bullet} \rightarrow \mathbf{GSSG} \tag{7.20}$$

GSH was measured in the brain, blood (Meral et al. 2007), liver, kidney (Moussa 2009; Yilmaz et al. 2016), hippocampus (Arendash et al. 2010) and reproductive tissue (Mailankot et al. 2009) of mammals after exposure to UHF-EMR. There was a decrease in these levels in all the organs. Although most studies show a reduction in the concentration of GSH, Praputpittaya et al. (2008) found no differences in the brains of rats exposed to 900 MHz UHF-EMR for 30 min a day for 1 week (Praputpittaya et al. 2008).

The decrease in GSH levels comes from its direct use in hydroxyl radical and singlet oxygen scavenging. In addition, the role of glutathione as co-substrate in the reactions of GPx and GST (glutathione S-transferase) causes a reduction in its concentration. The production of ROS and RNS induced by electromagnetic fields leads to the formation of organic peroxides, which are metabolized by some types of GST and GPx (Fig. 7.3).

7.6.1.5 Melatonin

Melatonin is produced largely by the pineal gland from serotonin. This happens mostly at night. The best-known function of this molecule is circadian rhythm control. Today, it is known that the hormone melatonin is a "scavenger" of free radicals. In addition, it stimulates the activity/expression of GPx and SOD (Piotr et al. 2003). Melatonin's potent antioxidant capacity comes from its ability to cross biological membranes. Treatment with melatonin prevented an increase in lipid peroxidation and in the activity of xanthine oxidase (XO) in rat brain 40 days after exposure to EMR but did not prevent a decrease of catalase activity and an increased protein carbonyl content (Sokolovic et al. 2008). In other studies, the administration of melatonin to mice reduces an UHF-EMR-induced increase of lipid peroxidation in the hippocampus and kidney (Oktem et al. 2005; Köylü et al. 2006). In that organ, there was also an increase in the concentration of NAG

(N-acetyl-β-D-glucosaminidase), a marker of damage to renal tubules. NAG levels after exposure to EMR decreased after administration of melatonin.

The hormone melatonin induces an increase in gene expression and in the activity of *SOD* and GPx that can then consume greater amounts of $O_2^{-\bullet}$ and H_2O_2, respectively, generated by UHF-EMR. The reduction of these ROS leads to reduced lipid peroxidation, protein carbonylation and kidney damage.

7.6.1.6 Zinc

Some researchers have found that certain metals have a pro-antioxidant function. An example is zinc. This transition metal is the thirtieth element of the periodic table and sits in the fifth period of that table. Unlike most of this period of transition metals, Zn has antioxidant activity rather than oxidant activity as observed in Fe, Cu, V, Ni, Co and Cr (Valko et al. 2005). Zinc stabilizes biological membranes and / or proteins, minimizing their susceptibility to attack by ROS (Bettger and O'Dell 1981; Powell 2000). This metal does not participate in redox reactions. Its antioxidant action is due to its binding to sulfhydryl groups of proteins, which protects them from ROS attack (Gibbs et al. 1985).

The antioxidant properties of zinc can be acute (described in the preceding paragraph) or chronic. This happens when an organism is treated with Zn for a long period, leading to increased synthesis of metallothioneins. These are metal-binding proteins, possessing large numbers of cysteines that bind to Zn, storing it (Powell 2000). Another situation in which the Zn has antioxidant activity occurs when it moves the Fe-binding site, inhibiting reactions that lead to the formation of ROS. In addition to the above mentioned antioxidant mechanisms, Zn stabilizes the structure of CuZnSOD, while Cu participates in the catalysis of reactions through oxidation and reduction where the superoxide is consumed (Ahmad 1995).

There are few studies relating non-ionizing electromagnetic radiation and quantification of metals in vivo. In one, Kiliçalp et al. (2009) investigated the action of 900 MHz UHF-EMR on the liver of *Cavia porcellus* (Kiliçalp et al. 2009). The Zn content did not change in the liver but decreased in the testicles. The Zn levels did not change in the liver due to its high quantity and activity of enzymatic and non-enzymatic antioxidants (other than zinc), which might have decreased the antioxidant levels instead of Zn. As the testis does not contain such a wealth of antioxidants, the Zn is used to defend against oxidative damage that can be caused by the generation of ROS induced by UHF-EMR.

7.6.2 Enzymatic

These defenses include mainly superoxide dismutase (SOD), catalase and glutathione peroxidase (GPx) (Hermes-Lima et al. 1998b; Hermes-Lima et al. 2001).

7.6.2.1 SOD

This enzyme catalyzes the dismutation of superoxide to hydrogen peroxide and molecular oxygen (reaction 7.21). It is widely distributed in aerobic organisms. Although there are four types of superoxide dismutase – FeSOD, MnSOD, CuZnSOD and EC-SOD – only the last three are present in mammals (Halliwell and Gutteridge 2007) . Regardless of type, all these superoxide dismutases have the same catalytic mechanism. MnSOD is present in the mitochondrial matrix, CuZn SOD occurs in the cytosol, nucleus, lysosome and mitochondrial intermembrane space, and EC-SOD is found in the extracellular medium (plasma, lymph and cerebrospinal fluid) (Ahmad 1995; Augusto 2006; Halliwell and Gutteridge 2007) (Figs. 7.2 and 7.3).

Reaction

$$\mathbf{O_2^{-} + O_2^{-} + 2H^{+} \xrightarrow[\text{}]{\substack{\text{FeSOD, MnSOD}\\ \text{CuZnSOD, EC-SOD}}} H_2O_2 + O_2} \qquad (7.21)$$

In most studies where animals were exposed to UHF-EMR, a decrease in SOD activity was observed. These results occurred independently in various organs of experimental models (Moustafa et al. 2001; Ozguner et al. 2005; Oktem et al. 2005; Ozgur et al. 2007; Hassan et al. 2010; Arendash et al. 2010; Elhag et al. 2007; Kesari and Behari 2009).

7.6.2.2 Catalase

Catalase decomposes hydrogen peroxide, forming water and oxygen (reaction 7.14). This enzyme is present in most aerobic organisms. Although catalase exists in other organelles, it is present mainly in peroxisomes where there is a greater concentration of hydrogen peroxide (Ahmad 1995; Halliwell and Gutteridge 2007) (Figs. 7.2 and 7.3).

Four heme groups constitute the majority of catalases. These groups are linked to the enzyme active site. This site is stabilized by NADPH that is connected to each subunit of catalase (Vainshtein et al. 1981).

Catalase activity decreased in most studies where organisms were exposed to UHF-EMR. Except for a few experiments (Kesari and Behari 2009; Ferreira et al. 2006), exposure for less than 10 days in rats or guinea pig – whether measured in the brain (Meral et al. 2007; Sokolovic et al. 2008), heart (Ozguner et al. 2005), kidney (Oktem et al. 2005) or blood (Hassan et al. 2010) – has always shown decreased activity for this enzyme. In 2014, Furtado-Filho et al. showed decreased catalase expression in rat liver after exposure to electromagnetic waves of 950 MHz.

Reaction

$$H_2O_2 + H_2O_2 \xrightarrow{\text{Catalase}} 2H_2O + O_2 \tag{7.22}$$

7.6.2.3 GPx

This enzyme breaks down not only hydrogen peroxide but also organic hydroperoxides, using glutathione as a reductant (reactions 7.23 and 7.24, respectively). There are four types of GPx: the classical Se-dependent GPx (located in the cytosol and mitochondrial matrix) depends on phospholipase A_2 to act on membrane hydroperoxides; PHGPx (located in the cytosol) uses phospholipid hydroperoxides as substrates without the need for phospholipase; GI-GPx (in the cytosol of cells of the gastrointestinal tract); and PLGPx (plasma) (Fig. 7.3). Even with different sizes in the active site of these enzymes, there is always an atom Se replacing the cysteine S (Ahmad 1995; Halliwell and Guteridge 2007) .

Reactions

$$H_2O_2 + 2GSH \xrightarrow[\text{(Classical GPx, PHGPx, GI-GPx, PLGPx)}]{} H_2O + GSSG \tag{7.23}$$

$$LOOH + 2GSH \xrightarrow[\text{(Classical GPx, PHGPx, GI-GPx, PLGPx)}]{} H_2O + LOH + GSSG \tag{7.24}$$

Although there are some studies that show no significant change in GPx activity in animals exposed to UHF-EMR (Ozgur et al. 2007; Ferreira et al. 2006; Irmak et al. 2002), other studies showed that the enzyme activity decreases significantly in liver, brain, sperm (Mailankot et al. 2009), blood (Moustafa et al. 2001; Hassan et al. 2010), heart (Ozguner et al. 2005) and kidney after continuous exposure to radiation mainly at a frequency of 900 MHz (Hassan et al. 2010).

The low total GPx activity can be explained by a decreased concentration of reduced glutathione. The concentration of this co-substrate decreases after exposure to UHF-EMR.

7.7 Oxidative/Nitrosative stress

The imbalance between the production of oxidants and the concentration of antioxidant defenses leading to cell damage is called oxidative/nitrosative stress (ONS). This can occur due to a high production of ROS/RNS by a reduction of antioxidant defenses or by both processes simultaneously (Sies 1993).

Oxidizing agents are formed in the normal process of metabolism and in various pathological conditions. They can be produced at very high rates, leading to oxidative/nitrosative stress and possibly to cell death and / or apoptosis. Oxidative/nitrosative stress is related to diseases such as cancer, hemochromatosis, rheumatoid arthritis, diabetes, metabolic syndrome, diabetic retinopathy, hepatic porphyria, neurodegenerative diseases, hypertension, cardiovascular diseases and aging (Hermes-Lima et al. 2004; Stadman 1992; Braughler and Hall 1989; Kleinveld et al. 1989; Kaul et al. 1993; Levine 1993; Reuter et al. 2010).

Recently, the ONS has been observed in mammalian organs when they are exposed to UHF-EMR (Kesari and Behari 2008; Reuter et al. 2010). As this radiation can induce the generation of ROS/RNS, it is likely that these animals become more susceptible to the initiation and / or worsening of various pathological conditions.

7.8 Conclusion

The main question addressed in this work was: Does UHF-EMR alter the metabolism of ROS and RNS in mammals? To answer this question we carried out an extensive literature review relating UHF radiation to oxidative/nitrosative damage and antioxidants. The majority of the studies reviewed show that UHF-EMR can cause changes in biomarkers of oxidative/nitosative damage. Of these, the products of lipid peroxidation show the greatest increase. Commonly, if reduced antioxidant defenses are observed, it is because they have been depleted (mainly GSH) by fighting against oxidative/nitrosative damage to macromolecules induced by ROS/RNS. Increased concentrations of aldehydes can lead to the aggravation of cancers, Alzheimer's disease, atherosclerosis and diabetes, as MDA is a biomarker of these diseases. Therefore, people with any of these diseases should avoid exposure to UHF-EMR.

References

Ahmad S (1995) Antioxidant mechanisms of enzymes and proteins. In: Ahmad S (ed) Oxidative stress and antioxidant defends in biology. Chapman and Hall, New York, pp 238–272

Altamura S, Muckenthaler MU (2009) Iron toxicity in diseases of aging: Alzheimer's disease. Parkinson's disease and atherosclerosis. J Alzheimers Dis 16:879–895

Anderson U, Leighton B, Young ME, Blomstrand E, Neswssholme EA (1998) Inactivation of aconitase and oxiglutarate dehidrogenase in skeletal muscle in vitro by superoxide anions and/or nitric oxide. Biochem Biophys Res Commun 249:512–516

Aragon G, Younossi ZM (2010) When and how to evaluate mildly elevated liver enzymes in apparently healthy patients. Cleve Clin J Med 77:195–2004

Arendash GW, Sanchez-Ramos J, Mori T, Mamcarz M, Lin X, Runfeldt M, Wang L, Zhang G, Sava V, Tan J, Cao C (2010) Electromagnetic field treatment protects against and reverses cognitive impairment in Alzheimer's disease mice. J Alzheimers Dis 19:191–210

Arthur JSC (2007) MAPK activation by radio waves. Biochem J 405:e5–e6
Augusto O (2006) Free radicals: good, bad and natural (in Portuguese), 1st edn. Office of Texts, São Paulo
Aweda MA, Meindinyo ROK, Gbenebitse SO, Ibitoye AZ (2011) Microwave radiation exposures affect cardiovascular system and antioxidants modify the effects. Adv Appl Sci Res 2:246–251
Aziz IA, El-Khozondar HJ, Shabat M, Elwasife K, Mohamed-Osman A (2010) Effect of electromagnetic field on body weight and blood indices in albino rats and the therapeutic action of vitamin C or E. Rom J Biophys 20:235–244
Barber T, Borrás E, Torres L, García C, Cabezuelo F, Lloret A, Pallardó FV, Viña JR (2000) Vitamin A deficiency causes oxidative damage to liver mitochondria in rats. Free Radic Biol Med 29:7–9
Beckman JS, Koppenol WH (1996) Nitric oxide, superoxide, and peroxinitrite: the good, the bad, and the ugly. Am J Phys 271:1424–1437
Belyaev IY, Hillert L, Protopopova M, Tamm C, Malmgren LO, Persson BR, Selivanova G, Harms-Ringdahl M (2005) 915 MHz microwaves and 50 Hz magnetic field affect chromatin conformation and 53BP1 foci in human lymphocytes from hypersensitive and healthy persons. Bioelectromagnetics 26:173–184
Bettger WJ, O'Dell BL (1981) A critical physiological role of zinc in the structure and function of biomembranes. Life Sci 28:1425–1438
Bilgici B, Akar A, Avci B, Tuncel OK (2013) Effect of 900 MHz radiofrequency radiation on oxidative stress in rat brain and serum. Electromagn Biol Med 32:20–29
Braughler JM, Hall DE (1989) Central nervous systems trauma and stroke: I. Biochemical considerations for oxygen radical formation and lipid peroxidation. Free Radic Biol Med 6:289–301
Bruno M, Brightman AO, Lawrence J, Werderitsh D, Morret DM, Morre DJ (1992) Stimulation of NADH oxidase activity from rat liver plasma membranes by growth factors and hormones is decreased or absent with hepatoma plasma membranes. Biochem J 284:625–628
Cadenas E (1995) Mechanism of oxygen activation and reactive oxygen species detoxification. In: Ahmad S (ed) Oxidative stress and antioxidant defenses in biology. Champman and Hall, New York, pp 1–61
Campisi A, Gulino M, Acquaviva R, Bellia P, Raciti G, Grasso R, Musumeci F, Vanella A, Triglia A (2010) Reactive oxygen species levels and DNA fragmentation on astrocytes in primary culture after acute exposure to low intensity microwave electromagnetic Field. Neurosci Lett 473:52–55
Castillo-Olivares AD, Castro IN, Medina MA (2000) Dual role of plasma membrane electron transport systems in defense. Crit Rev Biochem Mol Biol 35:197–220
Cerqueira FM, Gennari MH, Augusto O (2007) Dietary antioxidants: controversies and perspectives. Quim Nova 30:441–449
Céspedes O, Ueno S (2009) Effects of radio frequency magnetic fields on iron release from cage proteins. Bioelectromagnetics 30:336–342
Chehen Y, Millew A, Grisham MB (1995) Pathophysiology and oxygen metabolites. In: Ahmad S (ed) Oxidative stress and antioxidant defenses in biology. Chapmen and Hall, New York, pp 238–272
Cooke MS, Evans MD (2005) Reative oxygen species from DNA damage to disease. Sci Med 10:98–111
D'Andrea JA, Chou CK, Johnston SA, Adair ER (2003) Microwave effects on the nervous system. Bioelectromagnetics 6:S107–SS47
Darley-Usmar VM, Mason WC, Hogg N, Kalyanaraman B (1995) Lipid peroxidation and cardiovascular disease. In: Blake D, Winyard PG (eds) Immunopharmacology of free radical specie. Academic Press, London, pp 23–38
Dean RT, Fu S, Stocker R, Davies MJ (1997) Biochemistry and pathology of radical-mediated protein oxidation. Biochem J 324:1–18
Deneke SM, Fanburg BL (1989) Regulation of cellular glutatione. Am J Phys 257:L163–LL73

Desai NR, Kesari KK, Agarwal A (2009) Pathophysiology of cell phone radiation: oxidative stress and carcinogenesis with focus on male reproductive system. Reprod Biol Endocrinol 7:114

Di Mascio P, Murphy MM, Sies H (1991) Antioxidant defense sytems: the role of carotenoids, tocophenols, and thiols. Am J Clin Nutr 53:194–200

Dindic B, Sokolovic D, Krstic D, Petkovic D, Jovanovic J, Muratovic M (2010) Biochemical and histopathological effects of mobile phone exposure on rat hepatocytes and brain. Acta Med Medianae 49:37–41

Djordjevic B, Sokolovic D, Kocic G, Veljkovic A, Despotovic M, Basic J, Jevtovic-Stoimenov T, Sokolovic DM (2015) Bratisl Lek Listy 116:96–100

Eblem A (2011) Non-ionizing radiation: concepts, applications and risks. In: Engineering course safety. http://www.prorad.com.br/cursos/Cursos/rni.pdf. Accessed 4 Sep 2015

Elhag MA, Nabil GM, Attia AMM (2007) Effects of electromagnetic field produced by mobile phones on the oxidant and antioxidant status of rats. Pak J Biol Sci 10:84–84

Ferreira ALA, Matsubara LS (1997) Free radicals: concepts, related diseases, defense system and oxidative stress. Rev Assoc Med Bras 43:61–68

Ferreira RA, Knakievicz T, Pasquali MAB, Gelain DP, Dal-Pizzol F, Fernández CER, Salles AAA, Ferreira HB, Moreira JCF (2006) Ultra high frequency-electromagnetic field irradiation during pregnancy leads to increase in erythrocytes micronuclei incidence in rat offspring. Life Sci 80:43–50

Figura OL, Teixeira AA (2007) Food Physics. Physical properties: measurement and applications, 1st edn. Springer, New York

Fridovich I (1998) Oxygen toxicity: a radical explanation. J Exp Biol 201:1203–1209

Friedman J, Kraus S, Hauptman Y, Schiff Y, Seger R (2007) Mechanism of short-term ERK activation by electromagnetic fields at mobile phone frequencies. Biochem J 405:559–568

Furtado-Filho OV, Borba JB, Dallegrave A, Pizzolato TM,Henriques JAP, Moreira JCF, Saffi J (2014) Effect of 950 MHz UHF electromagnetic radiation on biomarkers of oxidative damage, metabolism of UFA and antioxidants in the livers of young rats of different ages. Int J Radiat Biol 90:159–168

Galleano M, Puntarulo S (1995) Role of antioxidants on the erythrcytes resistence to lipid peroxidation after acute iron overloads in rats. Biochim Biophys Acta 1271:321–326

Gibbs PNB, Gore MG, Jordan PM (1985) Investigation of the effect of metal ions on the reactivity of thiol groups in humans: aminolevulinate dehydratase. Biochem J 225:573–580

Hall G (2008) Maxwell's electromagnetic theory and special relativity. Philos Trans R Soc 366:1849–1860

Halliwell B (2006) Oxidative stress and neurodegeneration: where are we now? J Neurochem 97:1634–1658

Halliwell B, Gutteridge JMC (2007) Free radical in biology and medicine, 4th edn. New York, Oxford

Hassan N, Raafat BM, Aziz WA (2010) Modulator role of grape seed extract on erythrocyte hemolysis and oxidative stress induced by microwave irradiation in rats. Int J Integr Biol 10:106–111

Hermes-Lima M, Willmore WG, Storey KB (1995) Quantification of peroxidation in tissue extracts based on Fe (III) xylenol orange complex formation. Free Radic Biol Med 19:271–280

Hermes-Lima M, Storey JM, Storey KB (1998) Antioxidant defenses and metabolic depression. The hypothesis of preparation for oxidative stress in land snails. Comp Biochem Physiol B 120:437–448

Hermes-Lima M, Storey JM, Storey KB (2001) Antioxidant defenses and animal adaptation to oxygen availability during environmental stress. In: Storey KB, Storey JM (eds) Cell and molecular response to stress. Elsevier press, Amsterdam, pp 263–287

Hermes-Lima M, Ramos-Vasconcelos GR, Cardoso LA, Orr AL, Rivera PM, Drew KL (2004) Animal adaptability to oxidative stress: gastropod estivation and mammalian hibernation. In: Barnes BM, Carey HV (eds) Life in the cold: evolution, mechanisms, adaptation, and

applications – thelfth international Hibernation symposium, Biological Papers of University of Alaska. Institute of Arctic Bioogy, University of Alaska Fairbanks, Fairbanks, pp 585–593
Hershko A, Ciechanover A (1992) The ubiquitin system for protein degradation. Annu. Rev Biochem 61:761–807
Hess DT, Matsumoto A, Kim SO, Marshall HE, Stamler JS (2005) Protein S-nitrosylation purview and parameters. Nat Rev Mol Cell Biol 6:150–166
Irmak MK, Fadilhoglu E, Güleç M, Erdogan H, Yagmurca M, Akyol Ö (2002) Effects of electromagnetic radiation from a cellular telephone on the oxidant and antioxidant levels in rabbits. Cell Biochem Funct 20:279–283
Ischiropoulos H (1992) From physiology to pathology: maintenance of the critical balance by antioxidants. Arch Biochem Biophys 298:431–437
Júnior MFV, Martin MI (2006) Measurement of non-ionizing radiation in the cities of São Jose dos Campos-SP and Taubaté (in Portuguese). Annals the 12th Meeting of Undergraduate Research and Graduate of ITA – ENCITA XII, Brazil
Kaul N, Siveski-Iliskovic N, Hill M, Slezak J, Sinfal PK (1993) Free radical and the heart. J Pharmacol Toxicol Methods 30:55–67
Kesari KK, Behari J (2008) Whole body 900 MHz radiation exposure effect on enzyme activity in male wistar rats. In: Proceedings of the XXIXth URSI (International Union of Radio Science) General Assembly, Chicago, IL, USA, 7–16
Kesari KK, Behari J (2009) Fifty-gigahertz microwave exposure effect of radiations on rat brain. Appl Biochem Biotechnol 158:126–139
Kiliçalp D, Dede S, Deger Y, Aslan L (2009) Effects of green tea on mineral levels of liver and testis of Guinea Pigs electromagnetic field emitted by mobil phones. Asian J Anim Vet Adv 4:86–92
Kleinveld HA, Swaak AJG, Koster JF (1989) Interactions between oxygen free radicals and proteins. Scand J Rheumatol 18:341–352
Koppenol WH (2001) The Haber-Weiss cycle – 70 years later. Redox Rep 6:229–234
Köylü H, Naziroglu M, Delibas N (2006) Melatonin modulates 900 MHz microwave-induced lipid peroxidation change in rat brain. Toxicol Ind Health 22:211–216
Kristal BS, Yu BP (1992) An emerging hypotesis: synergistic induction of aging by free radicals and Maillard reactions. J Gerontol 47:B107–BB14
Lenaz G (2012) Mitochondria and reactive oxygen species. Which role in physiology and pathology? Adv Exp Med Biol 942:93–136
Levine RL (1993) Ischemia: from acidosis to oxidation. FASEB J 7:1242–1246
Mackul'Ak T, Prousek J, Švorc L (2011) Degradation of atrazine by Fenton and modified Fenton reactions. Monatsh Chem 142:561–567
Madamanchi NR, Vendrov A, Runge MS (2005) Oxidative stress and vascular disease. Arterioscler Thromb Vasc Biol 25:29–38
Mailankot M, Kunnath AP, Jayaleksmi H, Koduru B, Valsalan R (2009) Radiofrequency electromagnetic radiation (RF_EMR) from GSM (0.9/1.8 GHz) mobile phones induces oxidative stress and reduces sperm motility in rat. Clinics 64:561–565
Marban E, Koretsune Y, Kusuoca H (1994) Disruption of intracellular Ca^{+2} homeostasis in hearts reperfused after prolonged episodes of ischaemia. In: Cellular, biochemical, and molecular aspects of reperfusion injury, vol 723. Annals of the New York Academy of Science, New York, pp 38–58
Mccord JM, Day ED (1978) Superoxide-dependent production of hydroxyl catalyzed by iron-EDTA complex. FEBS Lett 86:139–142
Meister A, Anderson ME (1983) Glutathione. Annu Rev Biochem 52:711–760
Meral I, Mert H, Mert N, Dejer Y, Yoruk I, Yetkin A, Keskin S (2007) Effects of 900 MHz electromagnetic Field emitted from cellular phone on brain oxidative stress and some. Brain Res 1169:120–124
Moulder JB (2000) Cellular antennas (base station) and human health (in Portuguese). http://www.radiacao.com.br/traducaocelular.html. Acessed 21 Aug 2016

Moussa AS (2009) Oxidative stress in rats exposed to microwave radiation. Rom J Biophys 19:149–158

Moustafa YM, Moustafa RM, Belacy A, Abou-El-Ela SH, Ali FM (2001) Effects of acute exposure to the radiofrequency fields of phones on plasma lipid peroxide and antioxidase activities in human erythrocytes. J Pharm Biomed Anal 26:605–608

Nageswara RM, Vendrov A, Runge MS (2005) Oxidative stress and vascular disease. Arterioscler Thromb Vasc Biol 25:29–38

Oktem F, Ozguner F, Mollaoglu H, Koyu A, Uz E (2005) Oxidative damage in the kidney induced by 900-MHz-emitted mobile phone: protection by melatonin. Arch Med Res 36:350–355

Olgar Y, Hidisoglu E, Celen MC, Yamasan BE, Yargicoglu P, Ozdemir S (2015) 2.1 GHz electromagnetic field does not change contractility and intracellular Ca2+ transients but decreases β-adrenergic responsiveness through nitric oxide signaling in rat ventricular myocytes. Int J Radiat Biol 91:851–857

Ozguner F, Altinbas A, Ozaydin M, Dogan A, Vural H, Kisioglu AN, Cesur G, Yildirim NG (2005) Mobile phone-induced myocardial oxidative stress: protection by a novel antioxidant agent caffeic acid phenethyl ester. Toxicol Ind Health 21:223–230

Ozgur E, Güler G, Seyhan N (2007) Regenerative effects of (−)-epigallocatechin-gallate against hepatic oxidative stress resulted by mobile phone exposure. IFMBE Proc 16:214–217

Pacher P, Beckman JS, Liaudet L (2007) Nitric oxide and peroxynitrite in health and disease. Physiol Rev 87:315–424

Paulraj R, Behari J (2006) Single strand DNA breaks in rat brain cells exposed to microwave radiation. Mutat Res 596:76–80

Peng T, Lu X, Feng Q (2005) NADH oxidase signaling induces cyclooxygenase-2 expression during lipopolysaccharide stimulation in cardiomyocytes. FASEB J 19:293–295

Peters JM (1994) Proteasomes: protein degradation machines of the cell. Trends Biochem Sci 19:377–382

Phillips JL, Singh NP, Lai H (2009) Electromagnetic fields and DNA damage. Pathophysiology 16:79–88

Piotr DP, Murawska-Ciałowicz E, Jethon Z, Januszewska L, Podhorska-Okołów M, Surowiak P, Zawadzki M, Rabczyński J, Zabel M (2003) Melatonin stimulates the activity of protective antioxidative enzymes in myocardial cells of rats in the course of doxorubicin intoxication. J Pineal Res 35:183–187

Powell SR (2000) The antioxidant properties of zinc. J Nutr 30:1447–1454

Praputpittaya C, Pleumsamran J, Duangjai A (2008) Electromagnetic radiation from mobile phone causes no oxidative stress to the brain. Asian Biomed 2:507–510

Ragy MM (2015) Effect of exposure and withdrawal of 900-MHz-electromagnetic waves on brain, kidney and liver oxidative stress and some biochemical parameters in male rats. Electromagn Biol Med 34:279–284

Reuter S, Gupta SC, Chaturvedi MM, Aggarwal BB (2010) Oxidative stress, inflammation, and cancer: how are they linked? Free Radic Biol Med 49:1603–1616

Rosales MAB (2014) Stress nitrosative in the pathogenesis of diabetic retinopathy. Implications blood-retinal barrier external and possible therapeutic targets (in Portuguese). Doctoral thesis. UNICAMP, Brazil

Saffi J, Henriques JAP (2003) Repair of DNA in eukaryotic cells. In: Da Silva J, Erdtmann B, Henriques JAP (eds) Genetic toxicology. Alcance, Porto Alegre, pp 271–305

Sahin D, Ozgur E, Guler G, Tomruk A, Unlu I, Sepici-Dinçel A, Seyhan N (2016) The 2100 MHz radiofrequency radiation of a 3G-mobile phone and the DNA oxidative damage in brain. J Chem Neuroanat 75:94–98

Santos NCF (1998) Quantification of antioxidant activity of isonicotinoil pyridoxal hydrazone (PIH) against oxidative stress induced by iron ions. Dissertation, University of Brasilia, Brasilia, Brazil

Sengupta DL, Sarkar TK (2003) Maxwell, Hertz the Maxwellians, and the early history of electromagnetic waves. IEEE Antennas Propag Mag 45:13–19

Shan X, Aw TY, Jones DP (1990) Glutathione-dependent protection against oxidative. Injury Pharmacol 47:61–71

Sies H (1993) Strategies of antioxidants defense. Eur J Biochem 215:213–219

Silva DF, Barros WR, Almeida MD, Rego MA (2015) Exposure to non-ionizing electromagnetic radiation from mobile telephony and the association with psychiatric symptoms. Cad Saude Publica 31:2110–2126

Simko M, Hartwig C, Lantow M, Lupke M, Mattsson MO, Rahman Q (2006) Hsp 70 expression and free radical release after exposure to non-thermal radio-frequency electromagnetic fields and ultrafine particles in human Mono Mac 6 cells. Toxicol Lett 161:73–82

Sohal RS, Weindruch R (1996) Oxidative stress, caloric restriction and aging. Science 273:59–63

Sokolovic D, Djindjic B, Nikolic J, Bjelakovic G, Pavlovic D, Kocic G, Krstic D, Cvetkovic T, Pavlovic V (2008) Melatonin reduces oxidative stress induced by chronic exposure of microwave radiation from mobile phones in rat brain. J Radiat Res 49:579–586

Stadman ER (1992) Protein oxidation and aging. Science 257:1220–1224

Stadman ER, Levine RL (2000) Protein oxidation. Ann N Y Acad Sci 899:191–201

Tomruk A, Guler G, Dincel AS (2010) The influence of 1800 MHz GSM-like signals on hepatic oxidative DNA and lipid damage in nonpregnant, pregnant, and newly born rabbits. Cell Biochem Biophys 56:39–47

Türker Y, Naziroglu M, Gümral N, Çelik Ö, Saygin M, Çömlekçi S, Flores-Arce M (2011) Selenium and L-carnitine reduce oxidative stress in the heart of rat induced by 2.45-GHz radiation from wireless devices. Biol Trace Elem Res 143:1640–1650

Ushio-Fukai M (2006) Localizing NADPH oxidase–derived ROS. Sci STKE re8:1–6

Uttara B, Singh AV, Zamboni P, Mahajan RT (2009) Oxidative stress and neurodegenerative diseases: a review of upstream and downstream antioxidant therapeutic options. Curr Neuropharmacol 7:65–74

Vaca CE, Wilhelm J, Harms-Ringdahl M (1988) Interaction of lipid peroxidation products with DNA. Mutat Res 195:137–149

Vainshtein BK, Melik-Adamyan WR, Barynin VV, Vagin AA, Grebenko AI (1981) Three-dimensional structure of the enzyme catalase. Nature 293:411–412

Valko M, Morris H, Cronin MTD (2005) Metals, toxicity and oxidative stress. Curr Med Chem 12:1161–1208

Valko M, Leibfritz D, Moncol J, Cronin MTD, Mazur M, Telser J (2007) Free radicals and antioxidants in normal physiological functions and human disease. Int J Biochem Cell B 39:4–84

Vasconcelos SML, Goulart MOF, Moura JBF, Manfredini V, Benfato MS, Kubota LT (2007) Reactive oxygen species and nitrogen, antioxidants anddamage markers oxidative in human blood: key analytical methods for its determination (in Portuguese). Quim Nova 30:1323–1338

Verschaeve L, Maes A (1998) Genetic, carcinogenic and teratogenic effects of radiofrequency fields. Mutat Res 410:141–165

Yilmaz A, Tumkaya L, Akyildiz KA, Kalkan Y, Bodur AF, Sargin F, Efe H, Uydu HA, Yazici ZA (2016) Lasting hepatotoxic effects of prenatal mobile phone exposure. J Matern Fetal Neonatal Med 17:1–14

Yurekli AI, Ozkan M, Kalkan T, Saybasili H, Tuncel H, Atukeren P, Gumustas K, Seker S (2006) GSM base station electromagnetic radiation antioxidative stress in rats. Electromag Biol Med 25:177–188

Zmyslony M, Politanski P, Rajkowsk E, Szymczak W, Jajte J (2004) Acute exposure to 930 MHz CW electromagnetic radiation in vitro affects reactive oxygen species level in rat lymphocytes treated by iron ions. Bioelectromagnetics 25:325–328

Chapter 8
Lyse-It™: A Rapid Platform for Cellular Lysing and Tunable DNA/Protein Fragmentation

Tonya M. Santaus, Johan H. Melendez, Maraki Y. Negesse, Amanda Harvey, Melissa Cyr, Paula Ladd, and Chris D. Geddes

Abstract In this article, we review *Lyse-It™;* a method for the low cost, efficient, and rapid sample preparation of various bacteria and cell types. Currently, bacterial detection and identification is underpinned by two discreet processes, namely genomic isolation followed by its identification. While molecular detection strategies are available to detect DNA and proteins in time-frames ranging for minutes to hours, the bottleneck in all of these detection strategies lies in the initial sample preparation, i.e. in the cellular lysing, which can add several hours or more to the overall bacterial identification process. For intercellular components, such as DNA and proteins, there is typically a two-step process of breaking open cells (lysis) and then fragmenting the released DNA or degrading the proteins. Cellular lysing is typically a multi-step, multi-reagent process that is tedious, expensive, and time-consuming. Following cell lysis, the DNA and/or proteins must be extracted and purified before downstream analysis.

Keywords Bow-tie antennas • Microwave-induced lysing • DNA fragmentation • Protein degradation • Rapid lysing • Low-cost lysing • Microwave-based sample preparation

The Lyse-It™ technology, which eliminates many sample preparation hurdles, has been developed over several years by our research group through a combination of both theoretical design, using Finite Difference Time Domain (FDTD) analysis, coupled with significant experimental testing. In this chapter, we review how gram-negative and gram-positive bacteria such as *Salmonella* and *Bacillus Anthracis*, along with various food products can be lysed, readying their DNA and/or proteins for analysis in a low cost and rapid manner. Lyse-It™ offers the significant benefit

T.M. Santaus • J.H. Melendez • M.Y. Negesse • A. Harvey • M. Cyr
P. Ladd • C.D. Geddes (✉)
Institute of Fluorescence, Department of Chemistry and Biochemistry, University of Maryland Baltimore County, 701 E. Pratt St., Baltimore, MD 21202, USA
e-mail: geddes@umbc.edu

C.D. Geddes (ed.), *Microwave Effects on DNA and Proteins*,
DOI 10.1007/978-3-319-50289-2_8

of both cellular lysing and DNA fragmentation and protein degradation in a single rapid step, a combination not offered by other lysing approaches today.

8.1 Introduction

There is a long-standing need for rapid and sensitive detection platforms, including point-of-care tests, for a variety of bacteria.(Gaydos et al. 2010, 2003; Taylor et al. 2011; Van Der Pol et al. 2012) Many technologies such as Polymerase Chain Reaction (PCR) and others, have the potential to provide results in under 30 min (Gaydos et al. 2010, 2003; Taylor et al. 2011; Van Der Pol et al. 2012) but only after the target analyte has been extracted. Pathogen identification on these platforms can take a substantial amount of time, due to the upfront sample preparation, or specifically, sample cell lysis.(Melendez et al. 2013) Lysis, the process by which a bacterium is broken open to release its genetic material, is crucial for its detection and identification, and is typically undertaken by adding a chemical cocktail to the sample (lysis buffer) accompanied by heating. This process can be timely to complete and often involves expensive lysis buffers.(Melendez et al. 2013) Even in non-diagnostic settings, such as research laboratories, lysis is still performed by a number of researchers using the age-old lysing buffer technology.(Melendez et al. 2013; Tennant et al. 2011)

In addition to chemical lysis buffer methods, there exists a whole range of other approaches including bead beaters, sonicators, homogenizers, mortar and pestle grinders, French Press, freezer milling, and enzymatic digestion. In all of these examples, a two-step process is needed for both cellular lysing, followed by DNA fragmentation or protein degradation. For example, several homogenizers are available on the market today, such as the Fast-Prep™ instrument, by MP Bio, which simply blends the cells into a soup, and a secondary set of solutions are needed to digest the DNA. While this homogenizer can rapidly blend cells in under a minute, it retails for several thousand dollars, not including the other supplies needed for digestion (see http://www.mpbio.com/). Further, with regard to chemical lysis, several different chemical solutions are often needed to lyse gram-positive or gram-negative cells, followed by DNA fragmentation. Promega (www.promega.com) offers numerous kits for DNA lysis, fragmentation, and recovery. Kits are usually found for 100 μL to 10 mL solution volumes, and are usually priced $200–$1000. For all these chemical lysis kits, lysis is a lengthy procedure, typically many hours.

8.2 Development of Lyse-It™: Changing How We Think of Sample Preparation Today

The authors have developed a novel method, Lyse-It™, which has the potential to affect both diagnostic and research settings alike. Focused microwaves in small disposable sample chambers allow the user to lyse cellular samples with high efficiency, typically within 60 s or less, on a single platform at a cost less than other technologies available today. This lysing approach is generic to a whole range of cells unlike traditional lysing buffer approaches that all use multiple kits for different media.

8.2.1 Finite Different Time Domain (FDTD) Analysis

Constructs of the Lyse-It™ technology first began by using the Finite Difference Time Domain (FDTD) method. This method is a mathematical way of solving problems in electrodynamics and was first proposed by Kane Yee in 1966. It can be considered a resonant technique in which the characteristic dimensions of a domain can be set to a certain size. Spatial and temporal derivatives, specifically Ampere's and Faradays laws are employed in the finite difference approximations. To be brief, the algorithm can be summarized in five steps. First, the derivatives in Ampere's and Faradays laws are replaced with finite differences and the parameters are set so that electric and magnetic fields are staggered in space and time. The resulting equations are then solved to obtain equations that express the known past fields and the unknown future fields. The magnetic field and the electric field are then evaluated one time point ahead so that they become past fields. Finally, the evaluation of the magnetic and electric fields are repeated in one-time step fashions until the fields have been obtained over the desired time frame (Schneider 2016). For Lyse-it™, simulations of desired metal geometries have been performed in the microwave wavelength region, for the spatial focusing of microwaves into small sample volumes.

8.2.1.1 Determining the Geometric Lengths of the Microwave-Focusing Metal Bow-Tie Structures

Microwaves are a form of electromagnetic radiation with wavelengths ranging from 1 m to 1 mm. The goal of the FDTD simulations was to determine the optimal length for mm-sized equilateral triangles, which could focus 2.45 GHz, ~12.3 cm microwaves, the microwave wavelength used in household microwave-ovens, into a small sample volume for localized heating. Subsequently, triangle lengths ranging from 5 to 40 mm were analyzed considering both single and two-facing triangles (bow-ties), Fig. 8.1.(Aslan et al. 2007a, 2008) While triangles around 20–25 mm

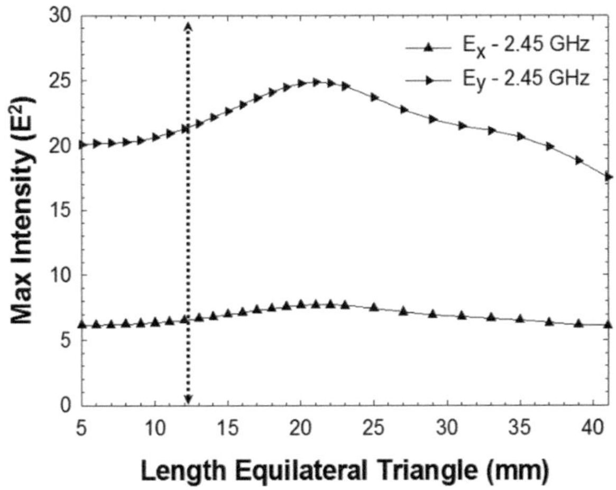

Fig. 8.1 FDTD simulations to determine the optimal length of the equilateral triangles. Each set of points represents the maximum intensities from the images in the Ix and Iy planes

were simulated to have the maximum intensities around the apexes located in the middle of the bow tie structure, Fig. 8.1, when tested these triangles rapidly failed, and often caused sparking within the cavity. Through testing different sized physical structures within a microwave cavity, we concluded that 12.3 mm equilateral triangles did not spark or arc, but did heat any water positioned at the apexes.

8.2.1.2 Microwave Field Distributions

Similar to how visible wavelength light can be focused at the apexes of nanometer scale metallic nano-triangles, (Hu and Weiss 2016; Hasan et al. 2016; Friedlander et al. 2015; Chen et al. 2004) we have used this as a starting point and subsequently designed and tested the use of 12.3 mm sized equilateral triangles to focus 12.3 cm microwaves using a conventional 2.45GHz transverse electric polarized total-field scattered field (TFSF) microwave cavity. By using FDTD to simulate the microwave field distribution (Ix + Iy) around the metallic geometries, (Aslan et al. 2008) we have been able to design our bow tie geometries for focusing microwaves not at all corners of the triangle, but now at the bow tie apex. Fig. 8.2 shows a typical FDTD image for a pair of 12.3 mm Aluminum triangles arranged in a bow-tie configuration with a 1 mm gap-size. As Fig. 8.2 shows, the microwave field is predominantly focused at the apex, noting that the field strength has been normalized to “1” in this image. Our extensive simulations have shown that for sizes ~λ/10, the microwave field strength at the apexes is significantly higher, so one would expect to heat water preferentially at these locations.

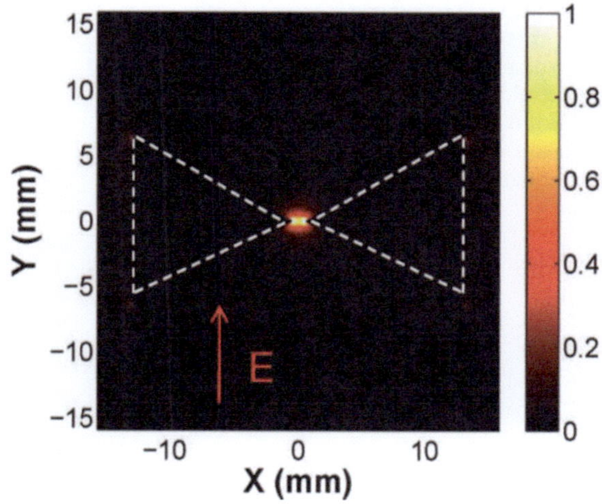

Fig. 8.2 FDTD Simulation of electric field distribution between equilateral Aluminum bow-ties

8.2.1.3 Gap Size

Following the microwave field simulations, the gap size (the distance between the triangle apexes) between the bow-tie structures was studied to explore the extent of field intensity at the apex. Fig. 8.3 shows the increase in electromagnetic field distribution for a 2.45 GHz microwave frequency incident upon the 2-D equilateral triangles in the bow-tie orientation. The max intensity is seen when the gap size is <1 mm, whereas large gap sizes tended to decrease the field intensity as shown in Fig. 8.4. Simulations show the field is about 40,000 times greater for bow ties versus a single triangle for small gap sizes <1 mm. In essence, small gap sizes allow for the rapid and more efficient heating of water located at the apexes as discussed below.

8.2.2 *Temperature Jump at the Bow Tie Apex*

Following the FDTD simulations and the initial testing of basic geometries for stability in a microwave cavity, we studied the temperature jump in real-time at the bow-tie apexes during microwave irradiation for a small volume of water placed at the apexes, Fig. 8.5 (Aslan et al. 2008). To undertake this, we developed a microwave-based thermal imaging microscope, which has been discussed elsewhere (Previte and Geddes 2007; Previte 2007). Fig. 8.5a, shows the increase in temperature between the bow tie. In approximately 2 s, the temperature has jumped ≈ 12 °C above the room temperature. When we thermally imaged a glass slide, the temperature only jumped ~1 °C in 2 s above room temperature. Fig. 8.5b shows thermal images taken in real time of the bow tie. Additional thermal imaging studies experimentally confirmed the gap size simulations of those in Fig. 8.4, in

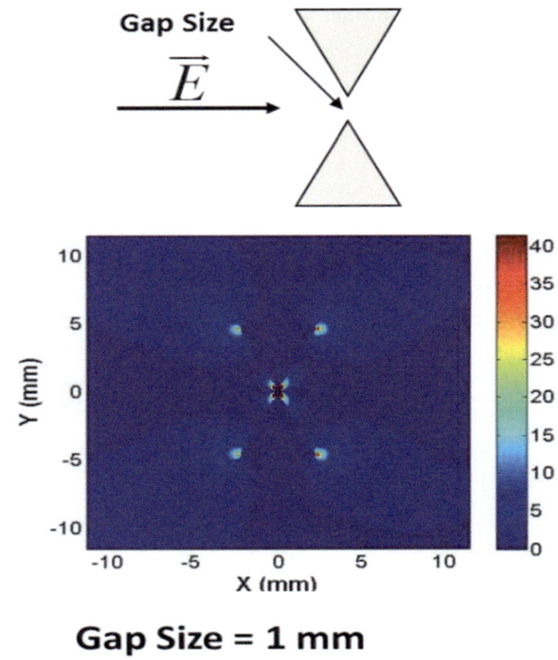

Fig. 8.3 FDTD simulation of the electromagnetic field distribution between equilateral aluminum bow-ties. The intensity of the field is maximum at the apex of the bow-tie structure

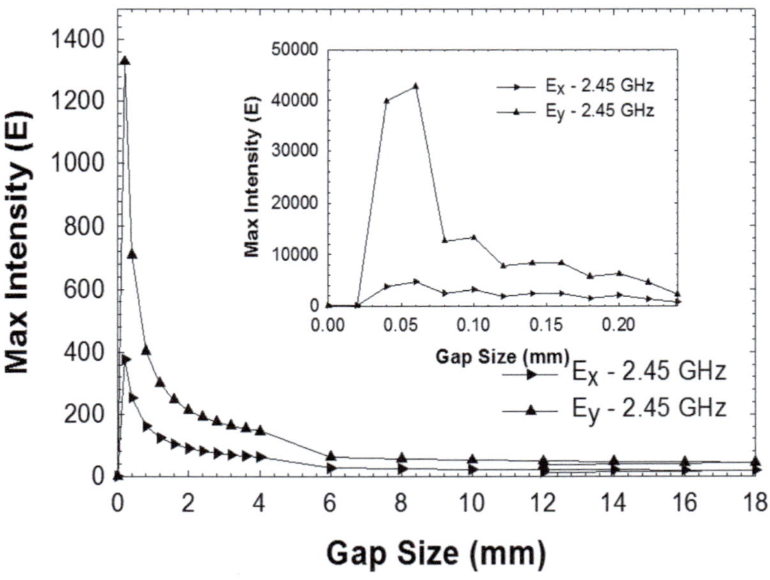

Fig. 8.4 FDTD simulation where the incident field is held constant and the gap size between equilateral Aluminum bow-ties is varied from 1 to 18 mm

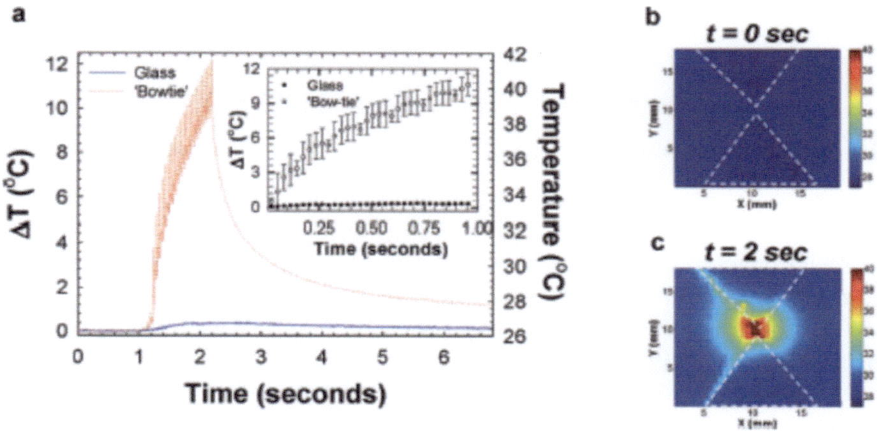

Fig. 8.5 Experimental temperature data taken with a thermal imaging camera for a microwave irradiation time up to 2 s. There is a substantial increase in temperature seen between the apex's of the metal bow-tie structures. (**a**) Temperature of water on a glass slide without bow-ties (*blue line*) versus temperature of water on glass slides equipped with metal bow-ties (*red line*). (**b**) Initial temperature reading pre-microwave irradiation. (**c**) Temperature reading/image after 2 s of microwave irradiation

that, smaller gap sizes lead to higher field intensities and therefore heating, for water positioned at the apexes (data not shown). For gap sizes $<<<$ 1 mm, we were able to thermally visualize super-heated water, >100 °C in only a few seconds of microwave exposure.

8.3 Gram-Negative Bacterial Lysing

Gram-negative bacteria do not retain the crystal violet stain, and thus appear pink under a microscope. The bacteria are characterized by their thin peptidoglycan cell wall in between an inner cytoplasmic membrane and outer membrane with periplasm between the membranes, as shown in Fig. 8.6 (Sigma-Aldrich 2016). The cell wall of these bacteria are much thinner than that of gram-positive bacteria. Gram-negative bacteria can be found in all parts of the world and usually do not forms spores. The three bacteria that we will be discussing are *Salmonella*, Chlamydia, and Gonorrhea.

8.3.1 Salmonella

Salmonella is a world-wide bacterium of concern. Non-typhoidal *Salmonella* produces vomiting, diarrhea, and fever in healthy humans, but for those who are

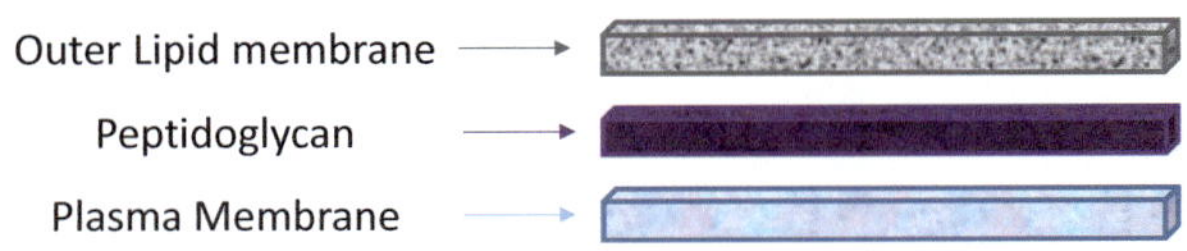

Fig. 8.6 Gram-negative bacterial cell wall architecture consisting of an outer lipid membrane, peptidoglycan layer, and a plasma membrane (Sigma-Aldrich 2016)

immunocompromised, the disease can be fatal; therefore, its detection is paramount (Tennant et al. 2011; Kennedy et al. 2004).

The Lyse-It™ technology was employed followed by culturing and Transmission Electron Microscopy (TEM) of the lysate (solution left after lysing) to understand the viability of cells post microwave treatment as well as test the utility of the lysate on a specific *ds*DNA detection platform, namely on a Microwave-Accelerated Metal-Enhanced Fluorescence (MAMEF) assay. The sample was first microwaved on high power (1.65 kW) for 13 s. The lysate was then plated to look for any growth. After incubation at 37 °C, there was no growth seen, data not shown, suggesting that *Salmonella* was not viable after lysing. In addition, agarose gels were performed on the lysates to observe the resultant DNA fragmentation patterns. The gel, stained with ethidium bromide, shows DNA fragments below 0.2 kb confirming that DNA was both extracted and fragmented during the microwave irradiation process (Fig. 8.7a). We found that different DNA fragmentation patterns could be observed depending on the length of the microwave heating time. Agarose gels of lysates from short microwave heating times, typically showed large DNA present on the gels, with the DNA sizes decreasing with both an increased heating time and microwave power. Transmission Electron micrograph (TEM) images were taken pre- and post-lysis to visually observe the disruption of the cell membrane. As we can see in Fig. 8.7b bottom, TEM images of lysed *Salmonella* show significant destruction of the cells, as compared to the unlysed sample.

Using the lysate of non-typhoidal *Salmonella*, a DNA specific MAMEF assay was performed to demonstrate that the lysates, specifically the DNA recovered, could be used on molecular DNA detection platforms, such as PCR and MAMEF. The MAMEF platform has been reviewed numerous times (Melendez et al. 2013; Tennant et al. 2011; Aslan et al. 2008; Joshi et al. 2014; Dragan et al. 2012; Zhang et al. 2011; Aslan and Geddes 2006; Aslan et al. 2007b; Aslan and Geddes 2008). The MAMEF anchor probe was attached to a silver island film surface contained within a well (Melendez et al. 2013; Tennant et al. 2011; Aslan et al. 2008; Joshi et al. 2014; Dragan et al. 2012; Zhang et al. 2011; Aslan and Geddes 2006; Aslan et al. 2007b; Aslan and Geddes 2008). The anchor probe, fluorescent probe, and *Salmonella* cells were microwave irradiated at 140 W for 30 s. The microwave produces a temperature gradient allowing the released and fragmented DNA to hybridize with the fluorescent probe and anchor probe. To confirm that *Salmonella* was detected, a 532-nm laser excited the fluorophore producing a fluorescent emission spectra. The resultant series-dilution intensity graph showed that a concentration as low as 9 colony forming units (9 CFU) could be detected with the

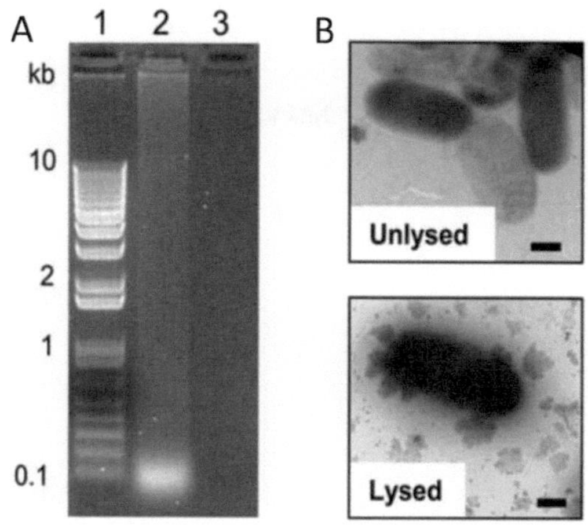

Fig. 8.7 (**a**) Agarose gel of *ds*DNA released from an overnight culture of *Salmonella* (1.9×10^{10} CFU/mL) with the use of gold bow ties, and microwave irradiation. Lane (1) 1 kb Plus DNA ladder; (2) lysed bacteria; (3) unlysed bacteria. (**b**) Transmission electron micrographs of unlysed and lysed *Salmonella* respectively

MAMEF assay specific to non-typhoidal *Salmonella* (Fig. 8.8a, b). Subsequently, by using a standard microwave oven in combination with the bow tie structure, we were able to lyse *Salmonella* cells in a very short time period, <30 s, where the lysate was viable for both the highly sensitive and specific detection of *Salmonella*.

8.3.2 Chlamydia trachomatis

Chlamydia, a sexually transmitted infection (STI) caused by the gram-negative *Chlamydia trachomatis,* is the most commonly diagnosed and reported STI to the Centers for Disease Control and Prevention (CDC). In 2008, there were 1.2 million reported cases (Zhang et al. 2011; Prevention 2016a). The current standard for the detection of chlamydia is the nucleic acid amplification test (NAAT); however, this test is time-consuming and not considered a point-of-care test. Therefore, we wanted to test to see if the lysing technology Lyse-it™ could be used to effectively and efficiently lyse the bacteria as well as fragment its DNA, so that it could be detected using a specific chlamydia MAMEF molecular assay.

A small aliquot of chlamydia sample was microwave irradiated on microscope slides with a standard 12.3 mm gold bow tie thermally vapor deposited onto the surface. A 2.45 GHz microwave was used and the sample irradiated for 10 s. Transmission electron micrographs were taken to demonstrate the disruption of the cell membranes (Fig. 8.9A). TEM images are shown on the same scale, where we can see the almost complete destruction of the Chlamydia cells. To confirm that the DNA was further released and fragmented during lysing, agarose gels stained with ethidium bromide were run on the resultant lysate. The gel confirmed that DNA was indeed fragmented down to sizes of 50 and 600 base pairs (Fig. 8.9B). Consistent with observations of *Salmonella* DNA, the size distribution of the Chlamydia DNA could also be *tuned*, where long microwave exposure times

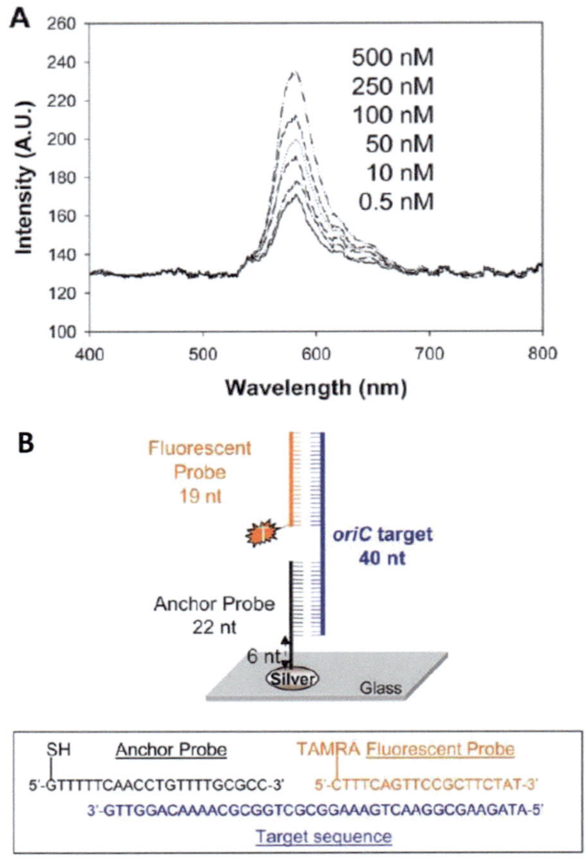

Fig. 8.8 (**a**) Increase in concentration of *Salmonella* DNA leads to an increase in fluorescent intensity on the MAMEF molecular assay. (**b**) MAMEF assay construct for targeting *Salmonella oriC*

resulted in progressively smaller DNA fragments over time. A MAMEF specific assay was also developed and tested for Chlamydia lysate, with the general protocol of attaching the anchor probe to a silver island film surface and microwave irradiating the anchor probe, lysate, and fluorescent probe to accelerate hybridization. As the concentration of DNA bound to the probes was increased, the fluorescent intensity from the assay increased (Fig. 8.10). Using this approach, the detection limit was about 100 colony forming units per ml (Zhang et al. 2011).

8.3.3 Neisseria gonorrhea

Gonorrhea is the second most prevalent sexually transmitted disease caused by the bacteria *Neisseria gonorrhoeae*. In 2014 the CDC reported that out of 100,000 men and women in, over 1000 people between the ages of 20 and 24 contracted the disease (Prevention 2016b). Like chlamydia, gonorrhea is also typically detected by

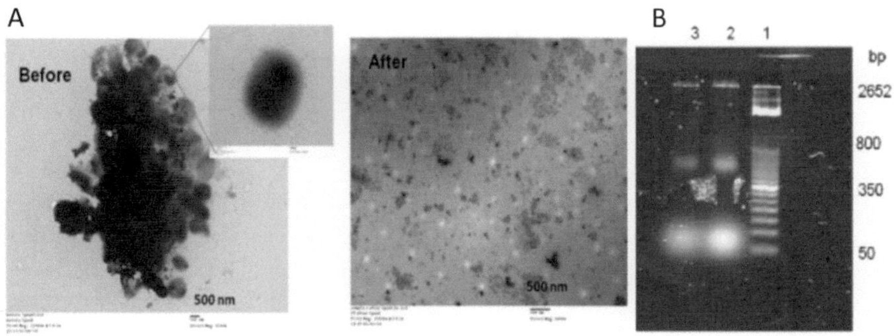

Fig. 8.9 (**a**) Transmission Electron Micrographs of Chlamydia pre- and post microwave irradiation. (**b**) Gel electrophoresis (2%) of lysed chlamydia. Lane 1: Ladder; 2: 40 μL lysed chlamydia sample; 3: 20 μL lysed sample

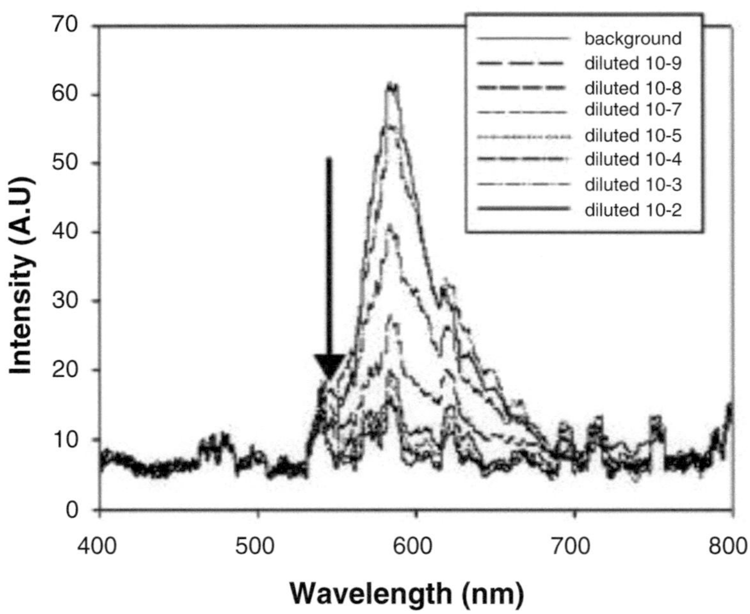

Fig. 8.10 MAMEF assay for Chlamydia. Dilutions in the stock concentration of chlamydia lead to a decrease in the fluorescent intensity

NAAT tests and Polymerase Chain Reaction (PCR). Using the Lyse-It™ technology, we have shown that gonorrhea cells were lysed and the DNA simultaneously fragmented for its downstream molecular detection in a very rapid time frame, as described below.

A 10^8 concentration of gonorrhea cells was microwave irradiated with the use of gold bow ties (Lyse-it™) at 30% and 50% power for 30, 60, and 90 s (corresponding to 8.1, 16.2, 24.3 kJ and 13.5, 27.0, and 40.5 kJ of energy in the

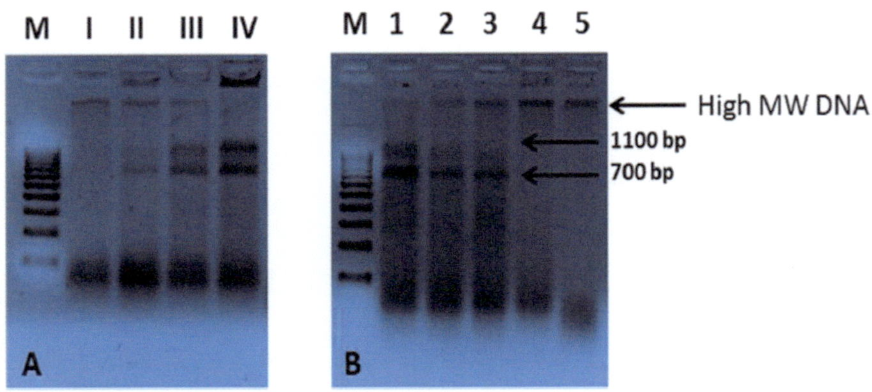

Fig. 8.11 Gel electrophoresis of gonorrhea. (**a**) Osmotic lysing Lane M: 100 bp molecular ruler, I: freshly prepared culture; II: 1 h incubation; III: 2 h incubation; IV: 3 h incubation (**b**) Conventional heating (Boiling) Lane M: 100 bp molecular ruler; 1: Pre-lyse sample; 2: 40 °C; 3: 50 °C; 4: 60 °C; 5: 70 °C

microwave cavity respectively) and was compared to conventional heating (the boiling method), as well as without the use of the bow ties. It was shown that by simple boiling at 70 °C, gonorrhea DNA fragments (Fig. 8.11). Following experimentation with conventional heating, trials comparing the use of the bow tie as compared to not using the bow tie were subsequently investigated. Microwave irradiation was performed at the powers and times described above. It was found that more DNA fragmentation is achieved with the use of the gold bow-ties (i.e. Lyse-it™) than without the use of bow-ties (Fig. 8.12) (Melendez et al. 2016). In addition, given that the same number of cells was used for all experiments, we see more DNA evident on the gels (i.e. darker bands) for longer heating times coupled with a higher microwave cavity power. In addition, we have recently been able to show that the recovered lysate using the Lyse-It™ technology, can be used with PCR, for the quantitative detection of Gonorrhea (Melendez et al. 2016).

8.4 Gram-Positive Bacterial Lysing

Gram-positive bacteria are inherently purple when stained with crystal violet. Gram-positive bacteria are characterized by a thick peptidoglycan wall above the plasma membrane (Fig. 8.13) (Sigma-Aldrich 2016). This thick peptidoglycan wall makes lysing these bacteria traditionally more difficult, time consuming and ultimately more expensive when using conventional lysing approaches. Similar to gram-negative bacteria, gram-positive bacteria can form spores and are found in all parts of the world. To demonstrate the significant utility of the Lyse-it™ technology to both rapidly lyse and fragment the DNA of Gram Positive bacteria,

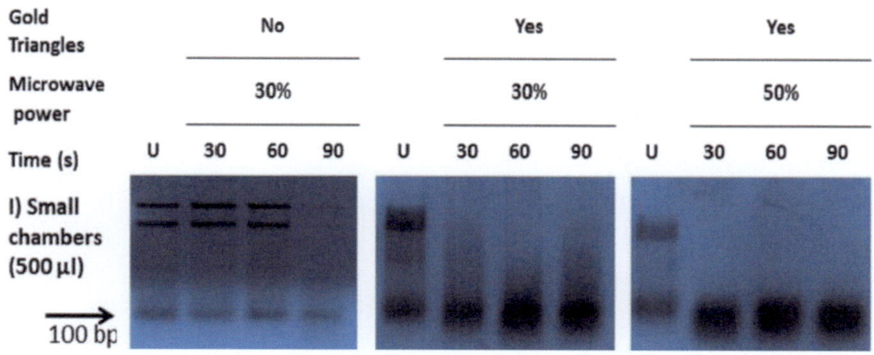

Fig. 8.12 Microwave irradiated gonorrhea with and without the use of gold bow ties for 30% and 50% power, where the entire cavity power was 900 W. From *left* to *right*: Gonorrhea microwave irradiated without bow ties prior to microwave irradiation (U) for 30, 60, and 90 s. *Middle* and *far right*: Gonorrhea microwaved with the use of gold bow ties for the same time/power points

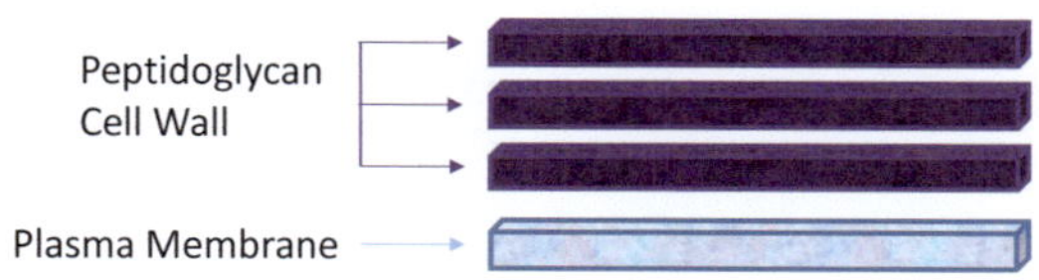

Fig. 8.13 Gram-positive bacterial cell wall composed of a thick peptidoglycan wall and a plasma membrane

we have studied three important organisms, namely *Bacillus anthracis*, *Listeria monocytogenes*, and *Clostridium difficile*.

8.4.1 Bacillus anthracis

Bacillus anthracis, the causative agent of anthrax, is a spore-forming gram-positive bacteria (Dragan et al. 2012; Aslan and Geddes 2008). Clinically there are three pathologies that anthrax can present with, depending on the route of infection. The routes of infection are cutaneous, gastrointestinal, and inhalational anthrax. Inhalational is the most serious form of infection and is considered a bioterrorism agent. Initially inhaled anthrax presents with flu like symptoms followed by respiratory distress, septic shock, and ultimately death in a very short time frame, usually <72 h (Dragan et al. 2012). Therefore, having a rapid sample preparation platform coupled with a rapid detection platform is of high importance.

The Lyse-It™ technology was applied to anthrax spores using bow ties and subjected to pulses of 20 s microwave irradiation. A comparison of the amount of disruption of the cell membrane was observed by TEM before irradiation, on a glass slide after heating in a microwave without and with bow ties, (Fig. 8.14). As shown

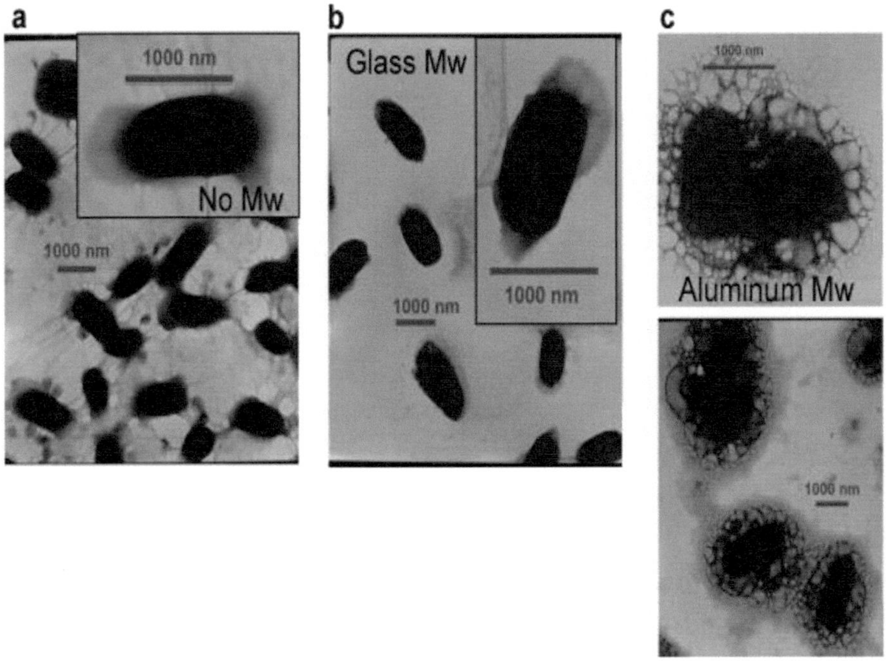

Fig. 8.14 TEM images of *B. anthracis* (**a**) before microwave irradiation, (**b**) on a glass slide after microwave irradiation and (**c**) with the use of bow ties, i.e. Lyse-It™ after microwave irradiation

in Fig. 8.14, the spores are fully intact prior to microwave irradiation. Even on the glass slide after microwave irradiation, there does not appear to be any significant structural changes in the spores, however, spores after irradiation with the bow ties show significant deformation. Interestingly, we had to reduce both the microwave power and exposure time in panel C, as compared to panel B, or else nothing except for cellular debris would have been observed, not unlike what was shown for the gram negative bacteria, c.f. Fig. 8.9. To similarly confirm the release, fragmentation, and efficacy of the lysed DNA, gel electrophoresis was performed (data not shown) and a MAMEF specific assay used to detect the DNA. The assay is specific for the protective antigen (PA) of anthrax (Dragan et al. 2012). Three assays were preformed where no microwave heating was involved, heating only on a glass slide, and heating with the bow ties (Fig. 8.15a–c). Without the use of microwave irradiation very little signal is detected. However, when microwave irradiation is used in conjunction with Lyse-It™, there is a significant increase in the fluorescent intensity (Aslan et al. 2008) of the MAMEF assay. Less than 100 copies of Anthrax could be detected in less than 1 min, including the sample preparation. Subsequently, the Lyse-It™ technology was not only able to lyse the Anthrax spores in a very rapid manner, but the fragmented DNA was additionally used for its successful specific genomic identification and quantitative detection on the MAMEF platform (Dragan et al. 2012).

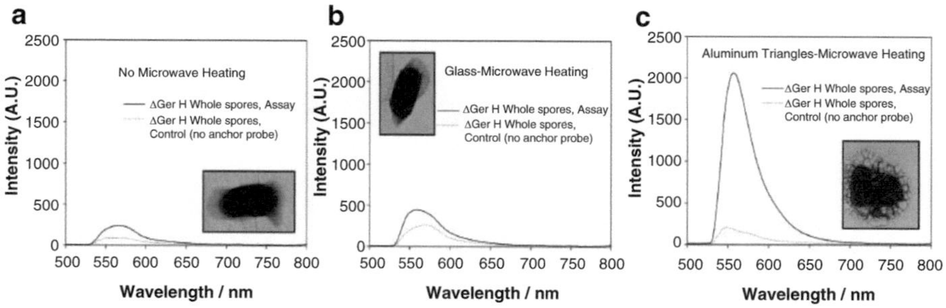

Fig. 8.15 MAMEF assay fluorescent intensity graphs for the detection of the PA gene of anthrax. There is an increase in intensity from no microwave heating (**a**) to microwave heating on glass (**b**) to microwave heating with bow ties (**c**), confirming the presence of released genomic Anthrax DNA

8.4.2 Listeria monocytogenes

Listeria is a rod-shaped, flagellated bacterium that can be found in soil and food processing plants around the world. The bacterium that causes disease in humans is *Listeria monocytogenes*. According to the Center for Disease Control and Prevention, those with immunocompromised systems, the elderly, and women who are pregnant are at the most risk for serious infection. Listeria has a fatality rate between 18 and 30%. The two most common detection methods are plate cultures and real-time PCR, where plate cultures typically take up to 48 h to perform, and PCR is inherently hindered by the sample matrix and the isolation of the DNA.

Similar to the other organisms tested with Lyse-it™, we see a decrease in the amount of growth of Listeria after microwave irradiation (Fig. 8.16). The recovered DNA was suitable for the genomic identification of the organism, with the recovered DNA ideally suited for the genomic identification of the organism.

8.4.3 Clostridium difficile

Clostridium difficile, more commonly abbreviated as *C. difficile* is a Gram-positive spore-forming bacterium that causes inflammation of the colon. People who are the most risk are those requiring long term antibiotics and the elderly. In 2011, there were half a million infections, 29,000 of which were fatal within 30 days of infection (Prevention 2016c). Current methods for detecting *C. difficile* are culturing from stool samples, cell cytotoxin neutralization assays (CCTA), and real-time PCR.

Subsequently, the Lyse-It™ technology was employed for the cellular lysis and DNA fragmentation of *C. difficile* with downstream detection on a MAMEF assay, similar to the other organisms tested. The samples were microwave irradiated at 80% power for between 8 and 16 s in a 1.65 kW microwave on slides supporting gold bow ties (Fig. 8.17) (Joshi et al. 2014). Cells microwaved at 80% power for

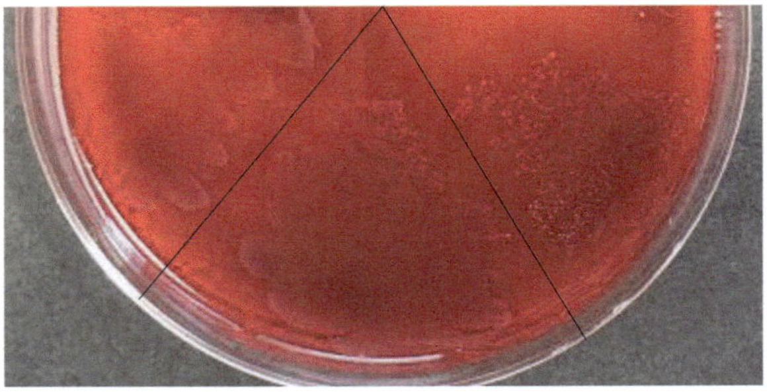

Fig. 8.16 *Listeria monocytogenes* on sheep blood agar. *Left*: 10^8 concentration of listeria grown at 37 °C pre-microwave irradiation. *Center*: Decrease in viable Listeria after microwave irradiation at 30% power for 60 s. *Right*: Significant decrease in Listeria colonies after 30% power, 90 s

Fig. 8.17 Gel electrophoresis of *C. difficile* microwave irradiated at 80% power for 8–16 s with bow ties. Lane M: 1 kb molecular DNA ladder, Lane 1: non-microwaved bacteria, Lanes 2–10: 8–16 s

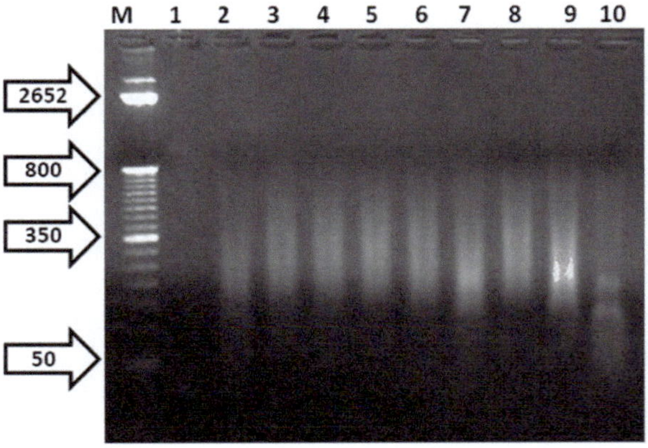

15 s resulted in the optimal release of DNA with fragments between 50 and 200 base pairs, ideal for MAMEF-based detection (Joshi et al. 2014). However, microwave irradiation for 16 s and above resulted in too small fragment sizes to be detected on the MAMEF assay, which was specific for *C. difficile* toxins A and B (Joshi et al. 2014). Toxins A and B from the *C. difficile* gene CD630 could be detected following microwave irradiation at 80% power for 15 s (Fig. 8.18). As the number of colony forming units increased, so did the intensity on the detection assay. Thus, 10 colony-forming units of the bacteria could be detected in stool using the Lyse-It™ technology in conjunction with a MAMEF specific assay (Joshi et al. 2014).

8.5 Lysing of Food Products

The Lyse-It™ technology is also able to disrupt plant and animal cells. Although both eukaryotic cells presenting many similarities, plant and animal cells differ in some aspects. The most important difference lies in the presence of a cell wall in

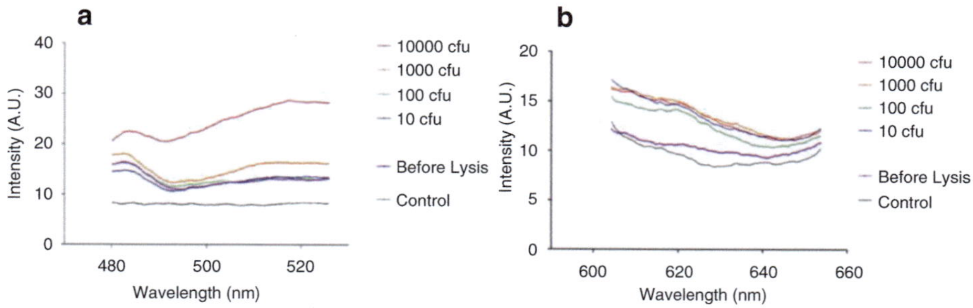

Fig. 8.18 MAMEF molecular DNA assay fluorescence intensities of *C. difficile* toxins A (**a**) and B (**b**) upon increasing colony forming units of the bacteria

plant cells. This cell wall functions like a stiff lattice providing structure and support to the cell. Because of its rigidity, lysing of plant cells is more challenging than that of animal cells. In the following section, we will address the effects of lysing on protein release and degradation.

8.5.1 Plant Cell Based Foods: Spinach and Cantaloupe

Spinach and cantaloupe are two foods that represent the plant cell group. The Lyse-It™ technology was employed to look at the differences in protein degradation with and without the use of gold bow tie structures as compared to conventional heating. Samples were aliquoted in to 2 mL and boiled for 1 min between 40 and 90 °C. The foods, in 1 mL aliquots, were microwave irradiated at 10% through 50% power in a 900 W microwave for 60 s. Following microwave irradiation, the samples were cooled to room temperature and run on SDS PAGE stained with SYPRO Ruby. The gels for cantaloupe lysing can be seen in Fig. 8.19. Complete protein degradation of the protein is seen in lane 20 corresponding to convention heating for 1 min at 90 °C. Protein degradation is seen without the use of bow ties at 50% power for 60 s (≈27.0 kJ). With the use of gold bow ties, the total amount of energy required to get the same extent of fragmentation is reduced to 21.6 kJ corresponding to ≈40% power for 60 s.

Spinach samples were also boiled and microwave irradiated with the same conditions as above. Analysis in comparing no bow tie structures to the use of the gold bow tie is currently underway. Preliminarily, protein degradation is seen with the use of conventional heating and microwave irradiation without the use of gold bow ties (Fig. 8.20). There is more protein degradation seen with conventional heating at 90 °C as compared to microwave irradiation. Protein degradation is seen in lanes 3 through 7 of Fig. 8.20, but not complete degradation. This is due to the temperature of the sample after microwave irradiation being lower than the 90 °C boiled sample. It is thought that with the use of gold bow ties, the temperature of the sample will be higher than without bow ties thus leading to more protein degradation.

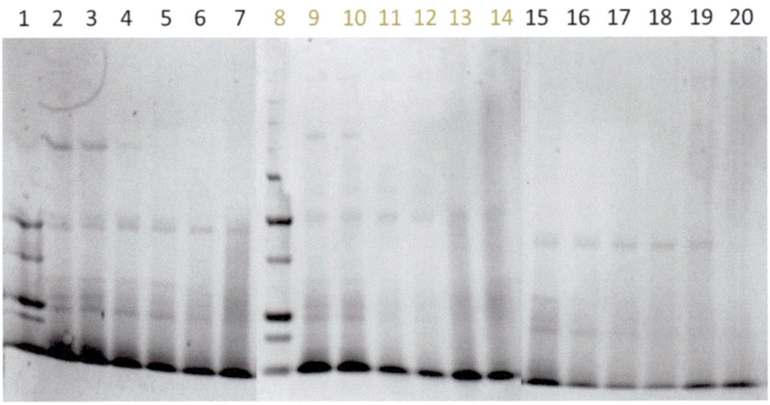

Fig. 8.19 Cantaloupe sample microwaved irradiated without bow ties (lanes 3–7), with gold bow-ties (lanes 10 through 14), and boiled 40 through 90 °C for 1 min (lanes 15–20 respectively). Lanes 1 and 8: 100 bp molecular ruler. Lanes 2 and 9: non-microwaved cantaloupe

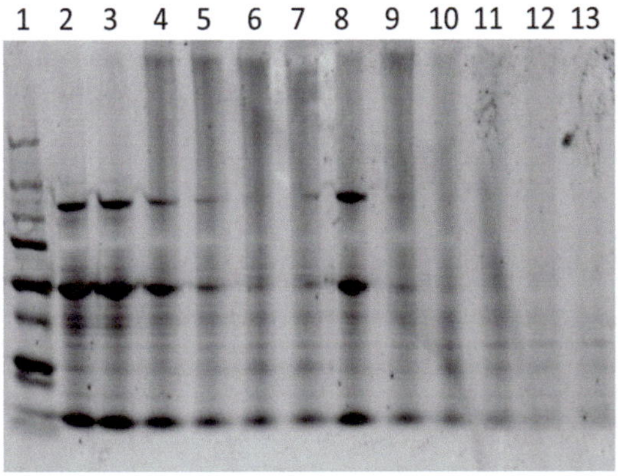

Fig. 8.20 Spinach samples microwave irradiated 10–50% power (lanes 3–7) and boiled 40–90 °C (lanes 8–13). Lane 1: Molecular ruler, Lane 2: Original spinach sample

8.5.2 Animal Cell Based Foods: Ground Beef and Beef Hot Dogs

Similar to the plant cell based foods, the Lyse-It™ technology was employed to analyze protein degradation from animal cells in particular from ground beef and beef hot dogs. Following the same steps as the plant foods, 2 grams of ground beef or beef hot dog were placed in to conical tubes for analysis. Samples were centrifuged to remove large pieces of food not able to be taken up by 1000 μL pipette. Tap water was added to the tubes and 1 mL of sample was microwaved from 10–50% for 60 s with and without gold bow ties. Conventional heating was also performed, where 2 mL of sample was boiled for 2 min at 40 °C through 90 °C.

Post microwave irradiation and conventional heating of ground beef, protein degradation is seen above 80 °C with boiling and from microwave irradiation

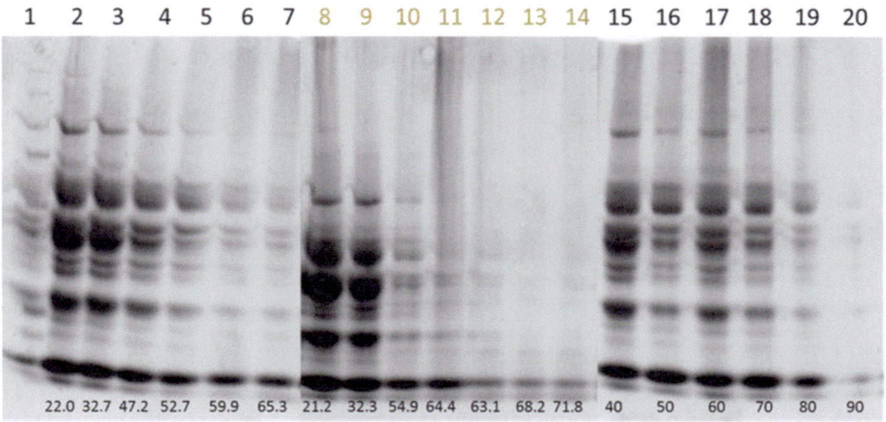

Fig. 8.21 Ground beef samples microwaved irradiated without bow-ties (lanes 3–7), with gold bow ties (lanes 10 through 14), and boiled 40 through 90 °C for 1 min (lanes 15–20). Lanes 1 and 8: 100 bp molecular ruler. Lanes 2 and 9: non-microwaved ground beef

(Fig. 8.21). Significant protein degradation is seen in lane 13 corresponding to 40% power at 60 s (21.6 kJ) with the use of gold bow ties. Significant protein degradation is *not seen* when metal bow ties are omitted. There is a decrease in the amount of intact proteins as the power of the microwave increases, i.e. protein that can be stained. To achieve protein degradation comparable to that seen with gold bow ties, the power or the time of microwave irradiation would have to be notably increased. Post microwave irradiation temperatures, seen at the bottom of Fig. 8.21, reveal that there is an increase in the sample temperature with the use of gold bow ties. This correlates to having increased thermal effects on the cells causing more cellular lysis and ultimately protein release and DNA fragmentation.

Consistent with the data obtained from ground beef, beef hot dog proteins degrade between 80 and 90 °C through conventional heating and more degradation is seen with the use of gold bow ties (Fig. 8.22). No significant protein degradation is seen without the use of gold bow ties. However, when gold bow ties are used, protein degradation can start to be seen at 40% power for 60 s (21.6 kJ). More degradation is seen at 50% power for 60 s (27.0 kJ). Subsequently, with the use of Lyse-It™, protein degradation from food products can readily be achieved and verified by SYPRO Ruby stained SDS PAGE analysis.

8.6 Conclusions

The Lyse-It™ technology has proven to be a valuable resource for shortening the sample preparation time required to lyse cells, fragment DNA, and degrade proteins. The lysing geometry, length and gap size have been systematically optimized through Finite Different Time Domain analysis along with experimentation to look

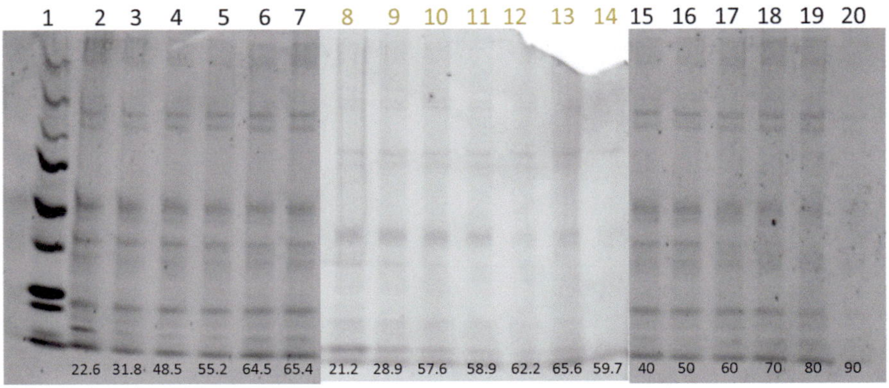

Fig. 8.22 Hot dog samples microwaved irradiated without bow-ties (lanes 3–7), with gold bow ties (lanes 10 through 14), and boiled 40 through 90 °C for 1 min (lanes 15–20). Lanes 1 and 8: 100 bp molecular ruler. Lanes 2 and 9: non-microwaved hot dog

at the effects of electric field and temperature distribution from the bow tie structure. Due to the cellular architecture of gram-positive versus gram-negative bacteria, an overall less microwave irradiation power and time, and thus energy, is required to lyse open gram-negative bacteria. Equipping standard microscope slides with metal bow tie structures allows for sequential cellular lysis, DNA fragmentation, and protein degradation. Cellular lysis was confirmed for gram-negative and gram-positive bacteria through Transmission Electron Microscopy. DNA fragmentation and protein degradation were studied on agarose gels and SDS PAGE respectively. Microwave irradiation with the Lyse-It™ technology can rapidly lyse bacterial cells and food substances providing sufficient DNA to be then used on both PCR and MAMEF molecular detection platforms.

Financial Disclosure Statement Professor Geddes is an inventor of a patent related to the technologies described in this book chapter review, for which he is entitled to receive royalties. One invention was issued to the University of Maryland Baltimore County (UMBC) (U.S. Patent No. 13/990,828). In addition, Professor Geddes is a partner in Lyse-It™ LLC, a Maryland based company (www.lyse-it.com) that has licensed the technology from UMBC.

References

Aslan K, Geddes CD (2006) Microwave-accelerated metal-enhanced fluorescence (MAMEF): application to ultra fast and sensitive clinical assays. J Fluoresc 16(1):3–8

Aslan K, Geddes C (2008) A review of an ultrafast and sensitive bioassay platform technology: microwave-accelerated metal-enhanced fluorescence. Plasmonics 3(2–3):89

Aslan K, Malyn SN, Bector G, Geddes CD (2007a) Microwave – accelerated metal – enhanced fluorescence: an ultra-fast and sensitive DNA sensing platform. Analyst 132:1122–1129

Aslan K, Zhang Y, Hibbs S, Baillie L, Previte MJR, Geddes CD (2007b) Microwave-accelerated metal-enhanced fluorescence: application to detection of genomic and exosporium anthrax DNA in <30 seconds. Analyst 132(11):1130–1138

Aslan K, Previte MJR, Zhang Y, Gallagher T, Baillie L, Geddes CD (2008) Extraction and detection of DNA from Bacillus anthracis spores and the vegetative cells within 1 min. Anal Chem 80(11):4125–4132

Chen M, Ho F, Lin M et al (2004) Localization of optical fields using a bow-tie nanoantenna. Scanning 6(25):I113–I117

Dragan AI, Albrecht MT, Pavlovic R, Keane-Myers AM, Geddes CD (2012) Ultra-fast pg/ml anthrax toxin (protective antigen) detection assay based on microwave-accelerated metal-enhanced fluorescence. Anal Biochem 425:54–61

Friedlander T, Mayo AE, Tlusty T et al (2015) Evolution of Bow-Tie architectures in biology. PLoS Comput Biol 11(3):e1004055

Gaydos CA, Quinn TC, Willis D, Weissfeld A, Hook EW, Martin DH, Ferrero DV, Schachter J (2003) Performance of the APTIMA Combo 2 assay for detection of Chlamydia trachomatis and Neisseria gonorrhoeae in female urine and endocervical swab specimens. J Clin Microbiol 41(1):304–309

Gaydos CA, Cartwight CP, Colaninno P, Welsch J, Holden J, Ho SY, Webb EM, Anderson C, Bertuzis R, Zhang L, Miller T, Leckie G, Abravaya KJR (2010) Performance of the Abbott Real-time CT/NG for the detection of Chlamydia trachomatis and Neisseria gonorrhoeae. J Clin Microbiol 48:3236–3243

Hasan D, Ho CP, Lee C (2016) Thermally tunable absorption-induced transparency by a Quasi 3D Bow-Tie nanostructure for nonplasmonic and volumetric refractive index sensing at Mid-IR. Adv Opt Mater 4(6):943–952

Hu S, Weiss SM (2016) Design of photonic crystal cavities for extreme light concentration. ACS Photonics 3(9):1647–1653

Joshi LT, Mali BL, Geddes CD, Baillie L (2014) Extraction and sensitive detection of toxins A and B from the human pathogen Clostridium difficile in 40 seconds using microwave-accelerated metal-enhanced fluorescence. PLoS One 9(8):e104334–e104334

Kennedy M, Villar R, Vugia DJ, Rabatsky-Ehr T, Farley MM, Pass M, Smith K, Smith P, Cieslak PR, Imhoff B, Griffin PM (2004) Hospitalizations and deaths due to Salmonella infections, FoodNet, 1996-1999. Clin Infect Dis: Off Publ Infect Dis Soc Am 38(Suppl 3):S142–S148

Melendez JH, Huppert JS, Jett-Goheen M, Hesse EA, Quinn N, Gaydos CA, Geddes CD (2013) Blind evaluation of the microwave-accelerated metal-enhanced fluorescence ultrarapid and sensitive Chlamydia trachomatis test by use of clinical samples. J Clin Microbiol 51 (9):2913–2920

Melendez JH, Santaus TM, Brinsley G, Kiang D, Mali B, Hardick J, Gaydos CA, Geddes CD (2016) Microwave-accelerated method for ultra-rapid extraction of Neisseria gonorrhoeae DNA for downstream detection. Anal Biochem 510:33–40

Prevention, Centre for Disease Control (2016a) Chlamydia. Center for Disease Control and Prevention, Atlanta

Prevention, Centre for Disease Control (2016b) Gonorrhea. Center for Disease Control and Prevention, Atlanta

Prevention, Center for Disease Control (2016c) Clostridium difficile. Center for Disease Control and Prevention, Atlanta

Previte MJR, Geddes CD (2007a) Microwave-triggered chemiluminescence with planar geometrical aluminum substrates: theory, simulation and experiment. J Fluoresc 17(3):279–287

Previte MJR, Geddes CD (2007b) Fluorescence microscopy in a microwave cavity. Opt Express 15(18):11640–11649

Schneider JB (2016) Understanding the finite different time domain method. http://www.eecs.wsu.edu/~schneidj/ufdtd/ufdtd.pdf

Sigma-Aldrich (2016) Peptidoglycans. Sigma-Aldrich, Darmstadt

Taylor SN, Van Der Pol B, Lillis R, Hook EW III, Lebar W, Davis T, Fuller D, Mena L, Fine P, Gaydos CA, Martin DH (2011) Clinical evaluation of the BD ProbeTecTM Chlamydia trachomatis (CT) Qx amplified DNA assay on the BD Viper TM system with XTRTM technology. Sex Transm Dis 38:603–609

Tennant SM, Yongxia Z, Galen JE, Geddes CD, Levine MM (2011) Ultra-fast and sensitive detection of non-typhoidal Salmonella using Microwave-Accelerated Metal-Enhanced Fluorescence ("MAMEF"). PLoS One 6(4):1–8

Van Der Pol B, Liesenfeld O, Williams JA, Taylor SN, Lillis RA, Body BA, Nye M, Eisenhut C, Hook EW III (2012) Performance of the Cobas CT/NG test compared to the Aptima AC2 and Viper CTQ/GCQ assays for detection of Chlamydia trachomatis and Neisseria gonorrhoeae. J Clin Microbiol 50:2244–2249

Zhang Y, Agreda P, Kelley S, Gaydos C, Geddes CD (2011) Development of a microwave—accelerated metal-enhanced fluorescence 40 Second, <100 cfu/mL point of care assay for the detection of Chlamydia Trachomatis. IEEE Trans Biomed Eng 58(3):781

Erratum to: Mobile Telephony EMFs Effects on Insect Ovarian Cells. The Necessity for Real Exposures Bioactivity Assessment. The Key Role of Polarization, and the “Ion Forced-Oscillation Mechanism”

Dimitris J. Panagopoulos

Erratum to:
Chapter 1 in: C.D. Geddes (ed.), *Microwave Effects on DNA and Proteins*, DOI 10.1007/978-3-319-50289-2_1

The original version of this chapter was inadvertently published with incorrect equation in page 28.

Current equation has long square root but the short square root is correct.

Incorrect:

$$E_{ave} = \sqrt{\frac{1}{2}\left(E_{o1}^2 + E_{o2}^2\right) \text{ or } E_o^2 = E_{o1}^2 + E_{o2}^2 \quad (= \text{constant})}$$

Correct:

$$E_{ave} = \sqrt{\frac{1}{2}\left(E_{o1}^2 + E_{o2}^2\right)} \text{ or } E_o^2 = E_{o1}^2 + E_{o2}^2 \quad (= \text{constant})$$

The original online version for this chapter can be found at
DOI 10.1007/978-3-319-50289-2_1

C.D. Geddes (ed.), *Microwave Effects on DNA and Proteins*,
DOI 10.1007/978-3-319-50289-2_9

Index

C.D. Geddes (ed.), *Microwave Effects on DNA and Proteins*,
DOI 10.1007/978-3-319-50289-2

MIX
Papier aus verantwortungsvollen Quellen
Paper from responsible sources
FSC® C105338
FSC
www.fsc.org

If you have any concerns about our products,
you can contact us on
ProductSafety@springernature.com

In case Publisher is established outside the EU,
the EU authorized representative is:
Springer Nature Customer Service Center GmbH
Europaplatz 3, 69115 Heidelberg, Germany

Printed by Libri Plureos GmbH
in Hamburg, Germany